Sedimentary Petrology
An Introduction

GEOSCIENCE TEXTS

SERIES EDITOR

A. HALLAM

Lapworth Professor of Geology
University of Birmingham

Sedimentary Petrology
An Introduction

M. E. TUCKER
BSc, PhD
Department of Geology
University of Newcastle Upon Tyne

A HALSTED PRESS BOOK

JOHN WILEY & SONS
NEW YORK TORONTO

To Vivienne

First published 1981

Published in the U.S.A and Canada by
Halsted Press.
a Division of John Wiley & Sons Inc., New
York

British Library
Cataloguing in Publication Data

Tucker, Maurice E
    Sedimentary Petrology. –
    (Geoscience texts; vol. 3).
    1. Rocks, Sedimentary
    I. Title II. Series
    552′.5 QE471

ISBN 0470-27160-4

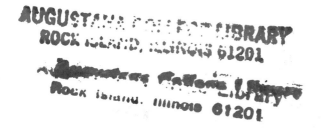

Contents

Preface

The study of sedimentary rocks, sedimentary petrology, goes back to the last century and beyond. It is only in the past few decades, however, that we have begun to understand and appreciate the processes by which these rocks are formed. Many of the latest steps forward have come from research on modern sediments and material from shallow and deep boreholes. The advent of sophisticated instruments, such as the electron microscope, has also been important. Some of the impetus for investigating sedimentary rocks has come from their economic importance: the fossil fuels coal and petroleum, and many essential minerals and raw materials are contained in these rocks. In spite of recent advances, there is still much that can be done just with a hammer in the field and simple microscope in the laboratory, and a pair of sharp eyes.

This book attempts to present a concise, up-to-date account of sedimentary petrology. In recent years, many texts have been published dealing more with the depositional environments and facies of sediments, with less attention being given to features of the rocks themselves. This book approaches the subject from the other direction, examining each rock group in turn, with discussions of composition, petrography, sedimentary structures and diagenesis. Depositional environments and facies are only briefly considered.

This book has been written with undergraduate students in mind. Because of this, references to the literature have been critically chosen. By and large, students do not want (or need) to consult original papers on a topic written in the early part of this century, or the last. Students require up-to-date information, the latest ideas, and reviews. Good review papers of course do cite the early literature so that the keen student can soon delve back and locate important papers. All the references cited in this book should be readily available in university and institute libraries.

ACKNOWLEDGEMENTS

Many friends and colleagues have willingly read early drafts of chapters and made very valuable and useful comments. I am particularly grateful to Hugh Battey, Colin Braithwaite, Paul Bridges, Trevor Elliott, John Hemingway, Mic Jones, Duncan Murchison, Andrew Parker, Tim Pharoah, Alastair Robertson, Colin Scrutton and Bruce Sellwood. I should also thank the many people (acknowledged in the text) who have supplied specimens, photographs or thin sections for text figures. I am indebted to Mrs. K. Sales of Newcastle University's Photographic Department for all her efforts in printing the photographs. My deepest gratitude must go to my wife, Vivienne, for doing much of the donkey-work (typing, etc. etc.), for giving up so many evenings and weekends without too much complaint, and for keeping the little horrors of Fig. 2.30 quiet during the day.

Maurice Tucker
Newcastle Upon Tyne

1
Introduction: Basic Concepts and Methodology

1.1 Introduction

Some 70% of the rocks at the Earth's surface are sedimentary in origin, and these include the familiar sandstones, limestones and shales, and the less common but equally well-known salt deposits, ironstones, coal and chert.

Sedimentary rocks of the geological record were deposited in the whole range of natural environments that exist today. The study of these modern environments and their sediments and processes contributes much to the understanding of their ancient equivalents. There are some sedimentary rock-types, however, for which there are no known modern analogues or their inferred depositional environments are only poorly represented at the present time.

Once deposited, sediments are subjected to processes of diagenesis, that is physical, chemical and biological processes which bring about compaction, cementation, recrystallization and other modifications to the original sediment.

There are many reasons for studying sedimentary rocks, not least because of the wealth of economic minerals and materials contained within them. The fossil fuels, oil and gas are derived from the maturation of organic matter in sediments and these then migrate to a suitable reservoir rock, mostly a porous sedimentary rock. The other fossil fuel, coal, is also contained within sedimentary sequences of course. Sedimentological and petrological techniques are increasingly used in the search for new reserves of these fuels and other natural resources. Sedimentary rocks supply much of the world's iron, potash, salt, building materials and many, many other essential raw materials.

Environments and processes of deposition and palaeogeography and palaeoclimatology can all be deduced from studies of sedimentary rocks. Such studies contribute much towards a knowledge and understanding of the Earth's geological history. Sedimentary rocks contain the record of life on Earth, in the form of fossils and these are the principal means of stratigraphic correlation in the Phanerozoic.

1.2 Basic concepts

1.2.1 CLASSIFICATION OF SEDIMENTARY ROCKS

Sedimentary rocks are formed through physical, chemical and biological processes. On the basis of the dominant process(es) operating, the common sediment lithologies can be grouped into four broad categories (Table 1.1). The terrigenous clastic sediments (also referred to as siliciclastic or epiclastic

1

Table 1.1 Principal groups of sedimentary rock

Terrigenous clastic sediments	Biogenic, biochemical and organic sediments	Chemical sediments	Volcaniclastic sediments
Conglomerates and breccias, sandstones, mudrocks.	limestones (and dolomites), cherts, phosphates, coal and oil shale.	evaporites, ironstones.	e.g. ignimbrites, tuffs and hyaloclastites.

deposits) are those consisting of fragments (clasts) of pre-existing rocks, which have been transported and deposited by physical processes. The conglomerates and breccias, sandstones and mudrocks, discussed in Chapters 2 and 3 belong to this group. Sediments largely of biogenic, biochemical and organic origin are the limestones, which may be altered to dolomite (Chapter 4), phosphate deposits (Chapter 7), coal and oil shale (Chapter 8) and cherts (Chapter 9). Sedimentary rocks largely of chemical origin, principally direct precipitation, are the evaporites (Chapter 5) and ironstones (Chapter 6). Volcaniclastic deposits (Chapter 10) constitute a fourth category and consist of lava and rock fragments derived from penecontemporaneous volcanic activity. Each of these various sedimentary rock-types can be divided further, usually on the basis of composition. In addition, many rock-types grade laterally or vertically into others through intermediate lithologies.

1.2.2   SEDIMENTARY ENVIRONMENTS AND FACIES

Sedimentary environments vary from those where erosion and transportation dominate, to those where deposition prevails. Most weathering and erosion, liberating sediment grains and ions in solution, take place in continental areas and climate, local geology and topography control the type and amount of material released. The main continental depositional environments are fluvial and glacial systems, lakes and the aeolian sand seas of deserts. Most shoreline environments, deltas, lagoons, tidal flats, sabkhas, beaches and barriers, and open marine environments, shallow shelves and epeiric seas, and bathyal-abyssal sites of pelagic, hemipelagic and turbidite sedimentation, are areas of net deposition, involving the whole range of sediment lithologies. Many of these sediments possess distinctive characteristics which can be used to recognize their equivalents in the geological record.

With sedimentary rocks, once they have been described and identified (the theme of this book), and their stratigraphic relationships elucidated then the concept of facies is applied. A *facies* is a body or packet of sedimentary rock with features that distinguish it from other facies. Features used to separate facies are sediment composition (lithology), grain size, texture, sedimentary structures, fossil content and colour. Lithofacies are defined on the basis of sedimentary characteristics, while biofacies rely on palaeontological differences. With detailed work, sub-facies can be recognized and microfacies if microscope studies are used to distinguish between rocks which in the field appear similar (often the case with limestones). Facies can be described in terms of (i) the sediment itself (e.g. cross-bedded sandstone facies), (ii) the depositional process (e.g. stream-flood facies) and (iii) the depositional

2

environment (e.g. tidal-flat facies). Only (i) is objective and, hopefully, unequivocal; (ii) and (iii) are both interpretative. Different facies frequently occur in association with each other. Repetitions of facies sequences are not uncommon and result from a variety of causes.

There are many factors which control and affect the sediments deposited and determine the sedimentary rock-type and facies produced. On a gross scale, overriding controls are (i) the depositional processes, (ii) the depositional environment, (iii) the tectonic context and (iv) the climate. Sediments can be deposited by a wide range of processes including the wind, flowing water as in streams, tidal currents and storm currents, waves, sediment + water flows such as turbidity currents and debris flows, the in-situ growth of animal skeletons as in reefs, and the direct precipitation of minerals, as in evaporites. The depositional processes leave their record in the sediment in the form of sedimentary structures and textures. Some depositional processes are typical of a particular environment, whereas others operate in several or many environments. Environments are defined on physical, chemical and biological parameters and they can be sites of erosion, non-deposition or sedimentation. Water depth, degree of agitation and salinity are important physical attributes of subaqueous environments and these affect and control the organisms living on or in the sediment or forming the sediment. Chemical factors such as Eh (redox potential) and pH (acidity-alkalinity) of surface and pore waters affect organisms and control mineral precipitation. The tectonic context is of paramount importance since this determines the depositional setting, whether it is for example a stable craton, intracratonic basin or rift, or geosyncline. Geosynclines, linear belts of thick, deformed sediments, are now interpreted in terms of plate tectonic models, so that deposition is inferred to have taken place in continental margin, ocean floor, trench and arc-related situations. Rates of subsidence and uplift, level of seismic activity and occurrence of volcanoes are also dependent on the tectonic context and are reflected in the sediments deposited. Climate is a major factor in subaerial weathering and erosion and this relates to the composition of terrigenous clastic sediments. Climate is instrumental in the formation of some lithologies, evaporites and limestones for example. Two other factors, controlled by climate and tectonic context, are sediment supply and organic productivity. Sediment supply is important in so far as low rates favour limestone, evaporite, phosphate and ironstone formation. High levels of organic productivity are important in the formation of limestones, phosphates, cherts, coal and oil shale.

Many attributes of a facies then are reflections of the depositional processes and environment. There is a finite number of environments so that similar facies and facies associations are produced wherever and whenever a particular environment existed in the geological past. Differences do arise of course, from variations in provenance (the source of the sediment), the nature of the fossil record at the time, and climatic and tectonic considerations. From studies of modern and ancient sedimentary environments, processes and facies, generalized facies models have been proposed to show the lateral and vertical relationships between facies. These models facilitate interpretations of sedimentary sequences and permit predictions of facies distributions and geometries. The importance of the vertical succession of facies was first

3

appreciated by Johannes Walther at the end of the 19th Century in his 'Law of the Correlation of Facies': different facies in a vertical sequence reflect environments which were originally adjacent to each other, providing there were no major breaks in sedimentation. Vertical changes in facies result from the effects of internal sedimentation processes and external processes. Familiar examples of the former are the progradation (building out) of deltas and tidal flats into deeper water, and the combing of a river across its floodplain. External processes are again chiefly tectonic movements, acting on a regional or global scale, and climatic changes. Both of these affect the relative position of sealevel, a major factor in facies development, and the supply of sediment, as noted above.

1.2.3 DIAGENESIS

Considerations of sedimentary rocks do not stop with environmental interpretations. There is a whole story to be told of events after deposition, that is during diagenesis. It is during diagenesis that an indurated rock is produced from an unconsolidated, loose sediment. Diagenetic processes begin immediately after deposition and continue until metamorphism takes over; this is when reactions are due to elevated temperatures (in excess of 150–200°C) and/or pressures. A distinction is made between early diagenetic events, taking place from sedimentation until shallow burial, and late diagenetic events occurring during deep burial and subsequent uplift.

Diagenetic processes which can be introduced here but which are considered further in later sections (2.8, 3.6 and 4.7) are compaction, recrystallization, dissolution, replacement, authigenesis and cementation. Compaction is a physical process arising from the mass of overlying sediment. Some grains and minerals deposited in a sediment or forming a sediment are unstable and during diagenesis they may recrystallize (i.e. their crystal fabric changes but the mineralogy is unaltered) or they may undergo dissolution and/or be replaced by other minerals. The effects of solution and replacement are common in limestones, sandstones and evaporites. The precipitation of new minerals within the pore spaces of a sediment is referred to as authigenesis, and if precipitated in sufficient quantity then cementation of the sediment results. Concretions and nodules such as commonly occur in mudrocks form through localized mineral precipitation. The ions for cementation are derived from porewaters and grain dissolution.

Diagenetic processes are important for several reasons. They can considerably modify a sediment, both in terms of its composition and texture, and in rare cases, original structures are completely destroyed. Diagenetic events also affect a sediment's porosity and permeability, properties which control a sediment's potential as a reservoir for oil, gas or water.

1.3 Methodology

The study of sedimentary rocks invariably begins in the field but after that there are several avenues which can be explored, depending on the objectives of the study and the interests of the investigator. Samples collected can be examined

4

on a macro-, micro- and nanno- scale. Sophisticated techniques and machines can be used to discover a sediment's mineralogy and geochemistry. Experiments can be devised to simulate the conditions of deposition. Data collected in the field or laboratory can be subjected to statistical tests and computer analysis. Account should be taken of any existing literature on the rocks being studied, and of descriptions of similar rocks and facies from other areas, together with their probable modern analogues. With all this information to hand, the rocks under consideration could then be interpreted with regard to origin, depositional process and environment, palaeogeography, diagenetic history, and possible economic significance and potential.

1.3.1 IN THE FIELD

The main point about fieldwork is being able to observe and record accurately what you see. With a little field experience and some background knowledge, you will soon know what to expect and what to look for in a certain type of sedimentary rock or particular facies. It is obviously important to appreciate the significance of the various sedimentary features you see, to know which are environmentally diagnostic for example, and also to know how they can be used to get maximum information: what to measure, what to photograph, what to collect. The field study of sedimentary rocks is discussed in Tucker (1981).

The study of sedimentary rocks in the field requires the initial identification of the lithology (often with the aid of a lens) in terms of composition, grain size, texture and fossil content. These attributes can be confirmed and quantified later in the laboratory. Sedimentary structures are usually described and measured in the field because of their size. It is relatively easy to see structures in hand-specimen or block, but those on the scale of a quarry or cliff face are easily overlooked. So observe on all scales. It is important to note the size and orientation of structures. Many sedimentary structures can be used for palaeocurrent analysis and these and others reflect the processes operating in the environment (Sections 2.3, 2.4, 3.2 and 4.6). Sedimentary structures should be described within their lithological context; many are related to grain size or composition for example.

One of the best methods of recording sedimentary rocks is to construct a log of the section. Basically, measure the thickness of each bed or facies interval, note its composition, grain size, colour, sedimentary structures, fossils and any other features. If a palaeocurrent measurement can be taken, record this too. A *graphic log* can be drawn up in the field using an appropriate vertical scale for the sediment thickness and a horizontal scale for the sediment grain size (Figs. 2.53, 2.55, 2.57 and 2.58). Different types of shading can be used for the various lithologies and symbols and abbreviations for the sedimentary structures and fossils (Bouma, 1962; Tucker, 1981). The value of such graphic logs lies in the immediate picture which is obtained of the vertical succession of facies. In logging a section the lateral extent and continuity of beds must be taken into account. Many beds are actually lenticular.

Although in the field study of sedimentary rocks, it is likely that a geological map will be at hand, some detailed mapping of small areas could well be required to ascertain the relationships between facies and affects of local structural complications.

In many cases the interpretation of sedimentary rocks hinges on the fieldwork, so much care and attention should be paid to it. Localities need to be visited several times; it is amazing how many new things you can see at an exposure on a second or third visit.

1.3.2 IN THE LABORATORY

A great deal can be done with sedimentary rocks in the laboratory and there are several books concerned with laboratory procedures (e.g. Müller, 1967; Bouma, 1969; Carver, 1971; Hutchison, 1974). Starting with a hand specimen, cutting and polishing a surface may reveal sedimentary structures poorly displayed or invisible in the field. With limestones, etching with acid and staining a surface may further enhance the structures. With unconsolidated sediments and sedimentary rocks which are readily disaggregated, sediment grain size can be measured though the use of sieves and sedimentation chambers (Section 2.2.1, Folk, 1974; Blatt *et al.*, 1980). The heavy minerals (Section 2.5.5) can be extracted from loose sediment by using heavy liquids such as bromoform or Thoulet's solution.

Much detailed work is undertaken on thin sections cut from sedimentary rocks or impregnated unconsolidated sediments. With limestones, acetate peels are frequently used and the staining of these and thin sections helps identify the

Table 1.2 Optical properties of common minerals in sedimentary rocks as observed with the petrological microscope. The properties of the heavy minerals are given in Fig. 2.5.

Mineral	Chemical formula	Crystal system	Colour	Cleavage	Relief
Quartz	SiO_2	trigonal	colourless	absent	v. low (+)
Microcline	$KAlSi_3O_8$	triclinic	colourless	present	low (−)
Orthoclase	$K(Na)AlSi_3O_8$	monoclinic	colourless	present	low (−)
Albite	$Na(Ca)AlSi_3O_8$	triclinic	colourless	present	low (−)
Muscovite	$KAl_2(OH)_2AlSi_3O_{10}$	monoclinic	colourless	1 direction	fair
Biotite	$K_2(Mg,Fe)_2(OH)_2AlSi_3O_{10}$	monoclinic	brown to green	1 direction	fair
Chlorite	$Mg_5(Al,Fe)(OH)_8(AlSi)_4O_{10}$	monoclinic	green	1 direction	fair
Kaolinite	$Al_2O_3.2SiO_2.2H_2O$	triclinic	colourless-yellow	1 direction	low (+)
Illite	$KAl_2(OH)_2[AlSi_3(O,OH)_{10}]$	monoclinic	colourless-yellow	—	low (+)
Montmorillonite	$MgCaAl_2O_35SiO_2.nH_2O$	monoclinic	colourless-pink	—	low (−)
Chamosite	$Fe_3^{2+}Al_2Si_2O_{10}.3H_2O$	monoclinic	green	—	moderate
Glauconite	$KMg(Fe,Al)(SiO_3)_6.3H_2O$	monoclinic	green	1 direction	moderate
Calcite	$CaCO_3$	trigonal	colourless	can be strong	low to high
Dolomite	$CaMg(CO_3)_2$	trigonal	colourless	can be strong	low to high
Siderite	$FeCO_3$	trigonal	colourless	strong	low to high
Gypsum	$CaSO_4.2H_2O$	monoclinic	colourless	strong	low
Anhydrite	$CaSO_4$	orthorhombic	colourless	3 directions	moderate
Halite	$NaCl$	cubic	colourless	strong	low
Collophane	$Ca_{10}(PO_4,CO_3)_6F_{2-3}$	a mineraloid	shades of brown	—	moderate
Pyrite	FeS_2	cubic	opaque	—	—
Hematite	Fe_2O_3	hexagonal	opaque	—	—
Magnetite	Fe_3O_4	cubic	opaque	—	—

carbonate minerals present. Stains can also be used for feldspars in terrigenous clastic sediments. There is a relatively small number of common minerals in a sedimentary rock and in many cases it is not necessary to examine their optical properties to identify them. The properties of the common sedimentary minerals are given in Table 1.2. The precise composition of many sedimentary rocks (the sandstones and limestones in particular), which enables them to be classified, is obtained from microscopic studies by the use of a point counter. Several hundred grains are identified as the thin section is systematically moved across the microscope stage. Grain sizes of indurated silt to sand-sized rocks are measured from a thin section or peel using a calibrated eye-piece graticule. Grain shape and orientation can also be assessed. Many aspects of diagenesis in sandstones, limestones and evaporites are deduced from thin section studies.

In recent years, much sedimentological work has been carried out on the scanning electron microscope (S.E.M.). This instrument allows examination of specimens at very high magnifications, features down to $0.1\mu m$ can be seen. The S.E.M. is especially useful for fine-grained sedimentary rocks, such as cherts, and for observing clay minerals and the cements of sandstones and limestones (Figs. 2.48, 4.38).

For mineral identification in fine-grained sediments and sedimentary rocks, X-ray diffraction (X.R.D.) is widely used. Clay minerals in mudrocks are invariably analysed in this way. It is becoming apparent that geochemical analyses of sedimentary rocks, especially limestones and shales, can give useful and vital information on the environment of deposition and path of diagenesis.

irefringence	Other Features	Form and occurrence	See section:
eak		as detrital grains (monocrystalline and polycrystalline types), cements and replacements: fibrous quartz (chalcedony), microquartz, megaquartz.	2.5.2, 2.8.2, 9.2
eak	grid-iron twinning	as detrital crystals, also authigenic.	2.5.3,
eak	simple twinning (Carlsbad)	Commonly altered to clays, so appearing	
eak	multiple twinning	dusty	2.8.4
rong	parallel extinction	common detrital minerals, occurring	2.2.4,
rong	parallel extinction	as flakes	3.4.3
eak		occur as detrital minerals, particularly	2.8.5,
eak	best identified through	in mudrocks, also as cement (as in sand-	3.4.1,
rong	x-ray diffraction since	stones) and replacements, such as of	
oderate	usually so fine-grained	feldspars and volcanic grains.	10.5
eak		ooids and mud in ironstones	6.4.3
oderate		forms syn-sedimentary grains	6.4.3
treme	can be distinguished	form grains, matrix, cement and replacements in	4.3, 4.7
treme	by staining	limestones, dolomites, sandstones (etc)	4.8, 2.8.3
treme	frequently stained brown	fine and coarse crystals in ironstones	6.4.2
eak		anhedral to euhedral crystals	
rong	parallel extinction	equant to lath-shaped crystals	5.2
—	may have fluid inclusions	often coarsely crystalline	5.3
tropic	if bone-	forms ooids, pellets, bones,	
weak	organic structure	some shells	7.2
—	yellow in reflected light	aggregates and cubic crystals, authigenic	6.4.4
—	red-grey in reflected light	cryptocrystalline, a pigment and replacement	6.4.1, 2.8.6
—	grey-black in reflected light	cryptocrystalline, detrital	6.4.1

7

Major and minor elements are mostly determined by atomic absorption spectrophotometry (A.A.) and X-ray fluorescence (X.R.F.). On the scale of individual grains and crystals, in limestone cements for example, the electron microprobe is used to determine trace elements on areas only a few microns across. A consideration of the isotopes of such elements as oxygen and carbon, measured with a mass spectrometer, is a powerful tool in the study of limestone and chert diagenesis. Finally, use of a cathodoluminoscope, which bombards a rock slice with electrons and causes luminescence, can reveal details of cements and overgrowths.

One further laboratory approach has been to carry out experiments to determine the conditions under which sedimentary structures, grain types, minerals, etc., were formed. Perhaps the best known are those involving laboratory channels or flumes, where the effects of water flowing over sediment have been monitored (Section 2.3.2), and the attempts to precipitate dolomite (4.8.1) and synthesize ooids (4.3.1).

Once the data on the sedimentary rocks under investigation have been gathered, then the interpretations can begin. Statistics and computers are being increasingly used for the evaluation and interpretation of these data. Mathematical geology is now an established branch of the Earth Sciences and reference should be made to the available textbooks on this subject (e.g. Griffiths, 1967; Harbaugh & Bonham-Carter, 1970; Davis, 1973; Till, 1974). Apart from the use of standard statistical procedures to test data for significance, correlation, regression and variance, more sophisticated statistical methods can be applied which allow more objective conclusions to be reached. For example, from petographic modal analyses of a large number of samples, cluster analysis and factor analysis can be used to pick out the principal facies-types and to determine relationships between the facies (Ekdale *et al.* (1976), Rao & Naqvi (1977) and Smosna & Warshaver (1979)). Markov chain analysis can be applied to recognize sequences in a stratigraphic succession. Coal measure sequences (Doveton, 1971), fluviatile sediments (Miall, 1973) and shelf limestones (Jones & Dixon, 1976) have been treated this way. Finally, using the computer, mathematical models can be devised to simulate sedimentation in a particular environment and to predict the facies distributions. This has for example been applied to sedimentation within a meandering stream (Bridge, 1975).

1.3.3 IN THE LIBRARY: SEDIMENTOLOGICAL READING

However good your field and lab work is, it must be supported with a knowledge of the literature on the subject. Publications on the petrology of sedimentary rocks go way back into the last century, but in fact most advances have come in the past two decades. There are many textbooks available which cover some aspects of the subject in more detail than can be given here or which deal with directly-related topics. General texts include Folk (1974), Füchtbauer (1974), Pettijohn (1975), Selley (1976), Greensmith (1978), Friedman & Sanders (1978) and Blatt, Middleton & Murray (1980). Limestones are described by Bathurst (1975) and Wilson (1975) and sandstones by Pettijohn, Potter & Siever (1973). Sedimentary processes, environments and facies are described by

Reineck & Singh (1973), Allen (1977), Selley (1978a), especially by Reading and others (1978) and Walker (1979). Chemical aspects of sediments are discussed by Degens (1965), Berner (1971) and Garrels & Mackenzie (1971) and diagenesis by Larsen & Chilingarian (1979). In addition, there are many collections of papers on a specific topic; here can be cited the series of special publications of the Society of Economic Paleontologists and Mineralogists (S.E.P.M.) and International Association of Sedimentologists (I.A.S.), and some memoirs of the American Association of Petroleum Geologists (A.A.P.G.). Of particular relevance are Memoirs 27 and 28 of A.A.P.G. (both by P. A. Scholle) on limestone and sandstone petrography respectively.

Most research papers, however, are published in the learned journals. The interested student should keep his eyes on the current journals for the latest information and ideas. Books soon go out of date(!). The two principal periodicals are the *Journal of Sedimentary Petrology* and *Sedimentology*, published by S.E.P.M. and I.A.S. respectively. Others devoted to sediments or containing many sedimentological papers are *Sedimentary Geology, Geology, Bulletin of the Geological Society of America, Bulletin of the American Association of Petroleum Geologists, Journal of Geology, Marine Geology* and *Palaeogeography Palaeoecology and Palaeoclimatology*. Perhaps a special mention should be made of the *Initial Reports of the Deep Sea Drilling Project* which contain a wealth of information on deepsea sediments and their diagenesis. In addition there are many other journals which often contain relevant articles; a regular perusal of the current periodicals in the library will spot these as they are published.

Finally, there are several periodicals which regularly cite all published papers. Indexes are provided so that all papers on a particular topic (or by a certain author) published in a particular year, can be located. Such journals are the *Bibliography and Index of Geology* published by the Geological Society of America, and *Geoabstracts Section E Sedimentology* published by the University of East Anglia, England.

2

Terrigenous clastic sediments
I: sandstones, conglomerates and breccias

2.1　　Introduction

Terrigenous clastic sediments are a diverse group of rocks, ranging from the fine-grained mudrocks, through the coarser grained sandstones to conglomerates and breccias. The sediments are largely composed of fragments, or clasts, derived from pre-existing igneous, metamorphic and sedimentary rocks. The clastic grains are released through mechanical and chemical weathering processes, and then transported to the depositional site. Mechanisms involved in the transportation include the wind, glaciers, river currents, waves, tidal currents and turbidity currents. The detrital grains may be rock fragments but the majority are individual crystals, chiefly of quartz and feldspar, abraded to various degrees. The finer breakdown products of the original rocks, formed during weathering and consisting mainly of clay minerals, are predominant in mudrocks and form the matrix to some sandstones and conglomerates. In a broad sense, the composition of clastic sediments is a reflection of the weathering processes, largely determined by the climate and geology of the source area. Source areas are generally upland, mountainous regions undergoing uplift, but detritus may also be supplied from erosion in lowland and coastal areas. Sediment composition is also affected by distance of sediment transport and by diagenetic processes.

Two important features of siliciclastic sediments are the sedimentary structures and textures. Many of these are produced by the depositional processes, while others are post-depositional or diagenetic in origin.

In this chapter, the coarser clastic sediments, the sandstones, conglomerates and breccias, are treated; the finer-grained terrigenous clastic sediments, the mudrocks, are discussed in Chapter 3.

2.2　　Sediment texture

The texture of a terrigenous clastic rock is largely a reflection of the depositional process and so many modern sediments have been studied to determine their textural characteristics. Studies of sediment texture involve considerations of grain size and grain-size parameters, grain morphology, grain surface texture and sediment fabric. On the basis of its textural attributes, a sediment can be considered in terms of its textural maturity.

2.2.1　　GRAIN SIZE AND GRAIN-SIZE PARAMETERS

The basic descriptive element of all sedimentary rocks is the grain size. Several grain-size scales have been proposed but one which is widely used and accepted is that of J. A. Udden based on a constant ratio of 2 between successive class boundaries together with terms for the classes by C. K. Wentworth (Table 2.1). The Udden–Wentworth grain-size scale divides sediments into seven grades: clay, silt, sand, granules, pebbles, cobbles and boulders, and subdivides sands into five classes and silts into four. From Table 2.1 the use of terms sand/sandstone, silt/siltstone and clay/claystone is clear. The term gravel is generally applied to loose sediments coarser than sand grade (2 mm), although there is usually some sand matrix; the term rudite or rudaceous rock is used for indurated gravels, and includes the conglomerates and breccias. For mudrock terminology refer to Chapter 4.

Table 2.1 Grain-size scale for sediments and sedimentary rocks (after J.A. Udden and C.K. Wentworth).

mm	phi	class terms		Sediment and rock terms implying grain size
256	−8	boulders		
128	−7	cobbles		
64	−6			gravel
32	−5			rudite
16	−4	pebbles		rudaceous seds.
8	−3			conglomerates
4	−2			breccias
2	−1	granules		
1	0		v. coarse	
0.5	1		coarse	sand
0.25	2	sand	medium	sandstones
0.125	3		fine	arenaceous seds.
0.0625	4		v. fine	arenites
0.0312	5		coarse	
0.0156	6	silt	medium	silt
0.0078	7		fine	siltstone
0.0039	8		v. fine	
		clay		clay
				claystone

Using millimetres as the units, the Udden–Wentworth scale is a geometric one (i.e. 1, 2, 4, 8, 16). W. C. Krumbein introduced an arithmetic scale (i.e. 1, 2, 3, 4, 5) of phi units (ϕ), where phi is the logarithmic transformation of the Udden–Wentworth scale: $\phi = -\log_2 S$ where S is the grain size in mm. For all serious work involving sediment grain size the phi scale is used since it has the advantage of making mathematical calculations much easier. For detailed work, class intervals in the sand field are taken at quarter phi intervals. Tables for the conversion of mm to phi units have been published by Page (1955), and a program for the conversion on a programmable calculator by Lindholm (1979).

When studying sandstones in the field, a first approximation of the grain size can be made with a hand-lens. The clast size of conglomerates and breccias can be measured directly with a tape measure. For accurate grain-size analysis, several laboratory methods are available. With poorly-cemented sandstones and unconsolidated sands, sieving is the most popular technique. Medium silt to small pebbles can be accommodated in sieves and it is the practice to use a similar sieving time, of around 15 mins for all samples, and similar weights, about 30 g or more for the coarser grades. Sedimentation methods, which measure the settling velocity of grains through a column of water, can be used for clay to sand grades. For well-cemented siltstones and sandstones (and limestones) thin sections have to be used and several hundred grain sizes measured with an eyepiece graticule and point counter. Details of these techniques together with their limitations are given in Müller (1967), Carver (1971), Folk (1974) and Blatt *et al.* (1980).

Once the grain-size distribution has been obtained then the sediment can be characterized by several parameters: mean grain size, mode, median grain size, sorting and skewness. A further parameter, kurtosis, has little geological significance. The parameters are often deduced from graphic presentation of the data, but an alternative is the method of moments, whereby the parameters are calculated directly from the size data. This latter method, outlined below, can be applied relatively easily with an electronic calculator.

For graphic representation, the histogram, smoothed frequency curve and cumulative frequency curve are plotted (Figs. 2.1, 2.2). It is the practice for grain size to decrease along the abscissa (x-axis) away from the origin. The histogram and smoothed frequency curve show the frequency of grains in each size class and usefully give an immediate impression of the grain-size distribution, particularly as to whether the distribution is unimodal or bimodal (Fig. 2.1). The cumulative frequency curve shows the percentage frequency of grains coarser than a particular value. When plotting cumulative frequencies it is best to use log probability paper which gives a straight line if the distribution is normal, i.e. Gaussian, generally the case with sediments. From the cumulative frequency plot are obtained the percentiles of the distribution, i.e. the grain

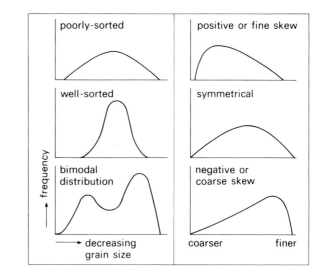

Fig. 2.1 Smoothed frequency distribution curves showing types of sorting and skewness.

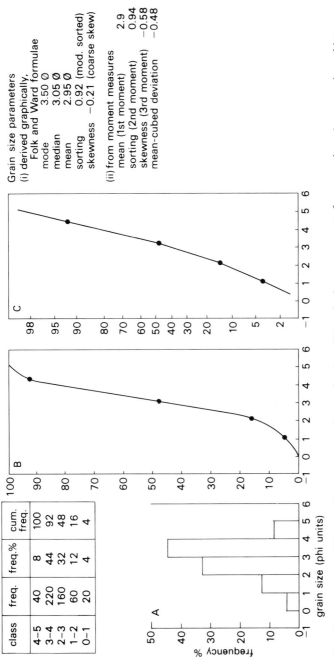

Grain size parameters
(i) derived graphically,
    Folk and Ward formulae
mode   3.50 Ø
median   3.05 Ø
mean   2.95 Ø
sorting   0.92 (mod. sorted)
skewness   −0.21 (coarse skew)

(ii) from moment measures
mean (1st moment)   2.9
sorting (2nd moment)   0.94
skewness (3rd moment)   −0.58
mean-cubed deviation   −0.48

class	freq.	freq.%	cum. freq.
4–5	40	8	100
3–4	220	44	92
2–3	160	32	48
1–2	60	12	16
0–1	20	4	4

Fig. 2.2 An example of the graphic presentation of grain size data (500 grain size measurements from a sandstone), presented as a histogram (A), and cumulative frequency curves plotted with an arithmetic scale (B) and log probability scale (C). Also given are the grain size parameters derived from the data (i) graphically using Folk & Ward's formulae of Table 2.2 and (ii) using the method of moments, formulae of Table 2.3.

13

sizes which correspond to particular percentage frequencies (percentiles), so that at the nth percentile, $n\%$ of the sample is coarser than that grain size.

The grain-size parameters are defined in Table 2.2. Nowadays the more precise parameters of Folk & Ward (1957) are used with grain size measured in phi. The simpler formulae of P. D. Trask for use with mm are also given.

The median grain size, simply the grain size at 50%, is not as useful as the *mean grain size* which is an average value, taking into account the grain sizes at the 16th, 50th and 84th percentiles (on the Folk & Ward scheme). The *mode* is the phi (or mm) value of the mid-point of the most abundant class interval.

Table 2.2 Formulae for the calculation of grain size parameters from a graphic presentation of the data in a cumulative frequency plot. With the Trask formulae, the percentile measure Pn is the grain size in millimetres at the nth percentage frequency, and with the Folk & Ward formulae, the percentile measure ϕ_n is the grain size in phi units at the nth percentage frequency.

Parameter	Trask formula	Folk & Ward formula
Median	$Md = P_{50}$	$Md = \phi_{50}$
Mean	$M = \dfrac{P_{25} + P_{75}}{2}$	$M = \dfrac{\phi_{16} + \phi_{50} + \phi_{84}}{3}$
Sorting	$So = \dfrac{P_{75}}{P_{25}}$	$\sigma\phi = \dfrac{\phi_{84} - \phi_{16}}{4} + \dfrac{\phi_{95} - \phi_5}{6.6}$
Skewness	$Sk = \dfrac{P_{25}P_{75}}{Md^2}$	$Sk = \dfrac{\phi_{16} + \phi_{84} - 2\phi_{50}}{2(\phi_{84} - \phi_{16})} + \dfrac{\phi_5 + \phi_{95} - 2\phi_{50}}{2(\phi_{95} - \phi_5)}$

Most sediments are unimodal, that is one class dominates but bimodal (Fig. 2.1) and even polymodal sediments are not uncommon, matrix-rich conglomerates for example, (Fig. 2.8). Where a grain-size distribution is perfectly normal and symmetrical, then the median, mean and mode values are the same. Trends in grain size over large areas can be used to infer the direction of sediment dispersal, with grain size decreasing away from the source area. Such down-current changes occur in fluviatile and deltaic systems, and in turbidites on submarine fans. From shorelines across shelves, the offshore decrease in grain size relates to a decrease in wave and current energy as the water depth increases.

Sorting is a measure of the standard deviation, i.e. spread of the grain-size distribution. It is one of the most useful parameters since it gives an indication of the effectiveness of the depositional medium in separating grains of different classes. Terms used to describe the sorting values obtained from the Folk & Ward formula are:

ϕ less than 0.35	very well sorted
0.35–0.5	well sorted
0.5–0.71	moderately well sorted
0.71–1.00	moderately sorted
1.0–2.0	poorly sorted
greater than 2.0	very poorly sorted

With thin sections of sandstones (and limestones) visual estimates of sorting can

14

be obtained by comparison with charts (Fig. 2.3). Also by reference to these charts, a rough estimate of sorting in a sandstone can be made in the field by use of a hand-lens.

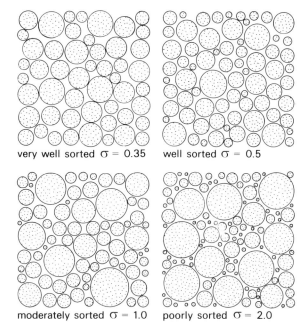

very well sorted σ = 0.35   well sorted σ = 0.5

Fig. 2.3 Charts for visual estimation of sorting (based on Pettijohn *et al.*, 1973).

moderately sorted σ = 1.0   poorly sorted σ = 2.0

Sorting of the sediment is determined by several factors. First there is the question of the sediment source: if a granite is providing the sediment, the grain sizes will be quite different from the sediment supplied by the reworking of a sandstone. The second factor is grain size itself: sorting is dependent on the grain size in that coarse sediments, gravels and conglomerates, and fine sediments, silts and clays, are generally more poorly sorted than sand-sized sediments which are more easily transported, and therefore sorted by wind and water. The third factor, which is where sorting is used for interpretation, is the depositional mechanism. Sediments which were deposited quickly, such as storm beds, or were deposited from viscous flows, such as mud flows, are generally poorly sorted; sediments which have been worked and reworked by the wind or water, the sandy deposits of deserts, beaches and shallow shelf seas for example, are much better sorted.

Skewness is a measure of the symmetry of the distribution and visually is best seen from the smoothed frequency curve (Fig. 2.1). If the distribution has a coarse 'tail', i.e. excess coarse material, then the sediment is said to be negatively skewed; if there is a fine 'tail', then the skew is positive. If the distribution is symmetrical, then there is no skew. Terms for skewness derived from the Folk & Ward formula are:

Sk greater than + 0.30	strongly fine-skewed
+ 0.30 to + 0.10	fine-skewed
+ 0.10 to − 0.10	near-symmetrical
− 0.10 to − 0.30	coarse-skewed
greater than −0.30	strongly coarse-skewed

Apart from being a useful descriptive term for a sediment sample, skewness is also a reflection of the depositional process. Beach sands for example tend to have a negative skew since fine components have been removed by the persistent wave action. River sands are often positively-skewed, since much silt and clay is not removed by the currents. Desert dune sands typically have a positive skew, but this is more complicated in origin, resulting from the inability of the wind to transport coarse particles from the source sediment into the sand dune area, and only the partial winnowing of the finest grains out of the dune sands.

Table 2.3 Formulae for the method of moments for grain size analyses. f is the percentage fraction in each class interval of the total weight of sediment (if a sieve analysis) or of the total number of grains (if data from a thin section), and $m\phi$ is the mid-point value of each class interval in phi units.

Mean (first moment) $= \bar{x} = (f_1 m\phi_1 + f_2 m\phi_2 \ldots + f_n m\phi_n)/100$, i.e. $\dfrac{\sum fm\phi}{100}$

Standard deviation $= \sigma = \sqrt{\dfrac{f_1(m\phi_1 - \bar{x})^2 + f_2(m\phi_2 - \bar{x})^2 \ldots + f_n(m\phi_n - \bar{x})^2}{100}}$, i.e. $\sqrt{\dfrac{\sum f(m\phi - \bar{x})^2}{100}}$
(second moment)

Moment coefficient of skewness $= \alpha_3 = \dfrac{\sum f(m\phi - \bar{x})^3}{100\sigma^3}$

Mean-cubed deviation $= \alpha_3\sigma^3 = \dfrac{\sum f(m\phi - \bar{x})^3}{100}$

For the *method of moments,* grain-size parameters are calculated from values of f: the percentage fraction in each class interval and $m\phi$: the mid-point value of each class interval in phi units, using the formulae of Table 2.3. the first moment gives the mean of the distribution ($\bar{x}$); the second moment gives the standard deviation, i.e. the sorting (σ), and the third moment gives the skewness (α_3). An additional parameter, the mean-cubed deviation ($\alpha_3\sigma^3$) is also frequently calculated.

In Fig. 2.2, grain-size parameters are derived from a graphic presentation of the tabulated data and from calculations using moment measures.

Interpretation and use of grain-size analyses

Grain-size analyses can be used to distinguish between sediments of different environments and facies and to give information on the depositional processes and flow conditions (see Middleton, 1976 for example). Many studies have attempted to distinguish between the sediments of modern depositional environments using the grain-size distribution. Scatter diagrams are usually constructed, such as sorting plotted against skewness. On this basis, it has been possible to distinguish between beach, dune and river sands (see Friedman, 1979 for a review). Grain-size analyses of sandstones alone should not be used in environmental interpretations, but combined with studies of sedimentary structures they can be most useful in facies description and analysis. Points to bear in mind though, are the possibilities of sand being reworked or supplied from adjacent or pre-existing environment, when there is a problem of inherited

16

characteristics, and the origin of the clay-grade material. The latter may well be depositional and so of consequence to the environmental interpretation, but fine-grained matrix can also be infiltrated, a breakdown product of labile grains and a diagenetic precipitate. To complicate the interpretation of grain-size analyses in terms of depositional process, several different processes may well have operated in one environment, and similar processes do take place in different environments.

2.2.2 GRAIN MORPHOLOGY

Three aspects of grain morphology are the shape, sphericity and roundness. The shape or form of a grain is measured by various ratios of the long, intermediate and short axes and descriptive terms for the four classes based on these ratios are oblate (tabular or disc-shaped), equant (cubic or spherical), bladed and prolate (rod-shaped) (Fig. 2.4). Sphericity is a measure of how closely the grain shape approaches that of a sphere. Roundness is concerned

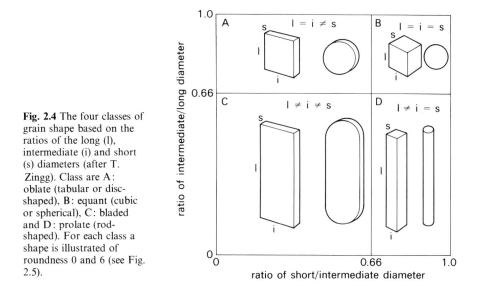

Fig. 2.4 The four classes of grain shape based on the ratios of the long (l), intermediate (i) and short (s) diameters (after T. Zingg). Class are A: oblate (tabular or disc-shaped), B: equant (cubic or spherical), C: bladed and D: prolate (rod-shaped). For each class a shape is illustrated of roundness 0 and 6 (see Fig. 2.5).

with the curvature of the corners of a grain and six classes from very angular to well rounded are usually distinguished (Fig. 2.5). Several formulae have been proposed for the calculation of sphericity and roundness (see for example Sneed & Folk, 1958; Dobkins & Folk, 1970). For environmental interpretations, the roundness measures are more significant than sphericity or form, and for most purposes the simple descriptive terms of Fig. 2.5 for roundness are sufficient.

The morphology of a grain is dependent on many factors; initially the mineralogy, nature of the source rock and degree of weathering, then on degree of abrasion during transport, and later on corrosion or solution during diagenesis. In a very general way, the degree of roundness increases with the duration of transportation or reworking. Beach and desert sands for example are typically better rounded than river or glacial outwash sands. As with grain-

17

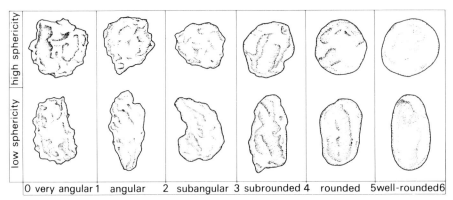

| 0 very angular | 1 angular | 2 subangular | 3 subrounded | 4 rounded | 5 well-rounded 6 |

Fig. 2.5 Categories of roundness for sediment grains. For each category a grain of low and high sphericity is shown (after Pettijohn *et al.*, 1973).

size analyses, care must be exercised in interpreting roundness values; the roundness characteristic of grains may be inherited and intense abrasion may lead to fracturing and angular grains.

2.2.3 GRAIN SURFACE TEXTURE

The surface of sand and coarser grains often has a distinct appearance. The dull, frosted nature of desert sand is an oft-cited example. With pebbles from a glacial deposit, striations are common on the surface. Crescentic impact marks occur on pebbles from beaches and river channels. Studies of the surface texture of modern sands with the scanning electron microscope have shown that there are features which may be environmentally diagnostic. Beach sand grains often show minute V-shaped percussion marks, the surfaces of desert sands possess 'upturned plates' and sands from glacial deposits have conchoidal fracture patterns and striations (Fig. 2.6; and Krinsley & Doornkamp, 1973). The surface texture of grains in sandstones must be interpreted with great caution since diagenetic processes can considerably modify the surface.

2.2.4 GRAIN FABRIC

The term fabric for grains in a sedimentary rock refers to their orientation and packing, and to the nature of the boundaries between them (Fig. 2.7). In many sandstones and conglomerates the sand grains and pebbles are aligned with their long axes in the same direction. This preferred orientation is a primary fabric of the rock (unless the rock has been tectonically deformed) and it is produced by the interaction of the flowing depositional medium (wind, ice, water) and the sediment.

Prolate pebbles in fluviatile and other water-lain deposits can be oriented both normal and parallel to the current direction. The normal-to-current orientation is produced by rolling of pebbles, while the parallel orientation arises from a sliding motion. In glacial sediments, the orientation of clasts is more commonly parallel to the direction of ice movement. One common fabric of oblate pebbles in water-lain deposits is *imbrication,* where the pebbles

18

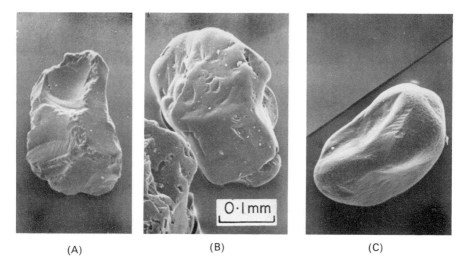

(A)   (B)   (C)

Fig. 2.6 Scanning electron micrographs of quartz sand grains from three modern environments. 2.6A Grain from glacial outwash deposit, Ottawa, Canada, showing conchoidal fractures and angular shape. 2.6B Grain from high energy beach, Sierra Leone, West Africa, showing rounded shape and smooth surface with small v-shaped percussion marks. 2.6C grain from desert sand sea, Saudi Arabia, showing frosted, pock-marked surface (due to 'upturned plates' which are visible at higher magnifications) and conchoidal fractures due to mechanical chipping.

overlap each other, dipping in an upstream direction. Elongate sand grains of sandstones can also show both parallel-to-current and normal-to-current preferred orientations, although in many sandstones the parallel orientation seems more prevalent (in turbidites for example, Parkash & Middleton, 1970). Grain orientation can be used as a palaeocurrent indicator (e.g. Shelton & Mack, 1970), particularly if sedimentary structures are poorly developed (see Section 2.4). Apart from detrital grains and pebbles, other components such as plant debris and fossils can show preferred orientations.

The *packing* of sediment grains is an important consideration since it affects porosity and permeability. Packing is largely dependent on the grain size, shape and sorting. Modern beach and dune sands, composed of well-sorted and rounded grains, have porosities from 25 to 65%. Where porosities are high, packing is loose and approaches the cubic packing of spheres (Fig. 2.7); lower porosities result from tighter packing in a rhombohedral arrangement. Poorly-sorted sediments have a closer packing and thus a lower porosity through the greater range of grain size and the filling of pore space between large grains by finer grains.

The common types of *grain contact* are: point contacts where the grains are just touching each other (giving the sediment a grain-supported fabric), concavo-convex contacts where one grain has penetrated another, and sutured contacts where there is a mutual stylolitic interpenetration of grains (Fig. 2.7). In addition where there is a lot of matrix, grains may not be in contact, but 'float' in the matrix (that is the grains are matrix-supported).

The fabric of sediments is another aspect of sediment texture which can give useful information on the depositional process. With conglomerates for example, pebbles floating in a matrix is a common feature of mudflows, glacial

19

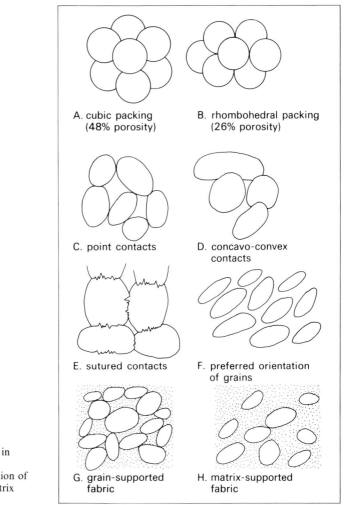

Fig. 2.7 Grain fabric in sediments: packing, contacts and orientation of grains, and grain-matrix relationships.

A. cubic packing (48% porosity)

B. rhombohedral packing (26% porosity)

C. point contacts

D. concavo-convex contacts

E. sutured contacts

F. preferred orientation of grains

G. grain-supported fabric

H. matrix-supported fabric

deposits (Fig. 2.8) and certain deepwater deposits such as debris flows. These contrast with conglomerates of river channel, stream flood and other more fluid, aqueous flows, where pebbles are in contact (clast-support) and there may be little matrix (Fig. 2.9). As with many aspects of sediment texture, fabric studies should be undertaken in conjunction with the sedimentary structures.

2.2.5   TEXTURAL MATURITY

Texturally immature sediments are those with much matrix, a poor sorting and angular grains; mature sediments are those where there is little matrix, moderate to good sorting, and subrounded to rounded grains; texturally supermature sandstones are those with no matrix, very good sorting and well-rounded grains. Primary porosity and permeability increase with increasing textural maturity, since the more mature the sediment is, the less matrix and more pore space it possesses.

20

Fig. 2.8 Matrix-support fabric; pebbles 'float' in matrix. Grain-size distribution in this polymictic conglomerate is polymodal. Tillite, Late Precambrian, northern Norway. Courtesy of T.C. Pharoah.

Fig. 2.9 Clast-support fabric; pebbles, mainly quartzite, are in contact. Late Precambrian, southern Norway.

Textural maturity in sandstones is largely a reflection of the depositional process, although it can be modified by diagenetic processes (Section 2.8.5). Where there has been minimal current activity, the sediments are generally texturally immature; persistent current or wind activity results in a more mature sandstone. Examples of texturally immature sediments include many fluviatile and glacial deposits; supermature sediments are typified by desert, beach and shallow marine sandstones.

2.3   Sedimentary structures

Sedimentary structures are the larger scale features of sedimentary rocks and include the familiar cross bedding, ripples, flute and load casts, dinosaur footprints and worm burrows. The majority of structures form by physical processes, before, during and after sedimentation, while others result from organic and chemical processes. Sedimentary structures, particularly those formed during sedimentation, have a variety of uses: for interpreting the depositional environment in terms of processes, water depth, wind strength etc.; for determining the way-up of a rock sequence in an area of complex folding, and for deducing the palaeocurrent pattern and palaeogeography. Many of the structures are on the scale of tenths to tens of metres and so are studied, recorded and measured in the field. In recent years, much experimental and theoretical work has been undertaken on the development of sedimentary structures, particularly those formed by the interaction of water and sediment. These experiments have been seeking the conditions of formation of the structures and they have allowed more meaningful interpretations to be made of these structures in the geological record.

Although there is no generally accepted classification of sedimentary structures, the four main groups are (i) erosional, (ii) depositional, (iii) post-depositional and (iv) biogenic structures.

Numerous photographs of sedimentary structures are given in Pettijohn & Potter (1964) and Conybeare & Crook (1968).

2.3.1   EROSIONAL SEDIMENTARY STRUCTURES

These structures are formed through erosion by aqueous and sediment-laden flows before deposition of the overlying bed, and by objects in transport striking the sediment surface. The most familiar are the sole structures, flute and groove casts, which occur on the underside of many beds deposited by turbidity currents; other structures of this group are scours and channels.

Flute marks have a distinctive appearance (Fig. 2.10) often described as spatulate or heel-shaped, consisting of a rounded or bulbous upstream end, which flares downstream and merges into the bedding plane. In section they are asymmetric, with the deepest part at the upstream end. They average 5 to 10 cm across and 10 to 20 cm in length, and occur in groups, all with a similar orientation and size. The formation of flutes is attributed to localized erosion by sand-laden currents, passing over a cohesive mud surface. The flute develops through a process of flow separation, whereby the current leaves the sediment surface at the upstream rim of the flute and a small eddy or roller, rotating in a horizontal plane, is trapped within the flute (Allen, 1971). While the current is moving swiftly the eddy keeps sediment out of the flute; deceleration of the current causes sediment to be deposited which infills the flute.

Flute marks are a characteristic structure of turbidites (Section 2.10.7 and Dzulynski & Walton, 1965; Pett & Walker, 1971) and they give a reliable indication of the flow direction.

Groove marks are linear ridges on the soles of sandstone beds which formed

Fig. 2.10 Flute marks on sole of turbidite. Flow from botton to top. Silurian, Southern Uplands, Scotland.

22

by the infilling of a groove cut into the underlying mudrock (Fig. 2.11). Groove marks may occur singly or many may be present on one undersurface, all parallel or deviating somewhat in orientation. It is generally held that grooves are formed through some tool, a fossil or mud clast for example, which was carried along by the current, gouging the groove into the mud. In rare cases the tool has been found at the end of a groove. Groove marks are common on the soles of turbidite beds but they can form elsewhere, on floodplains when a river breaks its banks for example, and on shallow marine shelves when there is a storm surge. Groove casts are a useful palaeocurrent indicator.

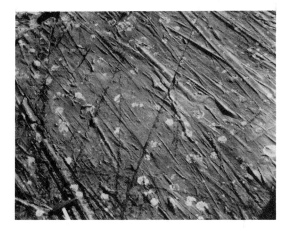

Fig. 2.11 Groove marks on sole of turbidite. Note that the orientation of the grooves varies through about 30°. Upper Carboniferous, Devon, England. Courtesy of T.C. Pharoah.

Impact marks are produced by objects striking the sediment surface as they are carried along by a current. Various types have been distinguished including prod, bounce, brush, skip and roll marks, depending on how the impact is thought to have taken place (Dzulynski & Walton, 1965). Objects making the marks are chiefly fossils and pebbles or lumps of sediment. Impact marks are common in turbidite sequences and tend to occur in more distal situations, where current flow has waned.

Channels and scours are found in sediments of practically all environments. Channel structures are generally on the scale of metres, in some cases kilometres, whereas scours are smaller scale erosional features, occurring within or on the bases of beds. They can both be recognized by the cutting of bedding planes and laminations in underlying sediments (Fig. 2.12). Scours are local structures, generally oval to elongate in plan view, with a smooth to irregular, concave-up shape in vertical section. Slightly coarser sediment or even pebbles may occur within the scours. Scour structures represent short-lived erosion events. Channels are more organized structures than scours and were often the pathways for sediment and water movement over considerable lengths of time. The larger channels can sometimes be mapped out on a regional scale, giving useful palaeogeographic information (e.g. Bluck & Kelling, 1963). Channels are generally infilled with coarser sediments than those beneath or laterally equivalent and frequently a thin lag deposit of pebbles and intraformational clasts is present in the base of the channel. Channels develop in many environments; those in fluviatile settings are especially well-known, but they also occur in glacial, deltaic, tidal flat, shelf-margin and slope and sub-

23

Fig. 2.12 Large channel structure (a major delta distributory channel), infilled with large-scale cross-bedded sandstone. The channel cuts down into horizontally bedded thin sandstones and shales which form a coarsening-upwards unit. The latter was generated through delta front progradation over a carbonate shelf deposit (a marine limestone at the base of this cycle). See Section 2.10.4. Upper Carboniferous, Northumberland, England.

marine fan locations. The margin of a large delta distributary channel is shown in Fig. 2.12.

2.3.2 DEPOSITIONAL SEDIMENTARY STRUCTURES

Sediment transport and aqueous flows

Much sediment is moved and transported in response to aqueous flows, such as the flow of a river, a tidal or storm current. A wide range of sedimentary structures arises from aqueous flow-sediment interaction (discussed below). Sediment is also transported by the action of waves and by the wind, to produce characteristic structures. In addition, sediment is moved downslope under the direct action of gravity, in mass-gravity transport; the most important types are sediment-gravity flows (sediment-fluid mixtures), in particular turbidity currents and debris flows (Section 2.10.7). Sediment is also transported by glaciers (Section 2.10.8).

Sediment in both water and air is transported either in suspension or as bed load. Suspended sediment is kept in suspension by the fluid turbulence. Coarser sediment is generally moved as bed load along the sediment surface by saltation (in short jumps) or by rolling and sliding.

To describe the strength of aqueous flows the mean flow velocity ($\bar{U}$) is most commonly used, but also the bed shear stress, denoted by τ_0, which is the average force per unit area exerted by the flow on the sediment surface. Flow depth (h) is a useful descriptive parameter, but less important than flow strength in terms of the sedimentary structures formed. Two basic types of aqueous flow are *laminar flow*, where layers or particles of flowing water move smoothly over one another with little diffusion or exchange, and *turbulent flow*, which develops at higher flow velocities and where eddies cause much diffusion; there are transverse components of motion and water moves as packets. Flow can be characterized in several ways. The *Reynolds number* is a ratio of inertial to viscous forces, given by $\bar{U}L/V$, where L is a characteristic length for the flow system and V is the kinematic viscosity. With any aqueous flow, the change from laminar to turbulent flow takes place at a critical Reynolds number. *Froude number* is a ratio of inertial to gravity forces and is given by $F = \bar{U}/\sqrt{gh}$

24

with g = acceleration due to gravity. A flow with Froude number less than unity is referred to as subcritical or tranquil flow; with Froude numbers greater than one the flow is referred to as supercritical or rapid flow. *Stream power* is given by $\bar{U}\tau_0$; the sediment transport rate is a function of the stream power. For aqueous flows, the relationship between grain size and the current velocity for sediment movement (critical erosion velocity) is approximately given by Hjulstrom's diagram (Fig. 2.13), while the relationship between grain size and bed shear stress for sediment movement is given in Fig. 2.14, based on the work of A. Shields. From Fig. 2.13, it can be seen that fine sand grades are more easily eroded than finer or coarser grades but once eroded, finer sediments are

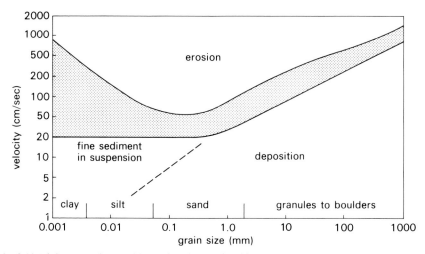

Fig. 2.13 Hjulstrom's diagram illustrating the relationship between grain size and the current velocity required for sediment movement (the critical erosion velocity). This relationship was deduced experimentally from flows of 1 metre depth. The stippled area indicates the scatter of the data and the increased width of this area at the finer grain sizes reflects the affects of sediment consolidation (higher velocities required if sediment more consolidated).

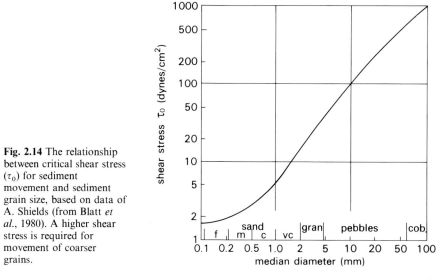

Fig. 2.14 The relationship between critical shear stress (τ_0) for sediment movement and sediment grain size, based on data of A. Shields (from Blatt *et al.*, 1980). A higher shear stress is required for movement of coarser grains.

25

held in suspension until the flow velocity is minimal. The higher erosion velocities required for silt and clay are a reflection of the cohesive forces between particles which have to be overcome. Detailed accounts of the mechanics of flow and the movement of grains in water, beyond the scope of this book, are given in Allen (1977), Friedman & Sanders (1978) and Blatt *et al.* (1980).

Once sediment is moving, then the nature of the sediment surface with its sedimentary structures or *bed forms* as they are called, is dependent on the flow conditions. Laboratory studies have shown that there is a definite sequence of bed forms (Fig. 2.15), related to increasing flow strength and grain size (see for example Simons *et al.,* 1965; Allen, 1968; Harms *et al.,* 1975). For some grain

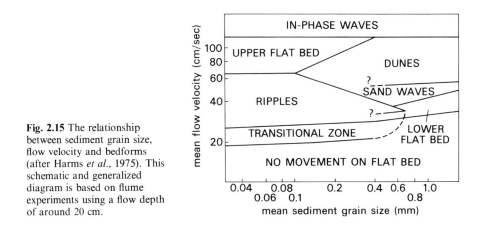

Fig. 2.15 The relationship between sediment grain size, flow velocity and bedforms (after Harms *et al.,* 1975). This schematic and generalized diagram is based on flume experiments using a flow depth of around 20 cm.

sizes and flow velocities there are overlaps of the bed form fields and transitional zones. For sediments finer than about 0.1 mm (down to about 0.03 mm), with increasing flow strength, an initially flat bed with no sediment movement, gives way to ripples and then to an upper flat bed. For sands of mean size from about 0.1 to 0.6 mm, the sequence is no movement, ripples, sand waves, dunes, upper flat bed. For sands coarser than about 0.6 mm, the sequence is no movement, lower flat bed, sand waves, dunes, upper flat bed. The characteristic feature of the upper flat bed is primary current lineation: low, flow-parallel ridges. At higher flow strengths still, in-phase waves develop, that is, sinusoidal undulations of the bed and antidunes, which are in phase with stationary or nearly stationary water-surface waves. Flow conditions under which ripples, sand waves and dunes are formed are referred to as *lower flow regime*; upper flat bed and antidunes form in the *upper flow regime*. The features of these bed forms and the internal sedimentary structures they give rise to are discussed in succeeding sections.

Bedding and lamination

The characteristic feature of sedimentary rocks, their stratification or bedding, is produced by changes in the pattern of sedimentation, usually changes in

26

sediment composition and/or grain size. Bedding is generally defined as a sedimentary layering thicker than 1 cm. Finer scale layering, only millimetres thick, is termed lamination. Lamination is frequently an internal structure of a bed (e.g. Fig. 2.16). The majority of beds have been deposited over a period of time ranging from hours or days, as in the case of turbidites and storm beds, to years, tens of years or even longer, as in the case of many marine shelf sandstones and limestones. Bedding planes themselves may represent considerable periods of time when there was little deposition. Sediment may well have been moved around during these intervals, but not buried. Subsequent modifications to bedding planes result from erosion, as the succeeding bed is deposited, deformation through loading and compaction, and solution due to overburden pressure. Tectonic events may also affect bedding planes.

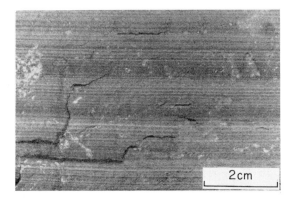

Fig. 2.16 Part of a bed showing well-developed lamination. Silt-grade quartz laminae alternate with clay and organic-rich laminae. Caithness Flagstones (lacustrine), Middle Devonian, N.E. Scotland.

Laminations arise from changes in grain size between laminae, size-grading within laminae, or changes in composition between laminae (e.g. Fig. 2.16). In many cases, each lamina is the result of a single depositional event. Although time-wise, this is often instantaneous, it can be much longer. Parallel lamination, also termed planar, flat or horizontal lamination, can be produced in several ways. In fine sands, silts and clays it is frequently formed through deposition from suspension, slow-moving sediment clouds or low density suspension currents. Such laminae occur in the upper parts of turbidite beds (Section 2.10.7) and in varves and rhythmically-laminated deposits of glacial and non-glacial lakes (Section 3.5). Laminae can also be formed by chemical precipitation, as in subaqueous evaporites (Section 5.2.2), and by phytoplankton blooms (Section 8.8).

Parallel lamination in sand-grade sediment, also termed flat bedding, is chiefly formed by turbulent flows at high flow velocities (see Fig. 2.15). The characteristic feature of this upper flat bed phase is the presence of *parting lineation* or *primary current lineation* on the surface of the laminae (Fig. 2.17). It consists of very low ridges, only a few grain diameters high, and hollows, which give the sandstones a visible fabric. The lamination results from turbulent eddies immediately adjacent to the sediment surface (Allen, 1964). In sands with a mean grain size coarser than 0.6 mm diameter, a flat bedding can be produced by bed load transport at low flow velocities (i.e. lower flat bed phase, Fig. 2.15).

Fig. 2.17 Parting lineation, also called primary current lineation, on the bedding surface of a parallel-laminated fluviatile sandstone. Carboniferous, N.E. England.

5 cm

Current ripples, sand waves, dunes and cross stratification

Current ripples, sand waves and dunes are downstream-migrating bed forms produced by unidirectional aqueous flows. Their formation depends on the flow strength and the sediment grain size (see Fig. 2.15 and earlier section). They are common in rivers, deltas and on shallow marine shelves, and in the geological record current ripples are commonly preserved, but sand waves and dunes rarely so. Ripples are small-scale bed forms with wavelengths (or spacings) of less than a few 10's of cm and heights of less than several cm. In profile they are asymmetric with a steeper, downstream-facing lee side and a gentle upstream-facing stoss side (Fig. 2.18). Ripples can be described by the wavelength to height ratio, referred to as the ripple index (Tanner, 1967). For current ripples this ranges between 8 and 20. Dunes (also called megaripples) are larger scale structures with wavelengths of 1 metre or more and heights of several 10's of centimetres. Dunes have a similar triangular profile and index to that of ripples. The shape of ripples and dunes is described as two-dimensional if the crests are straight, or three-dimensional, if the crests are sinuous, catenary, lunate or linguoid (Fig. 2.19). The shape of ripples and dunes is related to flow strength; with increasing flow velocity, ripples show the sequence straight-crested, sinuous, linguoid; for dunes the sequence is straight-crested, sinuous, catenary, lunate (Allen, 1968).

Sand waves are low, straight to sinuous-crested bed forms with well-defined lee faces and stoss sides which are either flat or covered with ripples. Sand waves

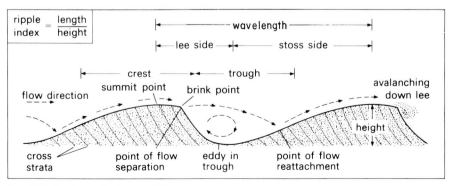

Fig. 2.18 Ripple terminology and flow pattern.

28

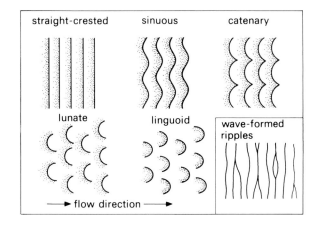

Fig. 2.19 Terminology for the shape of the crests of ripples and dunes formed by unidirectional currents. For comparison, the typical crest pattern of wave-formed ripples is also shown.

lunate   linguoid   wave-formed ripples

→ flow direction →

have wavelengths up to 100 m or more and the index is much higher than ripples or dunes. Sand waves in rivers are often referred to as a type of bar.

Ripples, sand waves and dunes migrate downstream through sediment being eroded from the stoss-side and carried to the crest from which it avalanches down the lee slope (Fig. 2.18). As a consequence of the stream flow over the bed form, flow separation takes place at the brink and an eddy develops within the trough. The backflow of the eddy may cause reworking of sediment at the toe of the lee slope, and in the case of large dunes, upstream directed ripples (backflow ripples) may develop. Also with large dunes, downstream-directed ripples may develop in the troughs downstream of the point of flow reattachment.

The downstream migration of ripples, sand waves and dunes under conditions of net sedimentation gives rise to *cross stratification,* a structure in the past loosely called current bedding or false bedding (Fig. 2.20). The cross strata, referred to as foresets, represent the former position of the ripple, dune or sand wave lee face. Two basic types of cross stratification, shown in Fig. 2.21, are (i) tabular cross strata, produced by two-dimensional bed forms and (ii) trough cross strata produced by three-dimensional structures. An individual bed of cross strata is termed a set; a group of similar sets is a coset.

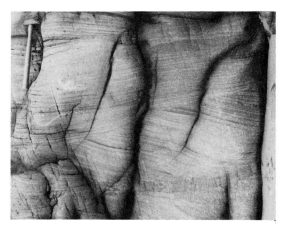

Fig. 2.20 Trough cross bedding; several sets of cross beds with tangential bases, formed through the migration of lunate dunes under conditions of net sedimentation in a fluvial channel. Old Red Sandstone, Devonian, Dyfed, Wales.

29

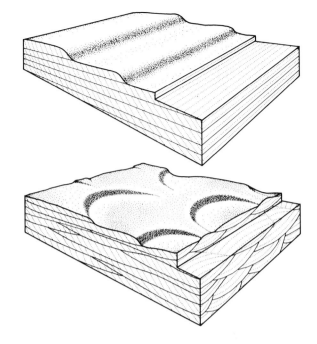

Fig. 2.21 Block diagrams showing the two common types of cross stratification: tabular and trough. Above: tabular cross stratification, chiefly formed through the migration of straight-crested (i.e. 2-dimensional) ripples, producing tabular cross lamination, and sandwaves, producing tabular cross bedding. Below: trough cross stratification, chiefly formed through migration of 3-dimensional bedforms, especially lunate and sinuous dunes, producing trough cross bedding (illustrated here) and linguoid ripples, producing trough cross lamination.

Tabular cross bedding generally consists of planar beds which have an angular contact with the basal surface of the set and an angle of dip of 30° or more. Sets range from a few decimetres to a metre or more in thickness. Tabular cross bedding is mostly deposited by migrating sand waves. Thicker tabular sets can be formed by the downstream migration of bars in river channels and by the growth of small deltas. In trough cross bedding (Fig. 2.20), the trough-shaped sets consist of scoop-shaped beds, with tangential bases, and dip angles reaching 25–30°. Trough cross bedding mostly forms through deposition by dunes, particularly sinuous and lunate types.

Tabular and trough cross lamination, with set heights of less than several cm, is mainly produced by straight-crested and linguoid ripples respectively. Where there is net deposition ripples build up as well as forward, so that a ripple 'climbs' up the stoss-side of the one downstream. This produces climbing-ripple cross lamination (also called ripple drift) and two types can be recognized. The most common is where the stoss-side of the ripples is represented by an erosion surface between cross-laminated sets. If there is very rapid sedimentation from suspension, then laminae can be formed on the stoss-side continuous with the foreset laminae (Fig. 2.22).

Where mud deposition is intermittent with ripple migration, thin streaks of mud occur between sets of cross lamination and mud is concentrated in ripple troughs. This type of structure is known as *flaser bedding,* while the term *lenticular bedding* is applied to isolated ripples, seen in section as cross-laminated lenticles, within mud or mudstone (Fig. 2.23). Both types are common in tidal flat sediments (Reineck & Singh, 1973) where mud is deposited during slack water periods; they also occur in delta front, prodelta and other situations where there are fluctuations in sediment supply and flow strength.

Cross bedding formed through the movement of dunes in a tidal regime is

30

Fig. 2.22 Two thin siliciclastic turbidite beds showing: sharp scoured base with convoluted and loaded sandstone balls and flame structures, and ripple cross lamination with preservation of stoss-side laminae (laminae dipping at low angle to right, current flow from right). Both beds show graded bedding, from fine sand at base to silt and clay at top. Upper Carboniferous, Devon, England.

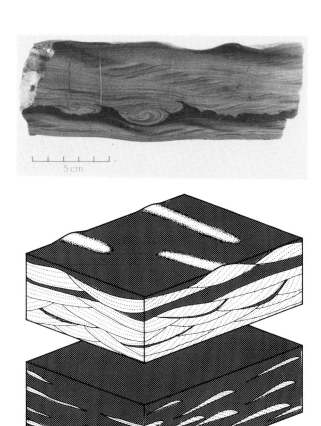

Fig. 2.23 Flaser bedding (above) and lenticular bedding (below).

occasionally 'herring-bone' in character (Fig. 2.24). The two opposing directions of sediment transport are reflections of the ebb and flood tidal currents.

One particular type of large-scale cross bedding known as epsilon cross bedding or lateral accretion surfaces (shown in Fig. 2.53) is formed by the lateral migration of point bars developed on the inside of meandering river and tidal channels. These surfaces are oriented normal to the current direction and

Fig. 2.24 Herring-bone cross bedding formed through tidal current reversals, in shallow-marine bioclastic limestone. Upper Jurassic, Dorset, England.

31

usually separate units of medium or small-scale cross stratification, formed by dunes and ripples migrating downstream on the point bar during the latter's episodic lateral movement. Examples of epsilon cross bedding have been described by Allen (1965a), Leeder (1973) and Elliott (1976).

Cross bedding can also occur in conglomerates. Many deposited in braided stream systems for example show a large-scale cross bedding produced by the downstream movement of gravel bars.

Antidunes and antidune cross bedding

Antidunes are low, undulating bed forms which develop at high flow strengths. A low-angle cross bedding, directed upstream, is developed through erosion on the downstream side of the structure, and deposition on the upstream-facing (stoss) side. Antidune cross bedding is rarely preserved, but it has been described from Ordovician turbidites of Quebec (Skipper & Bhattacharja, 1978). It can develop in base-surge tuffs (Sect. 10.3).

Wave-formed ripples and cross lamination

Wave-formed ripples are common in many shallow marine to intertidal, deltaic and lacustrine sediments, both sandstones and limestones. These ripples are characterized by a symmetrical profile and continuous straight-crests (Fig. 2.25). However, many are asymmetric, although these tend to have more

Fig. 2.25 Wave-formed ripples: ripple profiles are symmetrical and crests bifurcate. Lacustrine sandstone, Upper Triassic; Glamorgan, Wales.

persistent crestlines than current-produced forms. Wave-ripple crests are often rather pointed compared with the rounded troughs. Crests commonly bifurcate and this feature serves to distinguish wave-formed ripples from current ripples. The ripple index for wave-formed ripples (6–10) is lower than that for current ripples (8–20) in sediment of similar grain size. The wavelength of wave-formed ripples depends on the sediment grain size, diameter of water particle orbits in the waves and the water depth. Waves are able to move sediment at depths less than half the wavelength, a depth referred to as wave base.

The internal structure of wave-formed ripples is variable and often the laminae are not concordant with the ripple profile. Wave-ripple cross lamination can consist of a chevron to undulating pattern of laminae, or unidirectional foresets (Newton, 1968; Raaf et al., 1977).

The movement of sand by wind mainly occurs through saltation and surface creep of grains. Finer sediment is transported in suspension. Wind ripples are generally straight-crested asymmetrical forms with some crest bifurcation. Ripple wavelength and height depends on the grain size and wind strength, in particular on the length of the saltation path of the moving sand grains. Compared to current and wave ripples, the ripple index is generally much higher. The most important aeolian bed form is the dune, with a wavelength on the scale of 10's to 100's of metres and height of several metres. The two common types are the barchan, a lunate dune, and the seif, an elongated sand ridge. Others are parabolic, star-shaped and transverse dunes. Similar structures but on a larger scale, with wavelengths of kilometres and heights of 10's of metres are collectively referred to as draas (Wilson, 1972). The factors controlling the type of dune (or draa) are wind strength and pattern, the grain size and supply, and the nature of the underlying surface. In an analogous manner to subaqueous dunes, sand is moved up wind-facing slopes and on to the top of lee surfaces from where it avalanches down the slopes when the angle of repose (35°) is exceeded. It is often the case that on average the angles of dip of aeolian cross beds, 25–35°, are steeper than those of subaqueous origin. The set thickness in aeolian cross beds, commonly several metres thick, but reaching 30 m or more, is generally greater than that of subaqueous dunes and sand waves.

Movement of barchan dunes under conditions of net sedimentation gives rise to a large-scale trough cross bedding, dipping down-wind (Figs. 2.26, 2.27). Simple seif dunes on the other hand, generally elongated parallel to the wind direction, have an internal structure of cross beds dipping in opposite directions (Fig. 2.27). Complex seifs, larger-scale ridges with small parasitic dunes upon them, have a complex internal structure. Details of aeolian dunes and internal structures are given by McKee (1966), Glennie (1970), Biggarella (1972), Reineck & Singh (1973) and Hunter (1977). The physics of grain movement by wind has been studied by Bagnold (1941) and reviewed in Allen (1977).

Fig. 2.26 Aeolian cross bedding, characterized by greater set thickness and maximum angle of dip compared with cross bedding of subaqueous origin. The cross bedded set which forms most of the cliff is nearly 6 m high. Lower Permian, Cumbria, England.

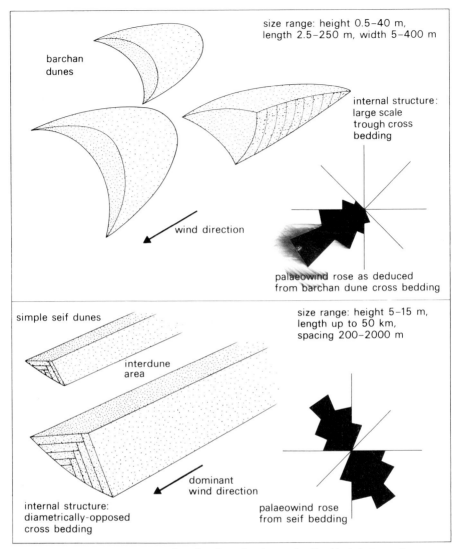

Fig. 2.27 The two principal types of aeolian dune, barchan and seif, with their internal structure and palaeowind pattern.

Graded bedding develops as a response to changes in flow conditions during sedimentation. Many different types of graded bedding have been described but two types of normal grading are (i) a gradual decrease in the whole grain size up through the bed, and (ii) a gradual upward decrease in size of the coarsest grains. Multiple grading, where there are several graded sub-units in a bed, also occurs. Coarse to fine graded bedding is generated through waning flow; thus graded bedding is a characteristic feature of turbidites deposited from decelerating density currents (Fig. 2.22; Kuenen & Migliorini, 1950). Decelerating flows giving graded beds also occur on floodplains, tidal flats and shallow marine platforms following storms.

Reverse or inverse grading, where grain size increases up through a bed, is not common. It occurs in beach lamination, through accelerating wave backwash, in certain deepwater sediment gravity flow deposits (Section 2.10.7) and in some volcaniclastic deposits (Sections 10.2, 10.3).

Massive bedding refers to beds without any apparent internal structure. Although many sedimentary rocks can appear massive or structureless, closer inspection may show that this is not the case. If the massive bedding is real, it could have formed through rapid deposition, a dumping of sediment as by grain flows and other re-sedimentation processes (Section 2.10.7). Alternatively, original internal depositional structures could have been destroyed by bioturbation, dewatering and recrystallization-replacement.

Other common depositional structures are mudcracks and rain spots. Mudcracks are common in fine grained sediments, which, if formed through desiccation, indicate subaerial exposure. Desiccation cracks are typically polygonal, with sharply-defined cracks. Cracks may also develop subaqueously as a result of synersis: contraction of the sediment through loss of water. It is thought that slight changes in water chemistry are responsible for this. Synersis cracks are common in lacustrine sediments (e.g. Donovan & Foster, 1972).

Rain spots are occasionally found in mudrocks of continental and shoreline environments.

2.3.3 POST-DEPOSITIONAL SEDIMENTARY STRUCTURES

In this group are included slump folds, convoluted bedding, load casts, sandstone dykes and dewatering structures.

Slump structures. Sediments deposited on slight to significant slopes are prone to slumping. Downslope movement can take place on a large or small scale, involving individual beds or thick packets of strata. Crumpling and folding of such beds are common (Fig. 2.28) and wholesale brecciation can result. The majority of slumps are initiated through earthquake shock. Slumped beds have been described by Helwig (1970) and Woodcock (1976).

Convoluted or contorted bedding develops in cross and planar laminated strata and consists of regular to irregular folds and contortions (Fig. 2.29). It is usually only the uppermost part of a bed which is affected, and the convolutions may be planed off, demonstrating their syn-sedimentary origin. The origin of convolute lamination is not fully understood, but likely causes are differential liquefaction and lateral intrastratal flow (dewatering processes, see below), and shearing of the sediment surface by currents. Convoluted

Fig. 2.28 Slump folds and thrusts formed through downslope mass movement of sediment. The bedding plane upon which the movement took place is clearly picked out. Note that the bedding is perfectly horizontal and undisturbed below this plane. Pleistocene, Israel. Courtesy of R.M. Tucker.

35

Fig. 2.29 Convoluted and deformed laminations in thin siliciclastic turbidite. Ordovician, Girvan, S.W. Scotland.

laminations are common in turbidite beds, but they also occur in fluviatile, tidal flat and other sediments.

Related to convolute bedding is *overturned cross bedding* where the upper parts of foresets are turned over, invariably in a consistent downstream direction. This has been attributed to frictional drag arising from the passage of a mass of saturated sand over the surface of the cross-bedded sand (e.g. Hendry & Stauffer, 1975).

Load casts are a common sole structure, seen as bulbous, downward-directed protuberances of a sandstone bed into underlying sediment, normally a mudrock (Fig. 2.30). Load casts show considerable variation in shape and size, and one common feature is the squeezing of mud up into the sand to form flame structures (Fig. 2.22). Lobes of sediment may become detached from the sand horizon to form load balls (*pseudonodules*). The structures are formed as a result of a vertical density contrast of more dense sand overlying less dense mud, so that the sand sinks down into the mud.

Related to load casts are ball-and-pillow structures where a sand bed lying within mudstone has broken up into pillow-shaped masses, still partly connected or free floating in the mud. It is thought that with many ball-and-pillow structures an earthquake shock was the triggering mechanism. High sedimentation rates also favour the development of this and the other deformation structures.

A number of structures in sandstones result from *dewatering*, the sudden loss of porewater causing the sediment to lose its strength. The two chief

Fig. 2.30 Large load casts on the underside of fluviatile sandstone; the mudrock of the underlying bed is visible between the loads. Late Precambrian, southern Norway.

36

processes are fluidization, whereby upward moving water produces a fluid drag on the grains, and liquefaction, whereby particles are shaken loose from each other through some applied stress, such as that associated with earthquakes. In sandstones, dewatering can give rise to a whole range of water escape structures. These include disruptions and contortions of bedding (as noted above); sandstone dykes cutting across primary structures and, if reaching the surface, forming a sand volcano; and dish structures, consisting of thin, concave-up laminae, often separated by pillars (Lowe, 1975). Water-escape structures have been described from sediments of most environments; recently described cases include Brenchley & Newall (1977), Johnson (1977) and Hiscott (1979).

2.3.4 BIOGENIC SEDIMENTARY STRUCTURES

The sedimentary structures formed by organisms are the trace fossils, also referred to as ichnofossils or lebenspuren (from the German). Trace fossils vary from discrete, well-organized structures which can be attributed to a particular organism and/or activity and are given Latin names, to vague bioturbation structures disrupting and even destroying primary features such as lamination and bedding (Fig. 2.31). With many trace fossils it is not known which type of organism was responsible, but the behaviour of the animal can be deduced. Similar trace fossils can be produced by quite different animals if they had similar modes of life.

The importance of trace fossils lies in the information they can give on the depositional environment, since in broad terms certain trace fossils or suites of trace fossils are characteristic of a particular environment, and often of a specific depth range (Seilacher, 1977). Trace fossils may often be the only evidence of life in a sediment if body fossils were not preserved. Sediments can

Fig. 2.31 Bioturbation and burrow structures in organic-rich calcareous siltstone deposited in a lagoon. In the upper part, original sedimentary structures have been completely disrupted and the sediment has a churned appearance; burrows penetrate down from the upper part into the laminated siltstones. Miocene, Iraq. Courtesy of M.G. Shawkat.

2 cm

37

be sub-divided on their trace fossils into 'ichnofacies'. Trace fossils are reviewed in Frey (1975) and relevant papers are contained in Crimes & Harper (1970, 1977).

Five principal groups of trace fossil are recognized: feeding structures, dwelling structures, crawling traces, resting traces and grazing traces (Fig. 2.32, and Seilacher, 1964; Frey, 1973). Some trace fossils are formed through a combination of activities.

Resting traces are formed by vagile epibenthic animals. They are impressions showing the broad shape of the animal. Impressions of star fish are a variety of resting trace occasionally found in the geological record.

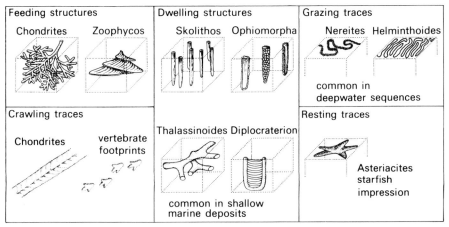

Fig. 2.32 Illustrations of the common types of trace fossil.

Crawling traces are made by any mobile animal, from a trilobite to a dinosaur. They are typical of predators, scavengers and some deposit feeders. The trails and trackways of moving animals are usually linear or sinuous, and uncomplicated by comparison with grazing and feeding traces. *Cruziana,* the trilobite foraging trail, is common on the surface or sole of many Lower Palaeozoic sandstone beds. Vertebrate footprints are relatively common throughout the Mesozoic and Cainozoic.

Grazing traces are mainly produced by mobile, deposit-feeding epibenthic organisms which feed at or near the sediment surface. The traces are epistratal, i.e. located on the bedding surface, and consist of curved, coiled or radiating furrows formed by the organisms systematically ingesting the sediment for food. Examples include *Helminthoides* and *Nereites* which are typical of deeper water sediments.

Feeding structures are intrastratal, i.e. formed below the sediment surface, and are formed by epibenthic and endobenthic deposit feeders, which live within a burrow system. The trace fossil typically consists of a network of filled burrows, branching or unbranched. *Chondrites* and *Zoophycos* are feeding structures found in many Phanerozoic shelf sediments.

Dwelling structures, mostly burrows, are mainly formed by sessile and semi-sessile endobenthic animals, particularly suspension feeders, predators and scavengers. The burrows may be simple vertical tubes, U-shaped or more

38

complicated burrow systems. Some U-shaped burrows, such as *Diplocraterion,* possess concave-up laminae (termed Spreite) formed through upward or downward movement of the animal in response to sedimentation or erosion. Many dwelling burrows are lined by pellets or mud, which serve to distinguish them from feeding structures. Examples include *Ophiomorpha* and *Thalassinoides,* crustacean burrows of intertidal–shallow subtidal environments, and *Skolithos,* a simple vertical tube common in intertidal zones.

One further type of trace fossil is the escape structure. In many situations organisms are occasionally smothered with sediment but they can regain their former position relative to the surface. In so doing, they leave behind characteristic escape structures of vertical tubes cutting and deflecting the sedimentary lamination. Borings are another type of trace fossil, although more common in limestones. They are distinguished from burrows by being made into a hard substrate, such as a hardground surface (Section 4.6.1), reworked concretion or carbonate skeleton.

2.4   Palaeocurrent analysis: the use of directional data

Many of the erosional and depositional sedimentary structures can be used to deduce the direction or trend of the currents that formed them. A knowledge of the *palaeocurrents* gives vital information on palaeogeography, sand body geometry and sediment provenance (Potter & Pettijohn, 1978), and in addition certain palaeocurrent patterns are restricted to a particular depositional environment (Selley, 1968).

Some sedimentary structures are vectorial, i.e. they give the direction in which the current was flowing; cross stratification, asymmetric ripples, flute marks and imbrication are structures of this type. Other structures give the trend or line of direction of the current; these include parting lineations, preferred orientations of grains and fossils, symmetrical ripples and groove marks.

Palaeocurrent data are collected in the field by simply measuring the orientation of the structure. However, if the strata are tilted to any extent (greater than 15°), the beds must be brought back to the horizontal by using a stereogram. This necessitates measurement of the dip and azimuth of the structure and dip and strike of the beds (for details see Potter & Pettijohn, 1978). Although the more data one has the more accurate is the palaeocurrent picture, some 20 to 30 readings for a locality or small area can be sufficient if one is dealing with a unimodal current system; for a bimodal or polymodal current system, more readings are desirable. To find the mean palaeocurrent direction, readings cannot simply be totalled and averaged since the data are directional. To obviate this, vector means can be calculated (Curray, 1956; Steinmetz, 1962) or constructed graphically (Raup & Miesch, 1957) but only if the distribution is unimodal. Statistical tests can be applied to palaeocurrent measurements to allow more precise comparisons between different sets of data (see Davis, 1973; Till, 1974). Palaeocurrent measurements are normally presented as a rose diagram as in Fig. 2.33, where readings are conventionally plotted in a 'current to' sense. If only a trend can be measured then the resulting rose diagram is symmetrical. Palaeocurrent patterns are of four types (Fig. 2.33): unimodal, bimodal-bipolar, bimodal-oblique and polymodal.

Palaeocurrent interpretation needs to be combined with studies and interpretations of the facies for maximum information. Measurements from fluviatile, deltaic and shallow marine deposits are most representative of the dominant flow direction(s) when cross bedding is used (Allen, 1966). Smaller scale structures such as ripples and cross laminations often reflect secondary

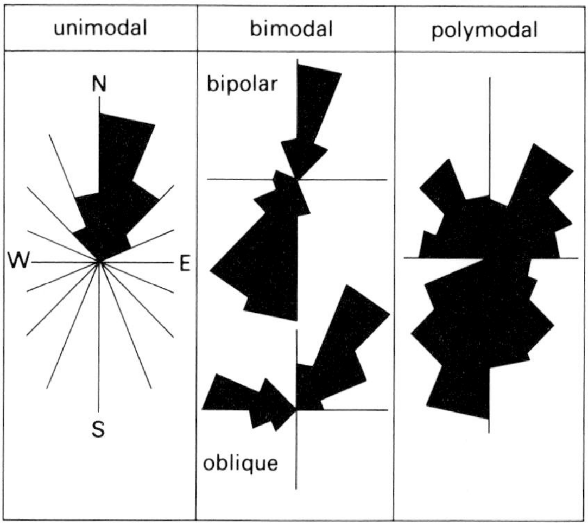

Fig. 2.33 The four common palaeocurrent patterns.

flows and so deviate from the direction of interest. With turbidites the flow direction is best given by the sole structures. A turbidity current is apt to wander when it is slowing down so that measurements from parting lineation, grain orientation and cross lamination within a turbidite bed often differ from those of sole structures. The palaeocurrent patterns of the principal depositional environments are given in Table 2.4.

2.5   Detrital components of siliciclastic sediments

Sandstones, conglomerates and breccias consist of: *detrital grains,* which form the framework of the sediments, fine-grained *matrix* located between grains, and authigenic minerals and *cements* precipitated after deposition of the sediment, during diagenesis (Section 2.8).

Practically any of the naturally-occurring minerals and fragments of any known rock-type can occur as grains in a clastic sediment. However, certain minerals and rock types are much more stable than others in the sedimentary environment so that in fact the number of common grain-types is relatively small. The abundance of a terrigenous mineral in a sedimentary rock is dependent on its availability, i.e. source-area geology, but also on its mechanical and chemical stability. With regards to chemical stability, minerals

40

Table 2.4 Depositional environments, directional structures and typical dispersal pattern.

Environment	Directional structures	Dispersal pattern
aeolian	large-scale cross bedding	unimodal if formed by barchan-type dunes and then indicates palaeowind direction; bimodal if formed by simple seif dunes and polymodal if complex seifs
fluvial	best to use cross bedding from dune bedforms; also parting lineation, ripples, scours, grain size changes, imbrication	palaeocurrents reflect palaeoslope and indicate provenance direction; unimodal pattern with small dispersion if low sinuosity rivers, unimodal with larger scatter if high sinuosity river or alluvial fan
deltaic	best to use cross bedding from dune bedforms; also channels, parting lineation, ripples	typically unimodal pattern directed offshore although marine processes (tidal and storm currents and waves) can complicate palaeocurrent pattern
shallow shelf	best to use cross bedding from dune bedforms; also ripples and scours	bimodal pattern common through tidal current reversals although tidal currents may be parallel or normal to shoreline; can be unimodal pattern if one tidal current dominates; polymodal and random patterns also occur
turbidite basin	best to use sole structures, especially flutes; also cross lamination, grain orientation, slump structures	unimodal pattern common, although may be downslope or along basin axis, or radial if on submarine fan. Contourites give palaeocurrent pattern parallel to slope

can be arranged into a series from the most to the least stable. With decreasing stability, the order is:

quartz, zircon, tourmaline
chert
muscovite
microcline
orthoclase
plagioclase
hornblende, biotite
pyroxene
olivine

Dissolution of minerals takes place at the site of weathering, particularly if the weathering processes are chemical rather than physical. The prevailing climate is important; mineral dissolution is more prevalent in hot and humid regions than in hot and semi-arid or polar areas. Dissolution of mineral grains also occurs during transportation if the grains are carried by water, and during diagenesis through the passage of pore fluids through the sediment, the so-called intrastratal solution. The mechanical stability of a mineral depends on the presence of cleavage planes and the mineral's hardness. Quartz, being relatively hard with no cleavage, is mechanically very stable and can survive considerable attrition during transportation. On the other hand, feldspars with their strong cleavage, and many rock fragments with their generally weak intercrystalline or intergranular bonds, are more easily broken down during sediment transport. Such unstable grains are called labile grains.

The detrital particles in terrigenous clastic rocks can be divided into six categories: (i) rock fragments (ii) quartz (iii) feldspar (iv) micas and clays (v) heavy minerals and (vi) other constituents. On the basis of their detrital components, sediments can be usefully considered in terms of their compositional maturity.

2.5.1 ROCK FRAGMENTS

These dominate conglomerates and breccias and tend to be the coarser particles in sandstones. As rock fragments get smaller they tend to break up into their constituent minerals or grains. The composition of the rock fragments depends basically on source-rock geology and durability of particles during transportation. In sandstones the fragments are commonly of (i) fine-grained sedimentary and metasedimentary rocks such as mudstone, shale, slate and pelite (e.g. Fig. 2.34), (ii) siliceous sedimentary rocks such as chert and siltstone

Fig. 2.34 Fragment of laminated shale, also other lithic and quartz grains in greywacke. Plane polarized light. Silurian; Southern Uplands, Scotland.

and (iii) igneous, in particular volcanic rocks. Compaction and diagenetic alteration may render fragments of shale, slate and igneous rocks indistinguishable from a fine-grained muddy matrix. Igneous rock fragments may be replaced by chlorite and zeolites. Many rock fragments are unstable (labile grains) but they are characteristic of certain sandstone types, in particular the litharenites and greywackes (Section 2.6).

With conglomerates and breccias, two types of clast occur: extraformational and intraformational clasts. Intraformational clasts, formed within the area of sedimentation, are chiefly lumps of mud, derived from erosional reworking of previously deposited muds. Extraformational clasts are derived from outside the area of sedimentation and can be of almost any rock-type, even very unstable varieties if the transport path is short. Conglomerates or breccias may contain a great variety of clasts particularly if the material is far-travelled. Such deposits are termed polymictic and examples include some glacial deposits and fluvial gravels. Conglomerates containing only one type of clast are referred to as monomictic (or oligomictic), and a local source is usually implied.

Rock fragments can often give very specific information on the provenance of a deposit if they can be tied down to a definite area. As an example of the use of lithic grains, Graham *et al.* (1976) were able to show that two greywacke sequences, some 500 km apart, in the Carboniferous of eastern U.S.A., had a common sedimentary/metamorphic provenance.

2.5.2 QUARTZ

The most common mineral in sandstone is quartz, the most stable of all minerals under sedimentary conditions. The average sandstone contains some 65% quartz, but some are practically 100% quartz. Many quartz grains in Mesozoic and Cainozoic sandstones are in their second or third cycle of sedimentation. The majority of quartz grains are derived from plutonic granitoid rocks, acid gneisses and schists. Various types of quartz can be distinguished: monocrystalline quartz grains are composed of a single crystal and polycrystalline grains of two or more crystals (Fig. 2.35). Polycrystalline

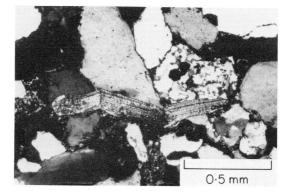

Fig. 2.35 Monocrystalline quartz grains with unit extinction, polycrystalline quartz grain with internal sutured contacts and flake of muscovite which has been bent through compaction. Crossed polars. Deltaic sandstone, Upper Carboniferous, N.W. England.

0·5 mm

grains can be divided into those with sutured, straight and irregular crystal contacts. Further subdivision of quartz grains can be made on the basis of extinction and inclusions. Crystals may possess unit extinction, where the whole crystal is extinguished uniformly under crossed polarizers, or undulose extinction, where extinction is not uniform but sweeps across the crystal as it is rotated (Fig. 2.36). Undulose extinction is usually a reflection of strain in the crystal lattice. Inclusions are common in quartz grains and are either vacuoles filled with fluid (appearing brown or black in transmitted light, silver in reflected light) or minute crystals of other minerals, especially rutile, mica, chlorite, magnetite and tourmaline.

With many quartz grains it is not possible to assign an origin, although there are differences in the proportions of the various quartz types in the common source rocks (Fig. 2.37 and Blatt, 1967; Basu *et al.,* 1975). Some features of the quartz grains are thought to be more diagnostic of one provenance than another. Quartz grains derived from volcanic igneous rocks for example are typically monocrystalline with unit extinction and no inclusions. Quartz from hydrothermal veins may be monocrystalline or coarsely polycrystalline but they characteristically possess numerous fluid-filled

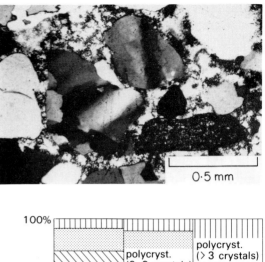

Fig. 2.36 Monocrystalline and polycrystalline quartz grain with undulose extinction, lithic grain of chert (lower right) and cement of microquartz. Crossed polars. Marine-shelf sandstone, Ordovician; Sierra Leone, West Africa.

0·5 mm

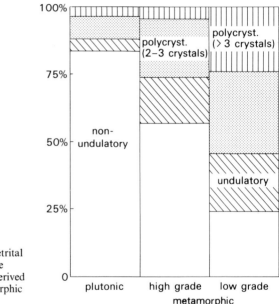

Fig. 2.37 Relative abundance of detrital monocrystalline and polycrystalline quartz grains in Holocene sands derived from known plutonic and metamorphic sources (from Basu *et al.*, 1975).

vacuoles. Pebbles in conglomerates are often of vein quartz, identified by their milky-white colour imparted by the fluid inclusions. Polycrystalline grains from a metamorphic source typically possess many crystals and they are often elongate, with a preferred crystallographic orientation. Undulose extinction, once thought to indicate a metamorphic source, in fact is common in igneous quartz crystals. Taken together, close examination of extinction properties and polycrystallinity can separate grains of plutonic and low-grade and high-grade metamorphic origin (Basu *et al.*, 1975).

Monocrystalline quartz grains with undulose extinction and polycrystalline quartz grains are both less stable than unstrained monocrystalline quartz. As a result of this, they are selectively removed during weathering, transportation and diagenesis, and sandstones contain a higher monocrystalline/polycrystalline quartz ratio and unstrained/strained quartz ratio, than the source igneous or metamorphic rock.

2.5.3 FELDSPARS

The feldspar content in sandstones averages between 10 and 15%, but in the arkoses (Section 2.7.2) it commonly reaches 50%. The mechanical stability of the feldspars is lower than that of quartz, since feldspars are softer and have a stronger cleavage. This leads to disintegration of feldspar crystals during transportation and in turbulent environments, so that on a broad scale for example, fluviatile sediments contain more feldspar than beach, shallow marine and aeolian sandstones (see Mack, 1978). The chemical stability of feldspars is also lower because of the ease with which they are hydrolysed. Chemical alteration typically involves replacement by clay minerals such as sericite (a variety of muscovite), kaolinite and illite (Section 2.8.5, Fig. 2.41). Incipient alteration gives the feldspars a dusty appearance; complete replacement produces clay mineral pseudomorphs after feldspar. Feldspar alteration takes place at the site of weathering, if it is dominantly chemical rather than physical weathering, and during diagenesis, either on burial or subsequent uplift.

Feldspar grains are derived from the same crystalline rocks as quartz. These are chiefly granites and gneisses where potash feldspars dominate over sodic plagioclase. The majority of feldspars in sediments are first cycle. Apart from suitable source rocks, the feldspar content of a sediment is largely controlled by the rate of erosion and climate. A humid climate in the source area promotes feldspar destruction because of the dominantly chemical weathering, whereas in an arid area fresh feldspars survive the dominantly physical weathering. Rapid erosion in an upland area will produce feldspar in spite of a humid climate.

2.5.4 MICAS AND CLAY MINERALS

These sheet silicates or phyllosilicates are particularly common in the matrix of sandstones and coarser clastics, and are the main component of mudrocks (Chapter 3). Biotite, chlorite and muscovite occur as large detrital flakes (e.g. Fig. 2.35), often concentrated along partings, laminae and bedding planes. Because of their flakey nature they are easily washed out of coarser sediments and so tend to accumulate with finer sands and silts; they are also easily removed from wind-blown sediments. Muscovite and biotite are derived from many igneous rocks but especially from metamorphic schists and phyllites. Chlorite comes largely from low-grade metamorphic rocks and the weathering and alteration of ferromagnesian minerals. Although biotite is more common in the source rocks than muscovite, the latter is more stable chemically and so is far more common.

Clay minerals in sandstones are both detrital and authigenic. The clay mineral types can rarely be identified with the petrological microscope (Section 3.4.1). All the chief clay mineral groups are represented in sandstones: kaolinite, illite, smectites and mixed-layer clays. Detrital clays reflect the source area geology, climate and weathering processes (Section 3.5). During diagenesis, clay minerals may be altered to other clays (Section 3.6), or they may form at the expense of other grains, in particular the feldspars. Chlorite often replaces labile rock fragments. Authigenic clay minerals may be precipitated around grains and in pore spaces (Section 2.8.5).

45

2.5.5   HEAVY MINERALS

These accessory grains are present in concentrations of less than 1%. They are chiefly silicates and oxides, many of which are very resistant to chemical weathering and mechanical abrasion. The common non-opaque heavy mineral grains are apatite, epidote, garnet, rutile, staurolite, tourmaline and zircon. Their principal features are given in Fig. 2.38. Ilmenite and magnetite are two common opaque detrital minerals. The specific gravity of heavy minerals, greater than 2.9, is higher than that of quartz and feldspar at 2.6. In view of this and their low concentration in the sediment, heavy mineral grains are separated from the crushed rock or loose sediment by using heavy liquids such as bromoform or Thoulet's solution, which have an intermediate specific gravity.

The study of heavy minerals can give some useful indications of provenance and of events in the source area. Certain heavy minerals, such as garnet, epidote and staurolite are derived from metamorphic terrains, whereas others, rutile, apatite and tourmaline for instance, indicate igneous source rocks. Major changes in the source-area geology, such as the uplift and unroofing of a granite, may be recorded in the heavy mineral assemblage of sandstones deposited in the region. Heavy mineral suites can be used to identify petrographic provinces within a formation, where, for example, sediment was supplied by two or more rivers draining areas of different geology.

Heavy minerals can be dissolved out during diagenesis through intrastratal solution. Because of this, older sandstones tend to have a less diverse heavy mineral suite.

As a result of their higher specific gravities heavy mineral grains tend to be smaller than the quartz grains with which they occur. In some instances the heavy minerals are concentrated in particular laminae or beds. This is a common feature of beach and other sediments where persistent winnowing takes place. Economic deposits, termed placers, may be formed in this manner. The monazite beach sand of Brazil, gold in fluviatile and beach sediments of the Yukon and Alaska, and ilmenite beach sands of Queensland are well known modern examples. In the gold deposits of Witwatersrand, S. Africa, the gold is principally located in early Proterozoic river channel conglomerates through syn-sedimentary concentration of detrital gold.

2.5.6   OTHER DETRITAL COMPONENTS

Carbonate grains in sandstones and coarser terrigenous clastics are mostly fossils or fragments of them, and non-skeletal grains such as ooids, peloids and intraclasts (see Chapter 4). Detrital grains of limestone and dolomite do occur, but unless there is an abundant supply, they are a very minor component. The importance of fossils in a sandstone or conglomerate lies in the stratigraphic and environmental information they may give. Other mineral grains in terrigenous clastics are skeletal phosphate (bone fragments), glauconite, chamosite and volcanic fragments. There is generally little disseminated organic matter in sandstones although carbonaceous plant fragments do occur in many sandstones.

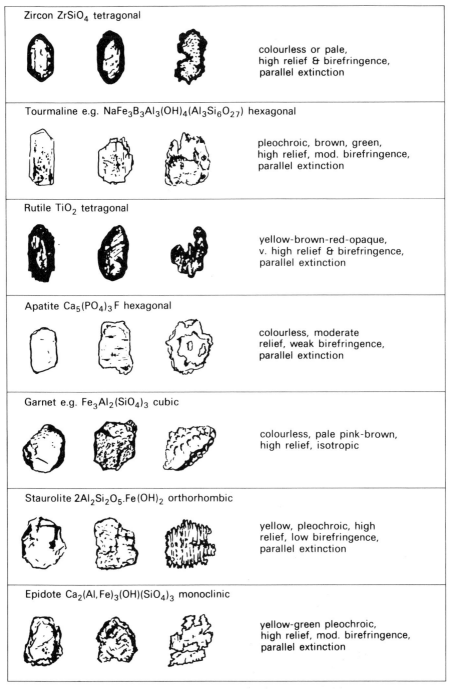

Zircon ZrSiO$_4$ tetragonal

colourless or pale,
high relief & birefringence,
parallel extinction

Tourmaline e.g. NaFe$_3$B$_3$Al$_3$(OH)$_4$(Al$_3$Si$_6$O$_{27}$) hexagonal

pleochroic, brown, green,
high relief, mod. birefringence,
parallel extinction

Rutile TiO$_2$ tetragonal

yellow-brown-red-opaque,
v. high relief & birefringence,
parallel extinction

Apatite Ca$_5$(PO$_4$)$_3$F hexagonal

colourless, moderate
relief, weak birefringence,
parallel extinction

Garnet e.g. Fe$_3$Al$_2$(SiO$_4$)$_3$ cubic

colourless, pale pink-brown,
high relief, isotropic

Staurolite 2Al$_2$Si$_2$O$_5$.Fe(OH)$_2$ orthorhombic

yellow, pleochroic, high
relief, low birefringence,
parallel extinction

Epidote Ca$_2$(Al,Fe)$_3$(OH)(SiO$_4$)$_3$ monoclinic

yellow-green pleochroic,
high relief, mod. birefringence,
parallel extinction

Fig. 2.38 Sketches of the seven most common heavy minerals (with the degree of weathering and/or dissolution increasing to the right) together with their optical properties (after Füchtbauer, 1974).

47

A compositionally immature sandstone contains many labile grains, i.e. unstable rock fragments and minerals, and much feldspar. Where rock fragments are of a more stable variety, there is some feldspar and much quartz, then the sediment is referred to as mature. For a sandstone composed almost entirely of quartz grains the term supermature is applied. Compositional maturity can be expressed by the ratio of quartz + chert grains to feldspars + rock fragments. This compositional maturity index is useful if comparisons between different sandstones are required.

Compositional maturity basically reflects the weathering processes in the source area and the degree and extent of reworking and transportation. Typically, compositionally immature sediments are located close to their source area or they have been rapidly transported and deposited with little reworking from a source area of limited physical and chemical weathering. Examples include many near-source fluviatile sediments, glacial and fluvio-glacial deposits. At the other extreme, supermature sediments are the end-product of intense weathering, where all unstable grains have been removed, or they are the result of intense abrasion and sediment reworking. The concepts of compositional maturity are considerably modified (i) where the source area itself consists of mature sediments, then the weathering mantle, soil and fluviatile sediments derived therefrom, will also be mature, and (ii) where sediment is supplied directly to a beach and near-shore area from adjacent igneous-metamorphic rocks, when immature sandstones will result.

2.6   Classification of terrigenous clastic sediments

Numerous classifications of terrigenous clastics have been proposed over the years with the majority based on two parameters: mineralogy and/or texture.

2.6.1   CLASSIFICATION OF SANDSTONES

Classifications of sandstones are petrographic, based on microscopic studies and requiring approximate, if not accurate, determinations of the modal composition. In spite of this, a rough estimate of a sandstone's composition can be made in the field through close scrutiny with a hand-lens. Most classifications use a triangular diagram with end members of quartz (+ chert), feldspar and rock fragments. The triangle is divided into various fields, and rocks with a modal analysis falling within a particular field are given a particular name.

An accepted and widely-used classification is that presented by Pettijohn *et al.,* (1973) and based on Dott (1964). In this scheme (Fig. 2.39), sandstones are divided into two major groups based on texture, that is whether the sandstones are composed of grains only, the arenites, or contain more than 15% matrix, forming the wackes. Of the arenites, the term *quartz arenite* is applied to those with 95% or more quartz grains, a rock-type formerly referred to as quartzite or orthoquartzite. *Arkosic arenite* refers to an arenite with more than 25% feldspar, which exceeds the rock fragments content, and *litharenite* is applied

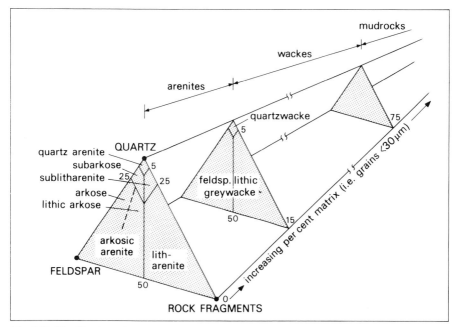

Fig. 2.39 Classification of sandstones (from Pettijohn *et al.*, 1973).

where the rock fragments content exceeds 25% and is greater than feldspar. The arkosic arenites can be divided into *arkoses* and *lithic arkoses*. Two rock-types transitional with quartz arenite are *subarkose* and *sublitharenite*. Specific names which have been applied to litharenite are *phyllarenite* where the rock fragments are chiefly of shale or slate, and *calclithite* where the rock fragments are of limestone.

The wackes are the transitional group between arenites and mudrocks. The most familiar is the *greywacke* and two types are distinguished: feldspathic and lithic greywacke. The term arkosic wacke is used for arkoses with a significant proportion of matrix. *Quartz wackes,* not a common rock-type, are dominant quartz plus some matrix.

This classification is primarily concerned with the mineralogy of the sediment and presence or absence of a matrix. It is independent of the depositional environment, although some lithologies are more common in certain environments. The nature of the cement in arenites is not taken into account. With regard to fine-grained interstitial material (matrix), a basic feature of the wackes, there is often a problem of origin. Some, perhaps most, is deposited along with the sediment grains. A part of the matrix, however, could be authigenic, a cement, and some a diagenetic alteration product of unstable grains. The latter is particularly the case with greywackes (Section 2.7.4).

With more detailed studies, particularly of the rock fragments and feldspars, it is often possible to subdivide the broad sandstone types into distinct *petrofacies*. These can then be related more precisely to different source areas and possibly to major tectonic events (e.g. Ingersoll, 1978; Dickinson *et al.*, 1979).

In addition to terrigenous clastic sandstones, there are many *hybrid sandstones*. These contain a non-clastic component derived from within the basin of deposition. The three main types are calcarenaceous, glauconitic and phosphatic sandstones. In glauconitic sandstones, the glauconite occurs as sand-sized pellets (Section 6.4.4). With phosphatic sandstones, the phosphate may be present as a cement, nodules, coprolites or bone fragments (Section 7.4).

Calcarenaceous sandstones contain up to 50% $CaCO_3$, present as carbonate grains. The latter are chiefly ooids, often with quartz nuclei, and carbonate skeletal fragments. Calcarenaceous sandstones are developed in carbonate-producing areas where there is a large influx of terrigenous clastics. They will often pass laterally into limestones or into purer sandstones towards the source of the siliclastic sediment. Terrigenous clastics cemented by calcite have been referred to as calcareous sandstones.

2.6.2 CLASSIFICATION OF CONGLOMERATES AND BRECCIAS

There are many ways of dividing up coarse clastic rocks. In the first place, on the basis of pebble shape, they can be split into *conglomerates* with subrounded to well-rounded clasts, and *breccias* with angular and subangular clasts. A division in terms of composition has already been mentioned: polymictic conglomerates (or breccias) contain a variety of pebbles and boulders, whereas mono- or oligo-mictic conglomerates consist of one pebble-type only. A division into intraformational and extraformational conglomerates divides those composed of clasts derived from within the basin of deposition, from those where pebbles are derived from beyond the area of sedimentation (Section 2.5.1). On sediment fabric, there are two types: clast-supported and matrix-supported conglomerates (Section 2.2.4). The latter have been termed diamictites.

Two particular types of breccia are slump breccias, consisting of broken and brecciated beds derived from downslope slumping (Section 2.3.3), and solution breccias, resulting from the dissolution of evaporites and the collapse of overlying strata (Section 5.5).

2.7   Petrography and origin of principal sandstone types

The four common types of sandstone are quartz arenite, arkose, litharenite and greywacke. They are frequently typical of a particular depositional environment, but because of the provenance control on sandstone composition, they are not restricted to that particular depositional setting.

2.7.1 QUARTZ ARENITES

Sandstones with at least 95% quartz grains (Fig. 2.40) are the most compositionally mature of all sandstones (Section 2.5.7). In addition, they frequently consist of well-rounded and well-sorted grains so that textural maturity is also very high. Cements are typically quartz overgrowths, but calcite is also common. Monocrystalline quartz grains dominate. Common heavy minerals are tourmaline, zircon and ilmenite.

Fig. 2.40 Quartz arenite. This sandstone is composed almost wholly of monocrystalline quartz grains, cemented by quartz overgrowths. Ganister, Upper Carboniferous, N.E. England.

0·5 mm

In many cases quartz arenites are the product of extensive periods of sediment reworking, so that all grains other than quartz have been broken down. The majority of the quartz grains are second cycle, derived from pre-existing sediments. Quartz arenites of this type, often deposited on shallow-marine shelves, are well developed in the Cambro-Ordovician of the Iapetus continental margins (N.W. Europe and eastern U.S.A.) and Cordilleran shelf of western U.S.A.. (e.g. James & Oaks, 1977). Other quartz arenites formed by extensive in-situ leaching, the ganisters (palaeosoils) of Carboniferous successions are examples.

2.7.2 ARKOSES

Rocks generally accepted as arkoses contain more than 25% feldspar, much quartz and some rock fragments (Fig. 2.41). Detrital micas are also present and some fine-grained matrix. The feldspar is chiefly potassium feldspar and much of this is microcline. The feldspar is usually fresh although some may be altered to kaolinite and sericite. Polycrystalline quartz and quartz/feldspar rock fragments are common. Arkoses are typically red or pink, through the feldspar's colour, but also through the presence of finely disseminated hematite, since many arkoses occur in red-bed sequences. Arkoses are derived from granites and gneisses and vary from in-situ weathering products, to stratified and cross-bedded arkoses where there has been substantial sediment transport.

Fig. 2.41 Arkose. Many of the feldspar grains (orthoclase, microcline and albite) show incipient replacement by sericite (minute bright crystals). Also present: quartz grains (clear by comparison with feldspar) with overgrowths, and hematite coatings on grains. Crossed polars. Torridonian Sandstone, Precambrian, N.W. Scotland.

0·5 mm

51

Arkose texture is typically poorly-sorted to well-sorted, with very angular to sub-rounded grains, the precise texture dependent on the degree of transportation. Grain-supported arkoses are cemented by calcite or quartz, while others are cemented by a matrix, often containing much kaolinite. The chemistry of arkoses is fairly uniform (Table 2.5). They are rich in Al_2O_3 and K_2O and there is an excess of Fe_2O_3 over FeO.

Arkoses are clearly derived from feldspar-rich rocks but apart from an appropriate provenance geology, climate and source area relief are also

Table 2.5 Average chemical composition of the three common sandstone types: quartz arenite, arkose, greywacke (from Pettijohn *et al.*, 1973). Points to note are the high SiO_2 and low everything else for quartz arenite, and the chemical differences between arkose and greywacke: the higher Fe_2O_3/FeO ratio and K_2O/Na_2O ratio in arkose relative to greywacke, and the higher Al_2O_3 and MgO contents of greywacke.

	quartz arenite	arkose	grey-wacke
SiO_2	95.4	77.1	66.7
Al_2O_3	1.1	8.7	13.5
Fe_2O_3	0.4	1.5	1.6
FeO	0.2	0.7	3.5
MgO	0.1	0.5	2.1
CaO	1.6	2.7	2.5
Na_2O	0.1	1.5	2.9
K_2O	0.2	2.8	2.0
TiO_2	0.2	0.3	0.6

important factors. Under humid conditions, feldspars weather to clay minerals, so that semi-arid and glacial climates favour arkose formation. If erosion is very rapid, however, i.e. the source area has a high relief, then arkoses can accumulate in spite of an adverse climate. Many arkoses were deposited in fluviatile environments.

2.7.3   LITHARENITES

These sandstones are characterized by a rock fragment content which is in excess of feldspar. They range widely in composition, both in terms of mineralogy and chemistry, depending largely on the types of rock fragment present. These are chiefly fragments of mudrock and their low grade metamorphic equivalents; other components are flakes of mica, some feldspar and much quartz of course (Fig. 2.42). There is little primary matrix, otherwise they are similar to greywackes in composition and in fact they have been referred to as subgreywackes. Cements are usually either calcite or quartz. Although they have a variable chemistry litharenites generally possess a high Al_2O_3 content from the dominant clay/mica-rich rock fragments, and low Na_2O and MgO.

Litharenites account for some 20 to 25% of all sandstones. Their immature composition implies high rates of sediment production from supracrustal sources followed by short transport distances. Many fluviatile and deltaic sandstones are litharenites.

Fig. 2.42 Litharenite. Lithic fragments in this case are limestone, with fossils; much silt-grade quartz is present, cement is calcite. Plane polarized light. Brockram, Lower Permian; Cumbria, England.

I mm

2.7.4 GREYWACKES

The characteristic feature of greywackes is the fine-grained matrix, which consists of an intergrowth of chlorite, sericite and silt-sized grains of quartz and feldspar (Fig. 2.43). Of the sand fraction, quartz dominates over rock fragments and feldspar. Many different rock fragments are usually present. Feldspar grains are chiefly sodic plagioclase and these are usually fresh in appearance. As indicated by the name, greywackes are dark grey or black rocks, often resembling dolerite.

The origin of the matrix has been referred to as the 'greywacke problem'. There are two possibilities: fine grained sediment deposited along with the sand fraction, and diagenetic alteration of unstable rock fragments. The evidence cited against a primary origin is the fact that modern deep sea turbidite sands do not contain much mud (e.g. Moore, 1974) and that in transportation and deposition by turbidity currents, one would expect a cleaner separation of mud and sand. Evidence for a diagenetic origin of the matrix through replacement of the labile components is based on the local presence of early calcite cements where matrix development has been inhibited (Brenchley, 1969) and the complete range of rock fragment preservation, from well-preserved, but with recrystallized borders, to barely discernible from matrix (Galloway, 1974).

On the whole, greywackes have a uniform chemical composition, which contrasts with that of arkoses (Table 2.5). Greywackes have high contents of

Fig. 2.43 Greywacke. Quartz and lithic grains are contained in a fine-grained matrix of chlorite and silt-grade quartz; lithic grains typically merge into the matrix. Crossed polars. Silurian, Southern Uplands, Scotland.

I mm

53

Al_2O_3, total Fe ($FeO + Fe_2O_3$), MgO and Na_2O. The high MgO and FeO values are reflections of the chloritic matrix and the high Na_2O is due to the dominant plagioclase. Greywackes differ from arkoses in the dominance of FeO over Fe_2O_3, MgO over CaO and Na_2O over K_2O.

Many greywackes were deposited by turbidity currents in basins of various types, often off continental margins and in association with volcanics. These greywackes can show all the typical turbidite features (Dzulynski & Walton, 1965; Section 2.10.7). Classic greywackes occur in the Lower Palaeozoic of Wales, Ireland and the Southern Uplands of Scotland, and in the Devonian-Carboniferous of S.W. England and Germany. In North America, they occur in the Mesozoic of California (Franciscan Formation), Washington and Alaska, and in the Palaeozoic of Newfoundland and the Appalachians. Although many greywackes are uniform in composition, in detail they do vary, especially in the nature of the rock fragments, so that distinct petrofacies can often be recognized. These variations can have important inferences on the tectonic situation at the time of deposition, since many greywackes were syn-orogenic. Many were deposited during periods of active vertical and horizontal plate motion, often coupled with island arc volcanism (Mitchell & Reading, 1969; Crook, 1974; Dickinson, 1974; Ingersoll, 1978).

2.8   Sandstone Diagenesis

As noted in Chapter 1, diagenesis has been divided into two broad stages: early diagenesis, for processes taking place from deposition and into the shallow burial realm, and late diagenesis for those processes affecting the sediments at deeper levels, and during and after uplift. The terms syndiagenesis, anadiagenesis and epidiagenesis have been used for early and burial diagenesis and diagenesis after uplift respectively (see Fairbridge, 1967 and Larsen & Chilingar, 1979, for reviews).

The important physical processes of diagenesis are compaction and pressure solution, both largely dependent on depth of burial. Chemical processes of diagenesis include the precipitation of minerals, often leading to the cementation of the sediments, the dissolution of unstable grains, and the replacement of grains by other minerals. Chemical processes take place in the medium of water, so that salinity, pH (a measure of the hydrogen ion concentration) and Eh (redox potential) of the water, and the ability of water to move through the sediment (dependent on the porosity and permeability) are of critical importance to diagenesis. In the early stages of diagenesis, lasting for some 1000 to 100000 years and affecting sediments to depths of around 1 to 100 m, pore waters are related to the depositional environment; connate waters if marine, fresh waters in the majority of continental sedimentary environments. These pore waters are soon modified by the breakdown of organic matter and bacterial activity. With marine sediments, for example, the initial stages of diagenesis take place in oxidizing pore waters, which with depth become reducing as oxygen is used up in bacterial processes.

During deep burial, pore waters are modified further by reactions with clay minerals, dissolution of unstable grains, precipitation of authigenic minerals and mixing with waters from other sources. Burial diagenesis operates over millions of years and affects sediments to depths of around 10000 m, where temperatures are in the region of 100°–200°C. Beyond this, processes of burial metamorphism take over. In general, pore waters in deeply-buried sediments are saline, neutral and alkaline. In comparison with seawater, these formation waters as they are called, have lower Na^+, Mg^{2+}, SO_4^{2-} and K^+ values relative to chlorine, but higher Ca^{2+}, Sr^{2+} and silica (Blatt *et al.*, 1980). Processes taking place in sedimentary rocks on uplift typically involve fresh ground waters with low Eh and acid pH. The extent of epidiagenetic processes depends largely on the porosity and permeability, which may well have been largely occluded during burial diagenesis.

There are many factors affecting sandstone diagenesis. The depositional environment, composition and texture of the sediment are primary factors and then pore-fluid migrations, the burial history and other factors affect the course of diagenesis. The principal processes are compaction and pressure solution, silica and calcite cementation, clay mineral and feldspar authigenesis and the formation of hematite coatings and impregnations. Recent reviews of sandstone diagenesis are contained in Larsen & Chilingar (1979) and Scholle & Schluger (1979).

2.8.1 COMPACTION AND PRESSURE SOLUTION

In the initial stages, compaction involves dewatering and a closer packing of grains. Further compaction through overburden pressure results in local fracturing and bending of weak grains (as in Fig. 2.35) and of especial importance, the solution of grains at points of contact. Pressure solution at grain contacts is minimal where the sediment is cemented early, before deep burial, or where there is much matrix, since in these cases the load is spread and the contact pressure reduced. Pressure solution can take place once a rock is fully cemented to produce irregular or sutured planes, known as stylolites. These cross-cut grains and cements, and a thin layer of insoluble material is concentrated along the stylolites. The importance of grain-contact pressure solution and stylolitization is that it could be the process by which material is supplied for cementation of the sediment.

2.8.2 SILICA CEMENTATION

One of the most common types of silica cement is the quartz overgrowth (Fig. 2.44). Silica cement is precipitated around the quartz grain and in optical continuity, so that the grain and cement extinguish together under crossed polarizers. The syntaxial overgrowth commonly gives the grain euhedral crystal faces. In many cases the shape of the original grain is delineated by a thin iron oxide-clay coating between the overgrowth and the grain. A thicker clay precipitate around the quartz grain, however, has often inhibited precipitation of a syntaxial overgrowth. In some cases the boundary between the grain and overgrowth cement cannot be discerned with the light microscope, and the whole rock has the appearance of a metamorphic orthoquartzite. The use of cathodoluminescence (Section 1.3.2) will often bring out the difference between grain and overgrowth (Sippel, 1968).

Fig. 2.44 Cementation by quartz overgrowths, in optical continuity with host grain. Crystal faces have developed in places. Crossed polars. Penrith Sandstone (aeolian), Lower Permian; Cumbria, England.

0·25 mm

The origin of the silica for this cementation has frequently been attributed to pressure solution. Pore solutions become enriched in silica which is then reprecipitated as overgrowths when supersaturation is achieved. Quartz overgrowths in sandstones without pressure solution effects may reflect significant upward migration of silica-rich solutions from more distant sites of pressure solution, or indicate another source of silica. Possible sources are dissolution of silica dust, other silicates and biogenic silica, and groundwater. Silica dust could be derived from grain abrasion, especially if it is an aeolian sandstone (e.g. Waugh, 1970). Dissolution of feldspars, amphiboles and pyroxenes would provide silica, as would the mineral transformations montmorillonite to illite and feldspar to kaolinite (e.g. Hawkins, 1978). In marine sediments, porewaters often contain significant concentrations of silica derived from the dissolution of diatoms, radiolaria and sponge spicules. These siliceous skeletons are composed of metastable amorphous opaline silica which has a higher solubility than quartz (Section 9.3 and Fig. 9.7). Because of this solubility difference and the increased solubility of opaline silica with increasing depth of burial, biogenic silica could be an important source. Groundwater is often supersaturated with respect to quartz and if this could be moved through a quartz sand in sufficient volume then cementation by overgrowth development could take place. Calculations by Sibley & Blatt (1976) from known rates of groundwater movement indicate that a fully-cemented sandstone could be produced in some 200 Ma.

Apart from quartz overgrowths, silica is also present as a cement in the form of microquartz, megaquartz, chalcedonic quartz and opaline silica (Section 9.2). In some cases, the presence of opal is related to the decomposition of volcanic particles.

One important feature arising from the early quartz cementation of sandstones is that they are then able to withstand better the effects of compaction and pressure solution during later burial. In this way a moderate porosity can be preserved which may be filled later with oil or gas.

2.8.3   CALCITE CEMENTATION

Calcite is one of the most common cements in sandstones, but other carbonate cements of more local importance are dolomite and siderite. The cement may vary from a uniform to patchy distribution, to local segregations and concretions. The two main types of calcite cement are poikilotopic crystals and drusy calcite spar. Poikilotopic crystals are large single crystals, up to several centimetres across, which envelope many sand grains (Fig. 2.45). Drusy calcite mosaics consist of equant crystals which fill the pores between grains, and typically show an increase in crystal size towards the centre of the original cavity (Section 4.7.2). As a result of calcite precipitation there is often a

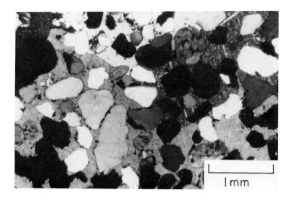

Fig. 2.45 Quartz grains cemented by large poikilotopic calcite crystals (twin planes visible in upper right calcite). Crossed polars. Yellow Sands (aeolian), Lower Permian; Durham, England.

1 mm

displacement of grains so that the grains appear to 'float' in the cement. Calcite may also be precipitated in cracks in grains and so force them to split. This is common with micas, but also occurs with quartz grains. Apart from filling pores, calcite and the other carbonates may also replace grains. Quartz grains cemented by calcite are often corroded and etched at their margins, to produce irregularly-shaped grains.

Calcite cements are common in grain-supported sandstones, such as quartz arenites, arkoses and litharenites. Calcite is frequently an early diagenetic cement and the first cement. In fact it is being precipitated in some modern river and desert sands and soils. The early precipitation of calcite inhibits later quartz overgrowth formation and feldspar alteration and can result in total loss of porosity and permeability. In other sandstones, calcite is a later precipitate, postdating quartz overgrowths and authigenic kaolinite (e.g. the Jurassic Brent Sands of the North Sea, Blanche & Whitaker, 1978). $CaCO_3$ precipitation, taking place when the solubility product is exceeded, invariably occurs through an increase in the activity of the carbonate ion. In the very shallow subsurface, this may happen through evaporation of vadose or near-surface phreatic ground water. At depths, carbonate precipitation can be brought about by an increase in the pH and/or temperature.

Dolomite cements vary from pore-filling microcrystalline rhombs to coarse anhedral mosaics and large poikilotopic crystals. They are often iron-rich (ferroan), indicating precipitation in reducing conditions below the groundwater table. Early dolomite precipitation may be related to near-surface evaporating conditions. Magnesium for later dolomite precipitation may be derived from clays, dissolution of magnesium-rich silicates, or dilution of seawater (Section 4.8). Siderite cements ($FeCO_3$) do occur in some sandstones, typically as micro-crystalline mosaics (Section 6.3 for conditions of formation of siderite).

Although in many sandstones, feldspars are altered to kaolinite and illite, feldspar overgrowths do occur on detrital feldspar grains (Fig. 2.46). They are most common on potash feldspars (e.g. Stablein & Dapples, 1977; Waugh, 1978), but they also occur on detrital albite grains (e.g. Hancock, 1978). Authigenic albite crystals not associated with detrital grains occasionally develop. For authigenic feldspar, alkaline porewaters rich in Na or K, Al and Si are necessary. These elements are largely derived from hydrolysis and

Fig. 2.46 Grain of microcline with authigenic overgrowth; crystal faces have developed. Crossed polars. Penrith Sandstone (aeolian), Lower Permian; Cumbria, England.

0·1mm

dissolution of less stable grains within the sediment. Feldspar overgrowth formation has been studied in Cainozoic continental sandstones of North America (Walker *et al.*, 1978), where the processes of hydrolysis and authigenesis take place above and below the water table, at very shallow depths of burial. Authigenic K-feldspars characterize the Cambro-Ordovician strata of the Iapetus shelf regions (Buyce & Friedman, 1975) and in this case the K, Si and Al ions are thought to have been supplied by decomposition of tephra (volcanic ash).

2.8.5   CLAY MINERAL AUTHIGENESIS

The importance of clay mineral precipitation within sandstones has only been appreciated in the last few years. Previously all clay in sandstones had been regarded as detrital. The precipitation of even small amounts of clay in a sandstone can have a great effect on the permeability and other properties of the rock and may seriously reduce its reservoir potential (e.g. Wilson & Pittman, 1977). Clay may also filter into a sandstone, carried down by porewaters from muddy interbeds (demonstrated by Walker *et al.* (1978) from desert alluvium). Extensive infiltration drastically alters the texture of the sediment and decreases the original textural and compositional maturity.

Illite and kaolinite are the most common authigenic clays in sandstones, but montmorillonite, mixed-layer illite-montmorillonite and mixed-layer montmorillonite-chlorite also occur (e.g. Almon *et al.*, 1976). Authigenic clay minerals occur as pore-filling cements and clay rims up to 50 μm thick around grains. The attenuation and absence of rims near and at grain contacts demonstrates their diagenetic origin. The precipitation of clay rims is usually an early or the first diagenetic event, often predating quartz overgrowths or calcite cementation. In some Cainozoic desert sediments this clay coating mostly consists of mixed-layer illite-montmorillonite, thought to have been precipitated from solutions that leached unstable minerals in the shallow burial environment. The clay rim can become impregnated with hematite (next section) or altered to other clays during diagenesis. Where clay rims are thick, they may inhibit later cementation and so preserve porosity. Illite in clay rims shows a variety of growth forms of fibres and whiskers (Fig. 2.47).

Within pores between grains, authigenic kaolinite characteristically forms 'books' or 'concertinas' of stacked pseudo-hexagonal plates (Figs. 2.48, 2.49).

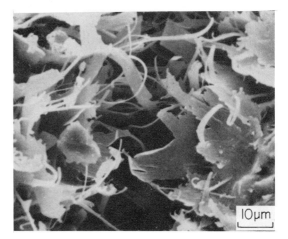

Fig. 2.47 Scanning electron micrograph of authigenic illite in the form of radially-arranged flakes and whiskers growing into pore space between two sand grains (left and right of picture). Rotliegend desert sandstone, Lower Permian; northern W. Germany. Courtesy of N.J. Hancock and Geological Society of London.

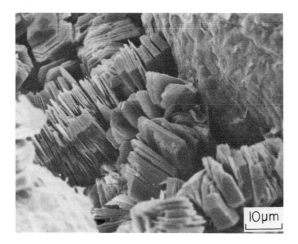

Fig. 2.48 Scanning electron micrograph of authigenic kaolinite, consisting of stacked pseudo-hexagonal platy crystals, between rounded sand grains. Rotliegend desert sandstone, Lower Permian; northern W. Germany. Courtesy of N.J. Hancock and Geological Society of London.

At depth, alkaline pore fluids and higher temperatures result in the replacement of kaolinite by illite. The typical kaolinite 'book' texture may be retained (pseudomorphed) by the illite.

For clay mineral authigenesis, alkaline pore fluids are required for illite, together with sufficient K, Si and Al. Kaolinite requires more acid porewaters and these can be produced by flushing of the sandstone by fresh water, either

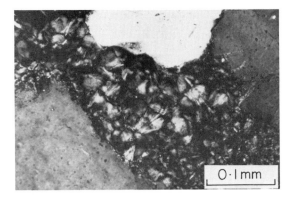

Fig. 2.49 Authigenic kaolinite in 'books' filling pore space between quartz grains. Crossed polars. Upper Carboniferous, Northumberland, England.

0·l mm

during an early burial stage if the sediments are continental, or if marine, during uplift after a burial phase. Kaolinite precipitation within marine sediments may also result from decomposing organic matter setting up a low pH. The ions for kaolinite and illite precipitation are largely derived from the alteration of labile detrital minerals, in particular feldspars.

Apart from clay mineral precipitation in pores and rims, much replacement of silicate minerals by clay can occur. The replacement may be irregular, peripheral or along fracture and cleavage planes. Feldspar grains and igneous and metamorphic rock fragments are often partially to completely replaced by clays. Detrital clays themselves may recrystallize and be replaced by other clays. The replacing clay minerals are commonly mixed-layer illite-montmorillonite and kaolinite, although in time these are often replaced by illite and chlorite (Section 3.6). One important effect of this replacement is the increase in the amount of interstitial matrix. This considerably affects the textural maturity of the sediment and gives rise to the greywacke-type texture (Section 2.7.4). The clay rarely preserves the shape of the grains it has replaced since compaction as a result of overburden pressure causes the clay to become squeezed between more rigid grains.

2.8.6 HEMATITE CEMENTATION AND PIGMENTATION

Many terrigenous clastic sediments are coloured red through the presence of hematite. In many cases these rocks were deposited in continental environments (deserts, rivers, floodplains, alluvial fans, etc.) and the term 'red beds' has been applied to them. The hematite typically occurs as a very thin coating around grains, but also stains red infiltrated or authigenic clay minerals and authigenic quartz and feldspar. It also develops within biotite cleavage planes and in some cases replaces the biotite (Turner & Archer, 1977). The hematite is chiefly amorphous or consists of micron-size crystals. These features of the hematite, together with the absence of hematite coatings at grain contacts (Fig. 2.50), indicate a diagenetic origin.

There has been much discussion on the source and origin of the hematite pigment in red beds. One view advocates a detrital origin, that amorphous iron compounds formed through moist tropical lateritic weathering in upland areas, are transported and deposited along with the sediments and then converted to hematite (van Houten, 1968; 1972). Many detrital sand grains have a yellow-brown stain of iron hydroxides, gained in the source area, and these can also age to hematite after deposition (Hubert & Reed, 1978). From studies of Recent to Tertiary desert (arid climate) sands, an alternative origin involves a purely diagenetic mechanism whereby the iron is supplied by intrastratal solution of detrital silicates such as hornblende, augite, olivine, chlorite and biotite, and magnetite (Walker, 1967; 1974; Walker & Honea, 1969). If the

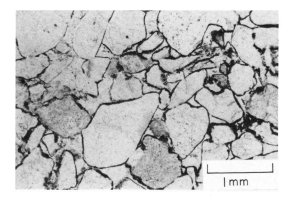

Fig. 2.50 Hematite coatings around sand grains; note that hematite is absent where grains are in contact. Plane polarized light. Torridonian Sandstone, Precambrian, N.W. Scotland.

diagenetic environment is oxidizing, then the iron is reprecipitated as hematite or a goethitic precursor oxide which converts to hematite on ageing. The length of time involved in the ageing process is the order of a million years; modern desert alluvium is a yellowish colour and the red colour gradually develops as the sediments get older. Only a small quantity of iron, 0.1%, is sufficient to impart a bright red colour to the sediments. The hematite develops above the water table and below, if the groundwater is alkaline and oxidizing. If reducing conditions prevail, the iron is present in the more soluble ferrous state and if incorporated into clays, rather than carried away in solution, it will impart a green colour to the sediments. (The chemistry of iron mineral formation is discussed in Section 6.3.)

Although in the majority of red beds, an intrastratal-diagenetic origin of the hematite is thought most likely, it is still difficult to exclude a detrital component, especially in matrix-rich sandstones.

Secondary alteration of the red colour takes place where reducing solutions penetrate into the sediments. This may be along more porous horizons or tectonic fractures. The colour of the sedimentary rock is then changed to green, grey or even white, if leaching is intense.

Other diagenetic minerals of less or only local importance are zeolites, sulphates and sulphides. Zeolites are formed chiefly from the breakdown of volcanic debris and so are common in volcaniclastic sediments (Section 10.5). Gypsum and anhydrite occur as cements in sandstones where there are evaporite beds in the sequence; otherwise they are rare. Celestite ($SrSO_4$) and barite ($BaSO_4$) rarely occur as cements. Pyrite occurs in many sandstones but only as an accessory diagenetic mineral.

Within any sandstone formation, the sequence of diagenetic events can be simple, involving only one mineral precipitate, or highly complex, involving many stages of precipitation and replacement. Factors controlling the path of diagenesis in sandstones are first the depositional environment, sediment composition and texture, and then later the porewater chemistry, depth of burial and timing of uplift. The relative timing of diagenetic events in sandstones is important in terms of the introduction of hydrocarbons. If a sandstone's porosity is occluded by early cementation, then it cannot act as an oil reservoir. Diagenetic processes take place in an aqueous medium so that the influx of oil terminates diagenesis and prevents further reactions. The affects of this are shown in the Jurassic Brent Sand of the North Sea where kaolinite is preserved in oil-bearing strata, but has been converted to illite in areas where the sandstone pores were filled with water (Hancock & Taylor, 1978).

2.9   Porosity and Permeability

Two important aspects of sedimentary rocks are their porosity and permeability. Porosity is a measure of the pore space and two types are defined:

$$\text{Absolute porosity } Pt = \frac{(\text{bulk volume} - \text{solid volume})}{\text{bulk volume}} \times 100$$

$$\text{Effective porosity } Pe = \frac{\text{interconnected pore volume}}{\text{bulk volume}} \times 100$$

Absolute porosity refers to the total void space, but since some of this will be within grains, effective porosity is more important. It is the latter which determines the reservoir properties of a rock, together with permeability, the ability of a sediment to transmit fluids. Porosity is a basic feature of a sediment or rock whereas permeability depends on the effective porosity, the shape and size of the pores and pore interconnections (throats), and on the properties of the fluid itself, i.e. capillary force, viscosity and pressure gradient. From Darcy's Law, permeability (K) depends on the rate (Q) at which fluid flows through a unit cross-section of rock: $Q = (K/\mu)(dp/dl)$ where μ is the fluid viscosity and dp/dl is the pressure gradient in the direction of flow.

With porosity, two major types are primary and secondary porosity. Primary porosity is developed as the sediment was deposited and includes inter- and intra-particle porosity. Secondary porosity develops during diagenesis by solution and dolomitization (the latter in limestones, Section 4.9), and through tectonic movements producing fractures in the rock.

Primary porosity in sandstones is principally interparticle porosity, dependent on the textural maturity of the sediment (Section 2.2.5), controlled largely by depositional processes and environments, and to a lesser extent on compositional maturity (Section 2.5.7), as shown in Fig. 2.51. In general, the primary porosity increases as the grain size increases, the sediment is better sorted and more loosely packed, and the clay content decreases (e.g. Beard & Weyl, 1973). The clean, well-sorted, loosely packed sands of beaches and aeolian dunes can have porosities in excess of 50% (Pryor, 1973), and they have high permeabilities too. Fine-grained sediments of high effective porosity, sorted siltstones for example, and chalks, often have low permeabilities since capillary forces prevent the flow of fluids through the small pore throats. The affect of compositional maturity relates to the breakdown of unstable grains; this can increase the porosity if the grains are simply dissolved out, but in most cases the porosity is reduced through the formation of clay minerals and other alteration products.

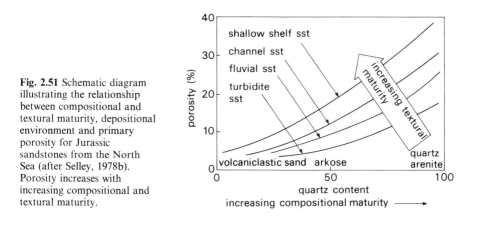

Fig. 2.51 Schematic diagram illustrating the relationship between compositional and textural maturity, depositional environment and primary porosity for Jurassic sandstones from the North Sea (after Selley, 1978b). Porosity increases with increasing compositional and textural maturity.

Once deposited, there are two processes in terrigenous clastics, apart from the breakdown of labile grains, which lead to a decrease in porosity: compaction and cementation. On a broad scale, these processes bring about a gradual decrease in porosity with increasing depth of burial (e.g. Selley, 1978b). Compaction takes place from a few metres below the sediment surface, and results in a closer packing of grains, and eventually at depths of 100's to 1000's of metres to pressure solution and interpenetration of grains (Section 2.8.1). Cementation, however, is the principal process of porosity loss in sandstones. Silica, calcite and clay can all be precipitated as cements (Sections 2.8.2–2.8.5), filling pores and decreasing both porosity and permeability. Most non-carbonate petroleum reservoirs occur in sandstones which have been only partially cemented and so retain much of their depositional porosity. Porosities necessary for good petroleum reservoirs are 20 to 35%; examples include the oil- and gas-bearing Jurassic Brent sands and Permian Rotliegendes of the North Sea, and the Cainozoic sandstones of southeast Texas and south Louisiana.

2.10 Depositional environments of sandstones and coarser clastics

The principal depositional sites of the coarser siliciclastic sediments are (i) fluviatile environments, (ii) deserts, (iii) lakes, (iv) deltas, (v) barriers, beaches, tidal flats and estuaries of marine shorelines, (vi) shallow marine shelves and epeiric seas, (vii) continental margins and deepwater basins and (viii) glacial environments.

Fluviatile environments are complex systems of erosion, sediment transport and deposition which give rise to a great variety of landforms. At the present time fluvial systems range from alluvial fans, through braided and low sinuosity stream networks to meandering (high sinuosity) rivers. Their sediments range from the coarsest conglomerates through sandstones to mudrocks. Fluviatile sandstones in general are usually sharp-based and cross-bedded (as in Fig. 2.20), with some flat bedding and cross lamination. They may be lenticular (the infills of stream channels) or laterally more persistent. Texturally and compositionally, fluviatile sandstones are generally immature to mature, although this depends on the sediment provenance and transport distance. Many are arkoses and litharenties; those derived from reworking of older sandstones are more quartzose. Fluviatile conglomerates are typically lenticular, often with cross bedding; many are polymictic; with both extra- and intra-formational clasts and have a pebble-support fabric. Fossils are not common in river sediments and mostly consist of plant material and skeletal fragments of freshwater and terrestrial animals, especially fish. Many fluviatile sediments were deposited under semi-arid climates and so are mostly red from the early diagenetic formation of hematite (Section 2.8.6). Soil horizons are common in fluviatile sequences: calcretes if a semi-arid climate and low water table (Section 4.10.1) and seatearths with rootlets, siderite nodules and even coals, if humid conditions prevailed and the water table was high (Sections 3.7.1; 8.7).

With detailed studies fluviatile sequences can be interpreted in terms of the fluvial system which operated: alluvial fan or low or high sinuosity river.

Alluvial fans are particularly common in semi-arid regions where there is periodic or infrequent heavy rainfall, but they also occur in humid settings. They are aprons of sediment occurring adjacent to upland areas, particularly those bounded by faults, with the fan apex located at the mouth of a canyon or wadi. Alluvial fans build out on to playas and lakes, floodplains of more permanent rivers, coastal plains and occasionally directly into the sea to form fan-deltas. Fans usually have a radius of 5 to 15 km., with their size depending on the area of the catchment basin. The surface of a fan is dissected by a network of channels radiating out from the fan apex. Compared with other fluviatile sediments, those of alluvial fans are generally coarser, consisting of much gravel and coarse sand, because of the short distance and time of transport. Deposition takes place from debris flows, stream floods and sheet floods. Debris flows (mud flows) are high density, high viscosity flows consisting of much fine sediment, together with clasts up to a boulder size. Debris flow deposits are laterally extensive beds without erosive bases and with a matrix-support fabric (Section 2.2.4). Stream floods are low viscosity flows, usually confined to channels. The deposits are cross-bedded pebbly sands and lenticular gravels with a pebble-support fabric and possibly with imbrication (Section 2.2.4). Sheet floods, shallow but extensive surface flows, deposit thin laterally-continuous beds of sand and gravel, which may show graded bedding, planar and cross stratification.

Relative to high sinuosity rivers, low sinuosity streams tend to occur where slopes are higher, sediment is coarser, and water discharge-sediment load is higher and more variable; these are factors controlled by climate, tectonics and source-area geology. Low sinuosity rivers are frequently braided: the flow is divided by sand and gravel bars, sand flats and islands into a number of smaller channels. Floodplains are poorly developed. Sediments deposited in braided rivers are chiefly sands, with coarse gravels in higher reaches. As result of the dominant channel and bar deposition, with little lateral migration, braided river sequences frequently form major elongate sand bodies consisting of lenticular cross-stratified units (Fig. 2.52). Tabular cross bedding, formed by bars and sand waves, and internal erosion surfaces between units are common. Palaeocurrents of low sinuosity rivers are unimodal with a low degree of variance.

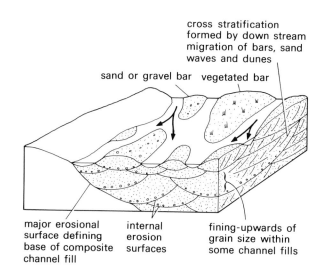

Fig. 2.52 Sketch showing part of a braided stream deposit.

cross stratification formed by down stream migration of bars, sand waves and dunes

sand or gravel bar   vegetated bar

major erosional surface defining base of composite channel fill

internal erosion surfaces

fining-upwards of grain size within some channel fills

High sinuosity or meandering streams possess distinct channel and overbank subenvironments (Fig. 2.53) and processes. The channel occupies only a small part of the alluvial plain but it migrates laterally through bank erosion and point bar sedimentation. Sand is moved as dunes on the channel floor and lower part of the point bar and as ripples on the upper part. The lateral accretion of point bars produces an erosive-based sandstone showing an upward decrease in grain size and scale of cross stratification (Fig. 2.53). The presence of a larger scale, epsilon cross bedding (Section 2.3.2) signifies this periodic lateral accretion. During floods silt and clay are deposited on the floodplain by overbank sedimentation. This results in a fine-grained member overlying the channel-point bar sandstone (Fig. 2.53). Some sand is deposited on levees adjacent to the channel when the river overtops its banks; sand is also deposited on floodplains as a result of crevassing, when a river breaches its banks. These sands occur as intercalations within the floodplain muds. The fining-upward sedimentary cycle produced through meandering stream deposition can be repeated many times in a fluviatile sequence. Palaeocurrents from meandering river sandstones have a greater dispersion than those from braided stream deposits.

65

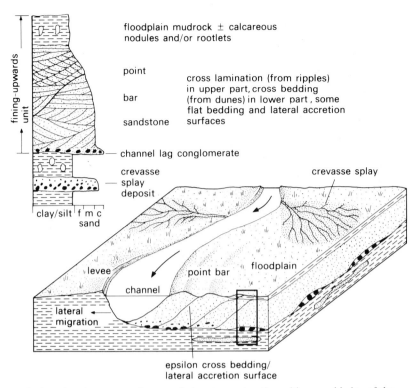

Fig. 2.53 The subenvironments of a meandering stream together with a graphic log of the sedimentary sequence produced through lateral migration of such a stream. Fluviatile fining-upward units are usually between about 2 and 20 m thick.

Reviews of fluviatile sediments have been presented by Allen (1965b), Collinson (1978) and Walker (1979). For descriptions of modern and ancient alluvial fan deposits see Bull (1972), Bluck (1967), Larsen & Steel (1978) and Heward (1978); for modern braided stream deposits see Williams & Rust (1969), Miall (1977) and Cant & Walker (1978), and for ancient ones see Smith (1970), Cant & Walker (1976) and Miall (1976); for ancient meandering stream deposits see Allen (1970) and Leeder (1973). Many papers on fluvial sediments are contained in Miall (1978).

2.10.2 DESERT ENVIRONMENTS

Deserts are areas of intense aridity, generally located in the subtropical belts (latitudes of 20 to 30°), although local topography and degree of continentality affect their development. Apart from areas of wind-blown sand, alluvial fans, ephemeral streams and desert lakes (playas) occur in a desert region, and there are vast areas of bare rock. Aeolian desert sands vary from a thin, impersistent cover to extensive sand seas.

Desert dune sands are typified by a grain size of fine to coarse sand (0.1 to 1.0 mm), good sorting and a positive skew (Section 2.2.1). Grains are well-rounded with a frosted surface (Section 2.2.3; Fig. 2.6). This textural maturity, produced by the frequent grain collisions, is matched by a compositional maturity: desert sands are mature to supermature, many being quartz arenites. Many desert sandstones are red through hematite pigmentation (Section 2.8.6) and fossils are absent apart from occasional vertebrate bones and footprints. The characteristic sedimentary structure of aeolian sands is large-scale, high-angle cross bedding, seen in Fig. 2.26 and described in Section 2.3.2. Thick sandstone sequences with this structure to the exclusion of others, and without finer grained interbeds, are likely of aeolian origin. Commonly associated with aeolian sandstones are wind-facetted pebbles (Dreikanters), waterlain deposits resulting from sheet and stream floods, and playa-lake sediments: mudrocks and evaporites.

Reviews of desert sediments, in particular aeolian sandstones, have been given by Glennie (1970) and Biggarella (1972). Some Permian and Triassic sandstones of Britain and N.W. Europe are interpreted as aeolian (e.g. Thompson, 1969; Glennie, 1972), as well as Mesozoic sandstones of the Colorado Plateau, U.S.A. Some 'classic' aeolian sandstones have been unsuccessfully re-interpreted as shallow-marine sand waves and tidal ridges (see discussion in Walker & Middleton, 1979).

2.10.3 LACUSTRINE ENVIRONMENTS

Sands and coarser sediments are being deposited along lake shorelines, in deltas where rivers drain into lakes, and on the deep lake-basin floors. Compared with their marine counterparts, beach sands and gravels of the lake shoreline are generally less well sorted and rounded since the level of wave activity is much less and there are no tides. Sediment distribution in a lake delta is similar to that of marine coastal deltas although the delta sequences produced are usually on a smaller scale. On the deep floors of lakes, river underflows deposit graded beds with scoured bases (Sturm & Matter, 1978), in a similar manner to turbidity currents of marine basins. The recognition of siliciclastic lake sediments is mainly dependent on the absence of a marine fauna and an association with certain minerals or rock-types which are restricted to lakes. Descriptions of modern and ancient lake sediments are contained in Matter & Tucker (1978).

2.10.4 DELTAIC ENVIRONMENTS

Deltas forming where rivers enter the sea are one of the main sites of siliciclastic deposition. Deltas are complex environments where the main factors controlling sediment distribution and delta morphology are the interplay between river regime, tides and wave action. Climate, water depth and subsidence rates are also important. In the study of deltas, a distinction is made between the *delta front*, from the sand bars at the mouths of distributary channels offshore to silty distal bars and the muddy prodelta, and the *delta plain*, the low lying area behind the delta front which includes the distributary channels, interdistributary bays, lakes, floodplains, marshes and swamps (Fig. 2.54). Two common delta types are the lobate delta, with numerous radiating and dividing distributary channels, and the birdfoot delta with just one or several major distributaries and well-developed interdistributary bays (Fig.

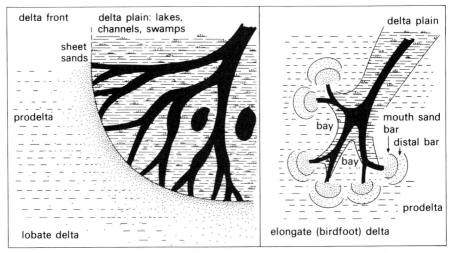

Fig. 2.54 The subenvironments of a lobate and elongate (birdfoot) delta. Progradation of a lobate delta gives rise to a laterally-extensive delta front sheet sand body, whereas an elongate sand body is generated by a birdfoot delta.

2.54). The Nile and pre-modern Mississippi deltas are examples of the lobate type while the modern Mississippi is a birdfoot delta.

Ancient deltaic sediments are usually thick sequences of sandstones and mudrocks, sometimes with interbeds of coal and limestone. The presence of marine fossils in some beds, usually in the limestones and some of the mudrocks, and non-marine fossils in others, is typical of deltaic sequences and indicates fluctuations in the position of the shoreline in response to sedimentation and sealevel changes. The characteristic appearance of deltaic sediments is as a vertical sequence of coarsening-upward units, produced by progradation (building out) of the delta front. Sands deposited at distributary mouths gradually build out over deeper water prodelta muds (Fig. 2.55). In a birdfoot delta, an elongate sand body is generated, whereas a more laterally-extensive sheet sand is produced through progradation of a lobate delta. The sand-infilled distributary channels themselves cut into the delta front deposits (Fig. 2.12). Delta plain deposits in the form of seatearths and coals, and possibly thin coarsening-upward units formed by the infilling of inter-distributary bays, occur above the delta front sequence. Mouth bar sands and channel sands show cross stratification and planar bedding while flaser and wavy cross bedding is common in distal bar and prodelta deposits. Many deltaic sandstones are lithic and arkosic arenites.

Repetition of delta advance sequences is brought about by delta switching and the formation of new delta lobes, and by external factors such as sealevel changes and tectonic movements. The thickness of each coarsening-up unit depends on the size and type of delta, water depth and subsidence rate, but it is generally in excess of 30 m. Deltaic sediments are common throughout the geological record but were particularly widespread during the Carboniferous, when extensive coal swamps developed on delta plains. Modern and ancient deltaic deposits have been discussed and reviewed in Morgan (1970), Broussard (1975), Coleman (1976) and Elliott (1978).

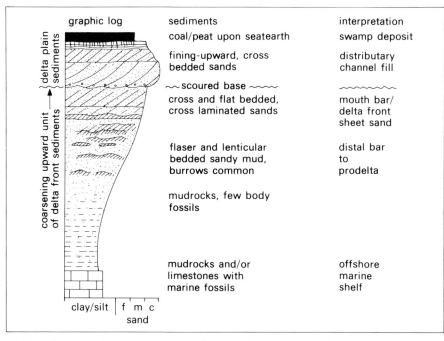

Fig. 2.55 Sketch graphic log and interpretation of typical sequence produced by delta progradation. Many variations can occur at the top of the sequence where the delta plain sediments occur, particularly if interdistributary bays are well-developed. Here the simple case of a delta distributary channel sand and swamp deposit is shown. The thickness of such a sequence varies considerably, but many are between 30 and 100 metres.

2.10.5 MARINE SHORELINE ENVIRONMENTS

Much siliciclastic sediment can be deposited in marine shoreline environments such as beaches, barrier islands, tidal flats and estuaries. Sediment supply, tidal ranges, wave action, sealevel and tectonic history, and climate all affect the sedimentation. Beaches and barriers are best developed in areas of moderate to high wave action, whereas tidal flats are more characteristic of areas with a high tidal range. Tidal processes dominate over wave activity in estuaries, although the river regime is also important.

Beaches are linear belts of sand along the coast whereas barriers are separated from the land by a lagoon. In many instances the lagoon is connected to the sea by tidal inlets. Well-studied barriers occur along the Texas and Georgia coasts, U.S.A., and the Dutch and German coasts of the North Sea. Offshore from beaches and barrier islands there is a decrease in grain size (Fig. 2.56). Symmetrical and asymmetrical wave-formed ripples and lunate dunes occur in the shallow subtidal (shoreface) off beaches and barriers, and planar surfaces formed by wave swash-backwash dominate the intertidal zone (foreshore). Cross stratification and planar bedding in truncated sets thus characterize beach and barrier sands, together with heavy mineral concentrations. Burrow structures are common in shoreface sediments. The sands are typically quartz arenites (i.e. supermature) in composition, and texturally supermature too. Barrier islands and beaches may migrate both

69

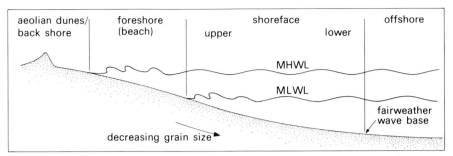

Fig. 2.56 Subenvironments from a sandy foreshore (beach) seawards to the offshore zone.

seawards and landwards, depending on the state of sealevel (whether it is static or rising), the supply of sand and the rates of subsidence. Coarsening-upward sequences are produced by seaward progradation (Fig. 2.57). For descriptions of modern and ancient beach and barrier sands see Davies *et al.* (1971), Schwartz (1973), Johnson (1975), Davis & Ethington (1976) and Davis (1978).

Tidal flats reach several kilometres in width and occur around lagoons, behind barriers, in estuaries and tidally-dominated deltas. There is usually a decrease in sediment grain size across the flat from sand in the low intertidal zone to silt and clay in the higher part. The characteristic sedimentary structures of the mid-upper tidal flat are various types of ripple, often showing interference patterns, and these give rise to flaser, wavy and lenticular bedding (Section 2.3.2). Cross-bedded sands with some herring-bone structure are deposited in the low tidal flat. Meandering tidal creeks dissect the tidal flats and their lateral migration produces sets of obliquely-inclined mud and sand laminae. Burrows and grazing trace fossils are common. Progradation of tidal

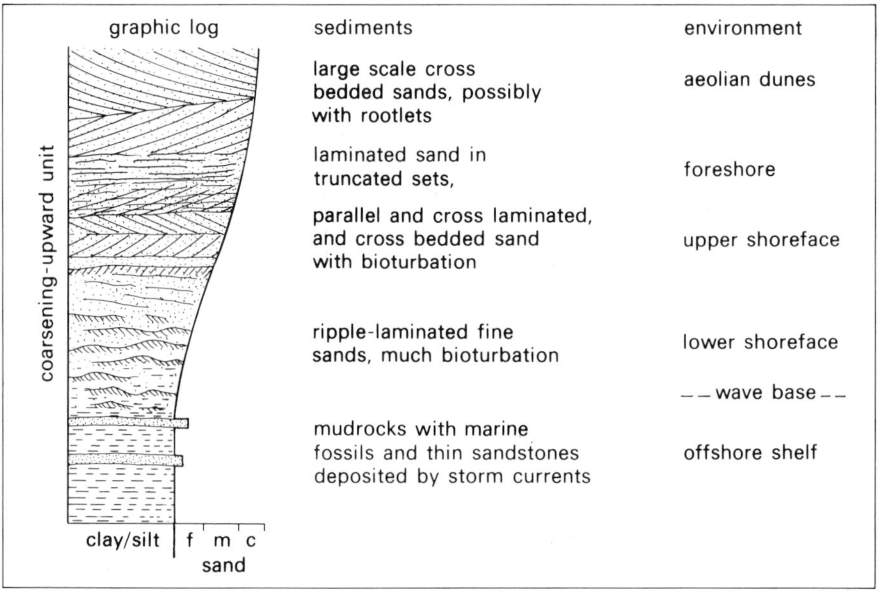

graphic log	sediments	environment
	large scale cross bedded sands, possibly with rootlets	aeolian dunes
	laminated sand in truncated sets,	foreshore
	parallel and cross laminated, and cross bedded sand with bioturbation	upper shoreface
	ripple-laminated fine sands, much bioturbation	lower shoreface
		— — wave base — —
	mudrocks with marine fossils and thin sandstones deposited by storm currents	offshore shelf

coarsening-upward unit

clay/silt | f m c
                sand

Fig. 2.57 Sketch graphic log and features of sequence produced by seaward progradation of beach (or barrier island). 10 m (or more) is a typical thickness of such a coarsening upward unit.

flat sediments typically forms a fining-upward sequence (Fig. 2.58), with the thickness determined by the palaeotidal range. Modern and ancient tidal flat deposits are discussed in Ginsburg (1975), Klein (1977) and Hobday & Eriksson (1977).

Most modern estuaries are drowned river valleys resulting from the post-glacial, Holocene sealevel rise. Estuarine sand bodies are located within and adjacent to the main channel and consist of sediment brought down by the river and supplied from the adjacent marine shelf. Mud flats and swamps also occur in estuaries. Modern estuarine sediments have been described by Lauff (1967), Howard & Frey (1973) and Greer (1975), but there are few truly estuarine ancient deposits; see Goldring *et al.* (1978) for an Eocene example from southern England.

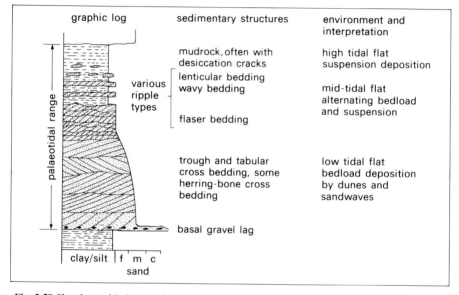

Fig. 2.58 Sketch graphic log and interpretation of fining-upward unit generated by progradation of a siliciclastic tidal flat. The thickness of the unit, usually several metres, indicates the palaeotidal range.

2.10.6 SHALLOW MARINE SHELVES AND EPEIRIC SEAS

In the shallow marine environment away from the coastline, sands are deposited in continental shelf seas, such as on the present-day eastern shelf of North America, and in epeiric, epicontinental seas, such as the North Sea. In some shelf and epeiric seas tidal currents dominate, while others are storm or wave dominated. Distinctions are made between fairweather and storm processes. Many of the sediments in modern shallow seas are relict; they were deposited before or during the Holocene transgression in glacial, fluvial and shoreline environments. Sands reworked from these relict sediments by tidal currents, storms and waves form sand waves, ripples, sand sheets and tidal ridges. These structures give rise to cross stratification and planar bedding.

Ancient shallow marine sands vary from thin-bedded, sharp-based sandstones with graded bedding and cross lamination, deposited by storm-

71

generated currents and interbedded with mudrocks, to thick sequences of tabular and trough cross-bedded sandstone where tidal ridges and sand waves existed. Palaeocurrent directions are often highly variable. Evidence of subaerial exposure is absent, but marine body and trace fossils can occur. As a result of the constant reworking, many shallow-marine sandstones are texturally and compositionally mature; many are quartz arenites. Shallow marine sands and sandstones have been reviewed by Swift (1976), Johnson (1978) and Walker (1979). Specific examples occur in the Late Precambrian of Scotland (Anderton, 1976) and Norway (Johnson, 1977), and the Lower Carboniferous of Ireland (Raaf et al., 1977). Storm beds are described by Goldring & Bridges (1973) and Brenchley et al. (1979).

2.10.7 CONTINENTAL MARGINS AND DEEPWATER BASINS

Continental margins and deepwater basins are the depositional sites of sandstones and conglomerates derived from adjacent slopes, platforms and shelves. Transport downslope is through sliding and slumping, and through sediment gravity flows: in particular turbidity currents, but also debris flows, and the less important grain flows and fluidized sediment flows. In addition oceanic bottom currents produced through thermo-haline density differences can transport and rework sediment. Apart from these resedimentation processes, deeper water environments are primarily the sites of pelagic and hemipelagic deposition: cherts (Section 9.3), pelagic limestones (Section 4.10.2) and mudrocks (Section 3.7.2). In fact, it is the association, often an interbedding, of sandstones and conglomerates with these deepwater deposits and their pelagic fossils which signifies resedimentation. There is also an absence of any typical shallow-water or subaerial sedimentary features of course, although shallow-water fossils and plant debris can be transported into deepwater. Some sedimentary structures and textures of the coarse clastics themselves indicate resedimentation.

Slides and slumps involve small to large masses of sediment, with more internal deformation (folding and brecciation) occurring in slumps. Turbidity currents are density currents of sediment and water in which the sediment is kept in suspension through the fluid turbulence. Some coarser sediment is transported as bed load. Slumps and turbidity currents today occasionally cause the breakage of submarine cables (Heezen & Hollister, 1971). Debris flows, like those on alluvial fans, are highly-concentrated sediment flows of high viscosity, usually consisting of fine-grained sediment which can support and thus transport large clasts. Grain flows arise from grain-to-grain collisions; fluidized sediment flows form through the upward escape of pore water. The sediment gravity flows usually develop from slumps and slides on a slope, and in many cases earthquakes are the trigger mechanism. Distinctive sediment types are produced on deposition.

Turbidites, deposited by decelerating turbidity currents, are the most important, and most common type of deepwater sandstone. They typically range from a few centimetres to a metre or more thick, and form sequences many hundreds or thousands of metres thick consisting of sandstones regularly alternating with hemipelagic mudrocks (as the turbidites of Fig. 4.61). Many turbidite sandstones possess well-developed sole structures: flutes (Fig. 2.10), grooves (Fig. 2.11) and other tool marks, as well as trace fossils and load casts (Dzulynski & Walton, 1965). Graded bedding is a characteristic internal structure of turbidites and horizontal and cross laminations are common (Fig. 2.22). Very rarely, antidune cross bedding is present (Section 2.3.2). Some turbidites show a definite sequence of internal structures, referred to as the Bouma sequence after Bouma, 1962 (Fig. 2.59). It comprises: a graded A

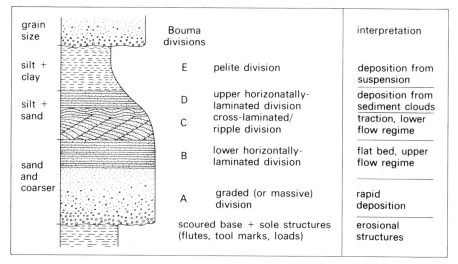

Fig. 2.59 Sequence of structures and Bouma divisions of a siliciclastic turbidite bed. Turbidites range in thickness from several centimetres to a metre or more.

division; a lower horizontally laminated B division, sometimes with parting lineation; a cross-laminated C division, frequently with stoss-side preservation as in Fig. 2.22; an upper horizontally-laminated D division and a pelitic E division. This sequence can be interpreted in terms of deposition from a waning flow: divisions A and B represent upper flow regime, C indicates lower flow regime, and D and E deposition from suspension (e.g. Walker, 1965). Cross laminations of the C division may be deformed into convolute laminations (Section 2.3.3). Incomplete Bouma sequences in turbidites, such as BCDE and CDE, are common. There are distinctive down current changes in the structures of turbidite beds (e.g. Walker, 1967; Enos, 1969). Those deposited closer to the source, proximal turbidites, are thicker, coarser-grained and poorly graded and have channels, scour marks and thin interbeds; distal turbidites, on the other hand, are thinner, finer-grained, parallel-sided and well-graded, with laminations, ripples and tool marks. Sole structures, internal structures and grain fabric can all be used for palaeocurrent analysis (Section 2.4).

The composition and texture of turbidites is largely a reflection of the material available for resedimentation. Many ancient turbidites are greywacke in composition (Section 2.7.4).

Of the other sediment gravity flow deposits, *debris flow deposits* are the more significant. They are chiefly matrix-supported conglomerates with little sorting of clasts. *Grain flow deposits* are thin, structureless sandstones, without size grading, or with some reverse grading in the lower part. *Fluidized sediment flow deposits* show poor size grading and an absence of depositional internal structures, but fluid escape structures such as sand volcanoes and dish and pillar structures are common (Section 2.3.3, Lowe, 1978).

In addition to the sediment gravity flows which generally move directly downslope to the basin floor, there are slope-parallel, contour-flowing bottom currents which can transport sediment. The deposits of these contourites are usually thin ($<$ 5 cm), laminated sandy siltstones (e.g. Stow & Lovell, 1979).

Sediment gravity flows deposit their clastic material in the lower reaches of submarine canyons, on submarine fans located at the foot of slopes, usually at the mouths of canyons, in deep ocean trenches, and on basin plains surrounding the fans. Proximal turbidites, often in channels, and the deposits of other sediment gravity flows occur in canyons and on submarine fans. Vertical changes in bed thickness and grain size in fan sequences result from migration and progradation of fan lobes (Walker & Mutti, 1973). Basin plain deposits are chiefly distal turbidites, with individual beds continuous for tens of kilometres.

Thick sequences of turbidites occurring within geosynclinal belts have been referred to as *flysch,* and in many cases flysch deposition took place during or just before the main orogenic deformational stage. With the better understanding now of geosynclines through plate tectonic theory and with deep sea drilling results, it is known that thick turbidite sequences can develop in many different tectonic situations: within trenches at destructive plate margins, at passive margins, in back-arc basins and within failed rifts for example. The composition and texture of these deepwater sandstones depend on the tectonic setting and on the amount of sediment supplied from the adjacent continent and/or volcanic sources (Reading, 1972). Reviews of deepwater clastic sedimentation include Walker (1973), Middleton & Hampton (1976) and Rupke (1978); compilations of relevant papers include Bouma & Brouwer (1964), Lajoie (1970) and Dott & Shaver (1974).

2.10.8   GLACIAL ENVIRONMENTS

Glacial environments include a wide range of depositional settings from continental to marine, subglacial to supraglacial and glaciofluvial to glaciolacustrine. Geomorphologically a number of different glacier types have been distinguished but a division of the thermal state into cold and temperate glaciers is more important in terms of glacial erosion and sedimentation. Cold glaciers are dry-based and have much rock debris incorporated into their basal parts; temperate glaciers are wet-based, contain less sediment, but move faster and so are more powerfully erosive.

Sediment deposited directly from a glacier, either subglacially while it is moving or when the glacier is stagnating and melting (ablating), is referred to as *till,* or *tillite* if indurated. Subglacial tills and tillites are mostly massive (Fig. 2.8) although some show a banding. Typical features are: an extensive lateral development, usually on a regional scale; a thickness of several to tens of metres; a lack of stratification; much matrix which supports clasts, some of which may show striations and facets; clasts of both local and exotic origin;

and a matrix which is largely comminuted rock fragments. Such sediments may be difficult to distinguish from debris flow deposits (Sections 2.10.1, 2.10.7).

Lenticular and stratified conglomerates and sandstones, often with cross bedding, are almost invariably interbedded or associated with tillites. These are waterlain deposits resulting from glaciofluvial processes. They may be deposited subglacially, as in eskers, in supraglacial streams and lakes during ablation, and in proglacial outwash plains (sandurs). Sediments of glacial lakes are sands and gravels where outwash streams construct deltas and, in deeper parts, rhythmically-laminated mudrocks (varves, Section 3.7.2), probably with scattered clasts dropped from rafted ice. Where ice sheets reach the sea, then till is deposited on the seafloor and can be reworked by marine processes and transported into deep-water through slumping and turbidity currents. Glaciomarine tills may contain fossils. Icebergs calved from ice shelves release sediment as they melt and deposit dropstones over a wide area.

There have been five major periods of glaciation: an early Proterozoic glaciation mainly documented from North America; a Late Precambrian glaciation which affected most continents; a Late Ordovician glaciation, evidence for which is particularly well displayed in North Africa; a Permo-Carboniferous glaciation which affected Gondwanaland, that is, S. Africa, S. America, India, Australia and Antarctica; and the Late Cainozoic, chiefly Pleistocene glaciation which began in the Oligocene-Miocene, and resulted in the thick glacial and glacial-associated drift that now covers much of northern Europe and Asia and North America. Modern and ancient glacial sediments have been reviewed recently by Wright & Moseley (1975) and Edwards (1978); many descriptions of pre-Pleistocene tillites are contained in Harland & Hambrey (1980).

2.10.9   FACIES SEQUENCES AND CONTROLS

Siliciclastic depositional environments give rise to characteristic sedimentary facies, as outlined above, with compositions, textures and sedimentary structures mainly dependent on sediment source and supply and depositional process(es). Tectonic context, position of sealevel and climate are major external factors affecting and controlling the pattern of sedimentation. The facies sequences developed in an area depend on the depositional processes and changes in the external factors. Vertical facies sequences can be generated through depositional processes alone, particularly through the progradation and lateral migration of environments (as propounded by Johannes Walther, Section 1.2.2) and through vertical aggradation. Deltas building out into deeper water: a lake or the sea (Fig. 2.55); rivers meandering across a floodplain (Fig. 2.53), beaches, barriers and tidal flats prograding seawards, and the construction of alluvial and submarine fans, all lead to lateral migrations of environments and subenvironments and the formation of characteristic depositional units and facies sequences. The infilling of a sedimentary basin with turbidites, bringing the depositional surface into the shallow marine environment, and sedimentation on a river floodplain are examples of vertical aggradation.

Many changes in siliciclastic facies sequences, however, are brought about by the external factors. Tectonic movements can have a drastic and varied effect. Examples include (i) the influx of coarse sediment through uplift in the source area; with rivers this could bring about a change from a meandering to braided character and in any siliciclastic sequence it could be marked by a change in sediment composition and/or texture and (ii) an increase in seismicity leading to more frequent sediment gravity flows, resedimentation and post-depositional sedimentary structures. Changes in climate can lead to increased aridity on the one hand, or a glaciation on the other. With increased aridity, river flow becomes more erratic and braiding may result; deserts and their aeolian sands can develop, and subaerial weathering which supplies much of the siliciclastic material for sedimentation, is more physical than chemical.

Sealevel changes, brought about by climatic and tectonic events have a profound effect on siliciclastic sedimentation. For example, with a rise in sealevel, deltaic cycles can be repeated, and clastic shorelines can be drowned and overlain by shelf or epeiric sea deposits; with a fall in sealevel turbidity currents may be more frequent as fluviatile sediments are deposited closer to shelf margins.

Shelf and epeiric seas and deepwater basins starved of siliciclastic sediment as a result of tectonic or climatic events may become sites of biogenic, particularly carbonate sedimentation or evaporite deposition if aridity is coupled with some restriction of circulation.

Extended discussions of facies and their controls are to be found in Selley (1976, 1978a), Reading (1978), Friedman & Sanders (1978), Walker (1979) and Blatt *et al.* (1980).

Further reading

Pettijohn, F.J., Potter, P.E. & Siever, R. (1973) *Sand & Sandstone*, pp. 618. Springer-Verlag, Berlin.

Reineck, H.E. & Singh, I.B. (1973) *Depositional Sedimentary Environments—with reference to terrigenous clastics*, pp. 439. Springer-Verlag, Berlin.

Folk, R.L. (1974) *Petrology of Sedimentary Rocks*, pp. 159. Hemphill Publishing Co., Austin, Texas.

Selley, R.C. (1976) *An Introduction to Sedimentology*, pp. 408. Academic Press, London.

Allen, J.R.L. (1977) *Physical Processes of Sedimentation*, pp. 248. Allen & Unwin, London.

Friedman, G.M. & Sanders, J.E. (1978) *Principles of Sedimentology*, pp. 792. John Wiley, New York.

Reading, H.G. (Ed.) (1978) *Sedimentary Environments and Facies*, pp. 557. Blackwell Scientific Publications, Oxford.

Selley, R.C. (1978) *Ancient Sedimentary Environments*, pp. 287. Chapman & Hall, London.

Dapples, E.C. (1979) Diagenesis of Sandstones. In: *Diagenesis of Sediments and Sedimentary Rocks* (Ed. by G. Larsen and G.V. Chilingarian), pp. 31–97. Elsevier, Amsterdam.

Scholle, P.A. (1979) A color illustrated guide to Constituents, Textures, Cements and Porosities of Sandstones and Associated Rocks. *Mem. Am. Ass. Petrol. Geol.* **28**, pp. 201.

Walker, R.G. (Ed.) (1979) *Facies Models*, pp. 211. *Geoscience Canada.*

3

Terrigenous Clastic Sediments
II: Mudrocks

3.1   Introduction

Mudrocks are the most abundant of all lithologies, constituting some 45 to 55%
of sedimentary rock sequences. However, because mudrocks are easily
weathered, they are frequently poorly exposed. In addition, as a result of their
fine grain size, their study often requires detailed laboratory analyses.
Mudrocks can be deposited in practically any environment although the major
depositional sites are river floodplains and lakes, large deltas, the more distal
areas of continental shelves and platforms, and the ocean floors. The main
constituents of mudrocks are clay minerals and silt-grade quartz. Since these
are largely detrital, the clay mineralogy to a greater or lesser extent reflects the
climate and geology of the source area.

In terms of grain size, clay refers to particles less than 4 μm in diameter,
whereas silt is between 4 and 62 μm (Table 2.1). Clay as applied to a mineral is a
hydrous aluminosilicate with a specific sheet structure (Section 3.4.1); the
typical size of clay minerals is less than 2 μm but they may reach 10 μm or more.
The term mud (also lutite) loosely refers to a mixture of clay and silt grade
material, with the former dominant. *Mudstone,* the indurated equivalent of mud
is a blocky, non-fissile rock, whereas *shale* is usually laminated and fissile
(fissility is the property of splitting into thin sheets). *Argillite* is used for a more
indurated mudrock and *slate* for one that possesses a cleavage. A sedimentary
rock of clay-grade material only is called a *claystone,* and one that contains
more silt-grade particles than clay is called a *siltstone.* As a general term for all
these rock-types mudrock is useful. Further than this, there is no generally-
accepted classification of mudrocks although several schemes have been put
forward recently (Picard, 1971; Lewan, 1978; Spears, 1980).

To describe mudrocks, particularly in the field, the terms mudstone, shale,
claystone and siltstone are best qualified by attributes such as colour, degree of
fissility, sedimentary structures, and mineral, organic and fossil content
(Table 3.1). Detailed studies in the laboratory, especially the use of X-ray
diffraction, may be required to determine the mineralogy.

3.2   Textures and structures of mudrocks

Fine terrigenous clastic sediments do not have the wide range of textures and
structures so typical of coarser clastic sediments (Chapter 2). This is mainly
because of the finer grain size and cohesive properties of mud. The grain-size
distribution of mudrocks is not easy to study and there are problems of
interpretation. The particle size of unconsolidated muds can be measured using

Table 3.1 Features to note in the description of mudrocks.

Attribute	Examples of descriptive adjectives
Colour	grey, green, red, brown, variegated
Degree of fissility	fissile, non-fissile, blocky, earthy flaggy, papery
Sedimentary structure	bedded, laminated, slumped, bioturbated or if no structures: massive
Mineral content	quartzose, illitic, kaolinitic, zeolitic, micaceous, calcareous dolomitic, gypsiferous
Organic content	organic-rich, bituminous, carbonaceous
Fossil content	fossiliferous, foraminiferal, ostracod, graptolitic

a sedimentation chamber or settling tube, but for many consolidated mudrocks an electron microscope is necessary. The interpretation of the grain-size data in terms of environmental conditions is complicated by the fact that clays are often deposited as floccules and that sediment-feeding organisms may generate pellets of mud.

A common texture is that of a preferred orientation of clay minerals and micas parallel to the bedding. This can be detected in thin section by areas of common extinction. The texture is largely the result of compaction and the associated dewatering which produces an alignment of clay flakes. Related to this is the property of *fissility*, possessed by shales. Fissility is the ability of mudrocks to split along smooth planes parallel to the stratification. The origin of fissility is not fully resolved. A major factor is the compaction-induced alignment of clay minerals, but also important is the presence of lamination. Fissility may be poorly developed or absent, as it is in mudstones, through bioturbation, the presence of much quartz silt or calcite, and flocculation of clays during sedimentation producing a random fabric which is retained on compaction. The degree of fissility shown by mudrocks may be related to weathering at outcrop.

One common sedimentary strcture of mudrocks is *lamination* (Fig. 3.1). Laminations are mainly due to variations in grain size and/or changes in composition. Size-graded laminations may be deposited from low density turbidity and suspension currents or from decelerating storm currents, in relatively short periods of time (hours or days). Other laminations may develop over much longer periods of time (months or years) if there is a seasonal or annual fluctuation in sediment supply and/or biological productivity. Organic laminae in mudrocks for example may be produced by seasonal algal blooms and the varved couplets of glacial lakes (Section 3.7.2) are taken to reflect the annual spring melting.

Siltstones may show the same sedimentary structures as finer grades of sandstone (Section 2.3.2). Small-scale current ripples occur and give rise to cross lamination. Planar beds with parting lineation (Fig. 2.17) can develop at higher stream powers. Symmetrical wave-formed ripples can also form. Where mud

(a)

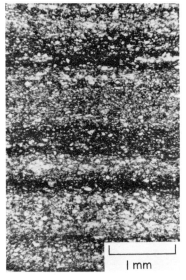

(b)

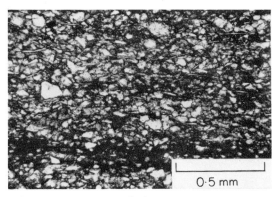

(c)

Fig. 3.1 Lamination in mudrocks. 3.1a Rhythmites, which are probably glacial varves, consisting of graded silt passing up into clay-grade material. Late Precambrian, Islay, Scotland. 3.1b Rhymthmites consisting of alternations of silt-grade quartz (in calcareous cement) and clay-organic matter; the result of seasonal deposition is a non-glacial lake. Caithness Flagstones Middle Devonian, N.E. Scotland (Fig. 2.16). 3.1c Detail of 3.1b showing subangular silt-grade quartz, muscovite flakes oriented parallel to lamination and organic-clay lamina (dark). All plane polarized light.

and sand are alternately deposited, through fluctuating current regimes and/or sediment supply, as in tidal flat, delta front and other environments, then flaser and lenticular bedding is produced (Fig. 2.23).

Some mudrocks do not possess sedimentary structures but are *massive*. This massive nature may be a result of the depositional mechanism. It is a common feature where the sediment was viscous, possessing little water at the time of deposition, as in mudflows and debris flows (Sections 2.10.1, 2.10.7). In other cases, mudrocks have become massive subsequent to initial deposition through bioturbation, mass sediment movement (slumping), dewatering, soil processes (pedogenesis) and root growth.

Other structures which occur in mudrocks and have been described in Chapter 2 include: the erosional structures cut into muds and preserved on the soles of sandstones (grooves and flutes); slump structures; desiccation cracks formed through subaerial exposure; syneresis cracks formed through subaqueous sediment shrinkage; biogenic structures and rain spot prints.

Many mudrocks contain *concretions* and *nodules* (Fig. 3.2). These are regular to irregular spherical, ellipsoidal to flattened bodies, commonly composed of calcite, siderite (Section 6.4.2), pyrite (Section 6.4.3), chert

Fig. 3.2 Calcareous nodules in red mudrock. Relatively deep water, marine deposit. Upper Devonian, Harz Mountains, Germany.

(Section 9.4) or calcium phosphate (Section 7.4), together with some original sediment. Nodules may grow at or just below the sediment surface at the time of deposition. These have been referred to as syngenetic and they may show evidence of attack from boring and encrusting organsims. Early diagenetic nodules are those formed in sediments which were still soft and uncompacted. These can be recognized from the presence of uncrushed fossils within the nodules and from the curving of laminae around the nodule, showing that compaction took place after growth. Late diagenetic or epigenetic nodules form after compaction of the enclosing sediments and with these, laminations in the sediment pass unaffected through the nodule. Nodules containing an internal network of cracks, particularly ones which widen towards the interior, are referred to as septarian nodules. The cracks form through shrinkage and indicate growth in water-laden sediment. The growth of nodules and concretions arises from the localized precipitation of cement from porewaters within the sediments. The Eh and pH of these porewaters is important in controlling nodule growth rates and mineralogy. Occasionally nodules may form around a nucleus, a fossil for example, as a result of the local chemical conditions. More commonly, nodules are without a nucleus, and form along definite horizons or within particular beds, reflecting a level at which supersaturation of porewaters was achieved. Discussions of the origin of nodules included Raiswell (1971) and Hudson (1978).

A poorly-understood structure occurring in mudrocks is *cone-in-cone structure*. It consists of nested cones of fibrous calcite, rarely ankerite or siderite, oriented normal to the bedding, the whole constituting a nodule or impersistent layer. There is often evidence that the structure has formed by

80

displacement. Cone-in-cone may be analogous to fibrous gypsum (Section 5.2.4), being formed through hydraulic jacking of the sediment, with the fibrous nature of the calcite reflecting growth under stress.

3.3 The colour of mudrocks

The colour of a mudrock, and indeed of any rock, is a function of its mineralogy and geochemistry. Colour can be very useful in field mapping to distinguish between various mudrock units or formations. The main controls on colour are the organic matter and pyrite content and the oxidation state of the iron. With an increasing amount of organic matter and pyrite, mudrocks take on a darker grey colour and eventually become black. Many marine and deltaic mudrocks are various shades of grey or even black as a result of finely disseminated organic matter and pyrite.

Red and purple colours are due to the presence of ferric oxide, hematite, occurring chiefly as grain coatings and intergrowths with clay particles. It is generally accepted that the red colour develops after deposition, through an ageing process of a hydrated iron oxide precursor, as is the case with red sandstones (Section 2.8.6). The origin of the precursor, however, whether it has formed through in-situ dissolution of metastable mafic grains or is detrital, is still a matter of discussion (Hubert & Reed, 1978; Van Houten, 1968, 1974; Walker, 1967, 1974; Walker & Honea, 1969). The impermeable nature of many red mudrocks might suggest a detrital source. Where iron-oxide coatings on grains are patchily developed or less intense then brown colours may result. Green mudrocks contain no hematite, organic matter or iron sulphides but result from ferrous iron within the lattices of illite and chlorite. The green colour may be original but in many instances it develops in mudrocks which were originally red through the reduction of hematite by migrating porewaters. The green colour may thus occur in more porous silty bands, in mudrocks close to porous sandstones, or adjacent to faults and joints. Many non-marine mudrock sequences of playa-lacustrine and floodplain origin are red in colour reflecting the dominantly oxidizing nature of the depositional and diagenetic environment.

Other colours in mudrocks result from a mixing of the colour-producing components. Olive and yellow mudrocks for example may owe their colour to a mixing of green clay minerals and organic matter.

3.4 Mineral constituents of mudrocks

3.4.1 CLAY MINERALS

Clay minerals are hydrous aluminosilicates with a sheet or layered structure; they are phyllosilicates, like the micas. The sheets of a clay mineral are of two basic types. One is a layer of silicon–oxygen tetrahedra with three of the oxygen atoms in each tetrahedron shared with adjacent tetrahedra and linked together to form a hexagonal network (Fig. 3.3). The basic unit is Si_2O_5 but within these silica layers aluminium may replace up to half the silicon atoms. The second type of layer consists of aluminium in octahedral coordination with O^{2-} and

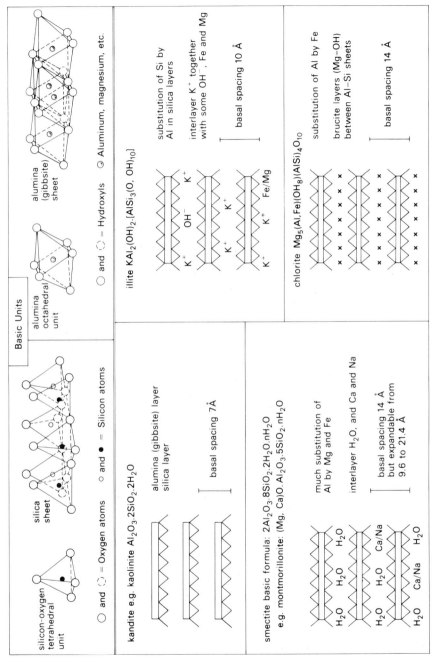

Fig. 3.3 Diagrams illustrating the structures of clay minerals.

82

OH$^-$ ions so that in effect the Al^{3+} ions are located between two sheets of O/OH ions (Fig. 3.3). In this type of layer, not all the Al (octahedral) positions may be occupied, or Mg^{2+}, Fe or other ions may substitute for the Al^{3+}. Layers of Al–O/OH in a clay mineral are referred to as gibbsite layers since the mineral gibbsite (Al(OH)$_3$) consists entirely of such layers. Similarly, layers of Mg–O/OH are referred to as brucite layers after the mineral brucite (Mg(OH)$_2$) composed solely of this structural unit. Clay minerals, then, consist of sheets of silica tetrahedra and aluminium or magnesium octahedra linked together by oxygen atoms common to both. The stacking arrangement of the sheets determines the clay mineral type, as does the replacement of Si and Al ions by other elements. Structurally, the two basic groups of clay minerals are the kandite group and smectite group.

Members of the kandite group have a two-layered structure consisting of a silica tetrahedral sheet linked to an alumina octahedral (gibbsite) sheet by common O/OH ions (Fig. 3.3). Replacement of Al and Si does not occur so that the structural formula is (OH)$_4$Al$_2$Si$_2$O$_5$. Members of the kandite group are *kaolinite*, by far the most important, the rare dickite and nacrite, which have a different lattice structure, and halloysite which consists of kaolinite layers separated by sheets of water. Related structurally to kaolinite are the alumino–ferrous silicate chamosite and the ferrous silicate greenalite (see Sect. 6.4.4). Kaolinite has a basal spacing, i.e. distance between one silica layer and the next, of 7 Å.

Members of the smectite group have a three-layered structure in which an alumina octahedral layer is sandwiched between two layers of silica tetrahedra (Fig. 3.3). The typical basal spacing is 14 Å but smectites have the ability to adsorb water molecules and this changes the basal spacing; it may vary from 9.6 Å (with no adsorbed water) to 21.4 Å. This feature of smectites, as a result of which they are often called 'expandable clays', is utilised in their X-ray identification. The common smectite is *montmorillonite;* it approximates to Al$_4$(Si$_4$O$_{10}$)$_2$(OH)$_4$. nH$_2$O but substitution of the Al^{3+} by Fe^{2+}, Mg^{2+} and Zn^{2+} can take place. A net negative charge resulting from such substitutions is balanced by other cations, especially Ca^{2+} and Na$^+$, which are contained in interlayer positions. Nontronite, saponite and stevensite are other smectites occasionally found in sediments; with nontronite Fe^{3+} replaces Al^{3+} in the octahedral layers and in saponite and stevensite Mg^{2+} substitution of Al^{3+} has taken place. Vermiculite has a structure similar to smectite, although it is less expansible, with all the octahedral positions occupied by Mg^{2+} and Fe^{2+}, and much substitution of Si^{4+} by Al^{3+}.

Illite, the most common of the clay minerals in sediments, is related to the mica muscovite. It has a three-layered structure, like the smectities, but Al^{3+} substitution for Si^{4+} in the tetrahedral layer results in a deficit of charge which is balanced by potassium ions in interlayer positions (Fig. 3.3). Some hydroxyl (OH$^-$), Fe^{2+} and Mg^{2+} ions also occur in illite. The basal spacing is about 10 Å.

Other clay minerals are chlorite, glauconite, sepiolite and palygorskite. *Chlorite,* like the smectites and illite, has a three-layered structure, but with a brucite (Mg–OH) layer between (Fig. 3.3). Substitution by Fe^{2+} occurs in chlorite (imparting the green colour) and the basal spacing is 14 Å. Glauconite

(Section 6.4.4) is related to illite and the micas, but contains Fe^{3+} substituting for Al^{3+}. Sepiolite and palygorskite are magnesium-rich alumino-silicates.

In addition to the four common clay minerals, illite, kaolinite, montmorillonite and chlorite, *mixed-layer clays* are also common. These consist of an interleaving of sheets of the common clays, in particular illite–montmorillonite and chlorite–montmorillonite. Specific names have been applied where there is a regular mixed-layering, corrensite for a chlorite–montmorillonite mixed-layer clay for example.

During weathering and diagenesis, interlayer cations can be leached out of the clay minerals by percolating waters. Such non-stoichiometric clays are termed degraded and in fact many illites and smectites in modern sediments are degraded, as well as some chlorites, biotites and muscovites.

Because of their fine crystal size and the presence of unsatisfied bonds, clay minerals are important in the process of ion exchange. Ions in aqueous solutions can be adsorbed on to and desorbed from clays, with the water chemistry controlling the exchange process. Some elements, such as iron (Section 6.2) can be transported by adsorption on clays.

The identification of clay minerals in a mudrock (or any other rock) is normally undertaken through X-ray diffraction of the less than 2 μm fraction of the sediment. The basal spacings of the clay minerals are deduced from the X-ray reflections for the untreated sample, and then the sample treated (a) with glycol, which causes expansion of the lattice of any smectites present and (b) with heat, which reduces the lattice spacing of smectite. The clay minerals can rarely be identified with the petrological microscope because of their fine crystal size. Occasionally where clays have been precipitated in cavities in sandstones (Section 2.8.5) or volcaniclastic rocks during diagenesis they may be large enough for recognition in thin section (Fig. 2.49). The clay minerals do differ in their optical properties; e.g. birefringence increases from kaolinite to montmorillonite to illite. They may also also differ in their crystal form; kaolinite often forms 'booklets'. A clearer picture of clay mineral crystal shape can be obtained from the scanning electron microscope which is now increasingly used for clay mineral identifications (Figs. 2.47; 2.48).

For further information on clay minerals and their structure see Grim (1968), Millot (1970), Weaver & Pollard (1973) Velde (1977), and Potter *et al.* (1980).

3.4.2   QUARTZ

Quartz in mudrocks is chiefly of silt-grade although coarser, sand-size, grains do occur, especially where the mudrocks grade laterally or vertically into sandstones. Quartz-silt is invariably angular in comparison with the typically more rounded quartz sand. Silt-grade quartz is derived from grain collisions in aqueous and aeolian media and from glacial grinding. Some quartz in mudrocks is diagenetic rather than detrital.

3.4.3   OTHER CONSTITUENTS

Feldspars are generally present in low concentrations in mudrocks in view of their lower mechanical and chemical stability relative to quartz (Section

84

2.5.3). Muscovite is common but biotite much less so. Calcite (or aragonite) may occur in the form of skeletal debris. Diagenetic calcite, as well as dolomite and siderite, occur as microscopic crystals, evenly disseminated or concentrated into nodules. Mudrocks may grade into calcareous mudrocks, referred to as marls. Pyrite occurs as cubes, framboids and nodules in dark, organic-rich mudrocks (Section 6.4.3). Other minerals occasionally present are glauconite, chamosite, hematite, gypsum-anhydrite and halite. Organic matter is common in mudrocks, particularly black shales (Section 3.7.2) and if present in sufficient concentrations it may give rise to oil shales (Section 8.8). Finely comminuted rock debris (rock 'flour') is a major constituent of fine terrigenous clastic sediments in glacial environments.

3.5 The formation and distribution of clay minerals in modern sediments

The major site of clay mineral formation is in the weathered mantles and soil profiles developed upon solid bedrock and unconsolidated sediments. Soils develop through physico-chemical and biological processes (pedogenesis) and possess distinct horizons. Many soils possess three horizons, A, B, C, and these can often be subdivided further. The upper or A horizon of a soil is chiefly a zone from which material has been transferred down to the B horizon. This process of eluviation transfers clay minerals, colloidal organic matter and ions in solution. The A horizon may contain much organic matter decomposing into humus. The B horizon is the zone of illuviation (i.e. accumulation) and so contains a higher clay content, precipitated sesquioxides of iron and manganese and carbonates. The C horizon consists of partly altered bedrock or sediment, passing down into fresh parent material. The thickness, development of horizons, and mineralogy of soils vary considerably, being largely dependent on climate, nature of the source material, topography, vegetation and time (for further information see soil science texts such as Hunt, 1972; Foth, 1978).

Practically all clay minerals can develop within soils through pedogenesis, and within weathering mantles. The clays form by the alteration and replacement of other silicate minerals such as feldspars and micas, the transformation of detrital clay minerals, and direct precipitation. The degree of leaching and the pH–Eh of the soil water, both largely determined by the climate, are the two main factors controlling clay mineral formation and stability. The nature and composition of the host rock or sediment is also important.

Where the degree of leaching is limited, as with many soils in temperate areas, then illite is the typical clay mineral formed. Chlorite also forms during intermediate stages of leaching in temperate soils but it is more easily oxidized and so occurs preferentially in acid soils. It also forms in soils of arid regions, both high and low latitude, where chemical processes are minimal. Montmorillonite is a product of intermediate leaching and weathering conditions, being common in temperate soils with good drainage and neutral pH, in gley (poorly drained) soils and in arid-zone soils which are highly alkaline. Mixed-layer clays mostly form through the leaching of pre-existing illite and mica. Kaolinite is characteristic of acid tropical soils where leaching is intensive. Further leaching of kaolinitic soils and the removal of silica gives rise to gibbsite and other aluminium hydroxides which form bauxite. Iron-rich soils of the humid tropics, the laterites or ferricretes are also formed through extreme weathering. They are composed of hydrated iron oxides and kaolinite.

Once formed, clay minerals are available for erosion, transportation and deposition. Many, perhaps most of the clay minerals in modern and ancient clayey sediments are derived from such sources and are thus detrital. There are other processes, however, whereby clays are formed. Clay minerals can be precipitated directly from water or porewaters in surficial sediments. Sepiolite and palygorskite for example are formed in this manner in saline lake and deep sea sediments. Clay minerals can form from the alteration of volcanic material (Sections 3.7.3, 10.5). Volcanic glass is metastable and in time it devitrifies and is replaced by smectite clays, chlorite and kaolinite, which can be reworked into younger sediments. Clay minerals can be precipitated within coarser terrigenous clastic sediments during diagenesis (Section 2.8.5) and the clays of mudrocks are themselves altered and replaced by other clays during diagenesis (Section 3.6). Clays formed through these non-pedogenic processes can amount to more than the detrital component in a muddy sediment.

The distribution of clay minerals in modern sediments is largely a reflection of the climate and weathering pattern of the source area. This is well illustrated by the clay mineralogy of the world's ocean-floor sediments, which also reflects contemporaneous volcanic activity (Griffin *et al.,* 1968; Rateev *et al.,* 1969). Kaolinite is dominant in low latitude areas, particularly off major rivers draining regions of tropical weathering, and illite is more common in ocean-floor muds of higher latitudes. The distribution of smectites, largely derived from alteration of volcanic material, is related to the active mid-ocean ridge systems and volcanic oceanic islands.

Two less important factors in the distribution of clay minerals in the oceans are grain size and clay mineral transformations. There are differences in the size of clay mineral flakes with kaolinite being the largest (up to $5\,\mu m$), illite typically 0.1 to $0.3\,\mu m$ and montmorillonite even finer grained. Montmorillonite frequently occurs as floccules or aggregates up to several μm in diameter. Clay mineral focculation arises from changes in water chemistry, with Eh–pH and salinity major factors. In relatively near-shore areas and on continental shelves, a zonation of clay on the basis of grain size can frequently be detected, with a more kaolinite-rich zone inshore of montmorillonite and illite (Baker, 1973). It has been suggested that clay minerals from fresh-water environments are modified and altered while suspended in seawater or soon after deposition on the seafloor (a process referred to as halmyrolysis). Postulated transformations include montmorillonite to illite, chlorite or palygorskite, and kaolinite to illite, chlorite or smectite. Although it is difficult to assess the importance of these clay mineral changes, the good correlation between clay mineralogy and climate/weathering of adjacent landmasses and ocean-floor volcanicity for modern ocean floor sediments, tends to suggest that such mineralogical transformations are not on a significant scale. They may be more important in areas of very slow sedimentation. Modifications to the chemistry of the clays do occur though.

Since clay minerals can be related to source areas, they can be used to monitor sediment dispersal patterns in estuaries and along continental shelves. Examples include studies by Knebel *et al.* (1977) and Baker (1973). In a vertical sense and allowing for clay diagenesis (see below), changes in the clay mineralogy of oceanic sediments can be related to climatic changes on adjacent continents. Jacobs (1974) for example has discussed variations in the clay mineralogy of Cainozoic seafloor muds of the Southern Ocean in terms of the impending Pleistocene glaciation of the Antarctic continent and its effect on weathering and erosion rates.

Environmental aspects of clay minerals have been reviewed by Keller (1970), Millot (1970) and Velde (1977).

3.6 Diagenesis of clay minerals and mudrocks

Clay minerals can be modified and altered during early and late diagenesis, and into metamorphism. The main physical post-depositional process affecting the mudrocks as a whole is compaction. Compaction in mudrocks expels water and reduces the thickness of the deposited sediment by a factor of up to ten. When muds are deposited they contain in the region of 70% to 90% water by volume. Compaction through overburden pressure soon removes much of the water so that at depths of 1 km or so, the mudrocks contain around 30% water (Fig. 3.4). Much of this is not free pore water but is contained in the lattice of the clay minerals and adsorbed on to the clays. Further compaction through water loss requires temperatures approaching 100°C and these are attained through burial at depths in the region of 2–4 km (Fig. 3.4). Dehydration of clays then takes place, accompanied by some changes in the clay mineralogy (see below). Final

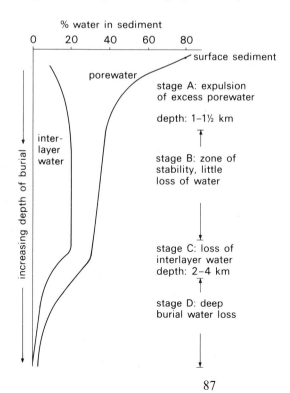

Fig. 3.4 Diagram illustrating the stages of water loss from muddy sediments with increasing depth of burial.

87

compaction to give a mudrock with only a few per cent water requires a much longer period of overburden pressure with elevated temperatures. Further information on compaction of argillaceous sediments is given by Rieke & Chilingarian (1974).

Changes in the clay mineralogy during diagenesis take place principally through the rise in temperature accompanying increased depth of burial. Studies of long cores through thick mudrock sequences (e.g. Burst, 1959; Powers, 1967; Dunoyer de Segonzac, 1970; Hower *et al.*, 1976), show that the main change is an alteration of smectites to illite via mixed-layer clays of smectite–illite (Fig. 3.5). This alteration involves the incorporation of K^+ ions into the smectite structure and loss of interlayer water. The process is largely temperature dependent and the temperature at which smectite begins to disappear is in the order of 70 to 95°C, that is at depths of 2–3 km, in areas of average geothermal gradient. At slightly higher temperatures and greater depths, kaolinite is replaced by illite and chlorite.

There is a change in the clay mineralogy of mudrocks through geological time (Fig. 3.6). Mudrocks back to the Lower Palaeozoic contain a variety of

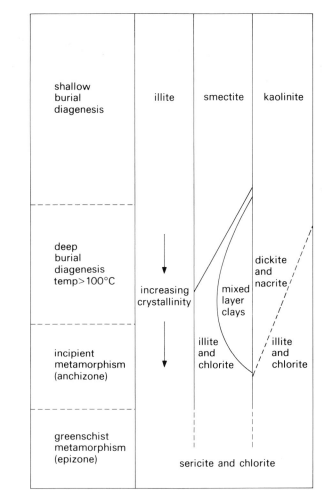

Fig. 3.5 Diagram illustrating the changes of clay minerals with increasing depth of burial and into metamorphism.

88

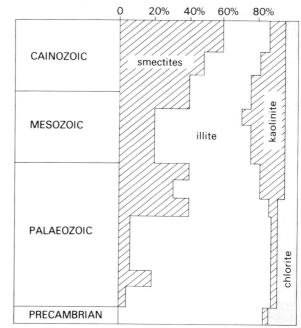

Fig. 3.6 The distribution of clay minerals through time (from Dunoyer de Segonzac, 1970).

clay minerals, but Lower Palaeozoic and Precambrian mudrocks are dominated by illite and chlorite. The more stable composition of the older sediments probably arises from diagenetic alteration through burial, as noted above, with their greater age, allowing more time for diagenetic reactions, a less important factor. A further factor which may be significant is the evolution of land plants during the late Devonian and into the Carboniferous, and the effects these plants must have had on soil-forming and weathering processes (such as enhanced leaching through organic decomposition).

Passing into the realm of incipient metamorphism (catagenesis or anchimetamorphism) clay minerals are further altered and replaced (Fig. 3.5). The phyllosilicates pyrophyllite and laumontite may develop at the expense of the clay minerals. Although smectites, mixed-layer clays and kaolinite do not survive into metamorphism, illite and chlorite do. With increasing degree of incipient and low-grade metamorphism, there is an increase in the order or crystallinity of the illite lattice and changes in the chemical composition. Illite is then replaced by sericite, a finely crystalline variety of muscovite. Studies of clay mineralogy can thus indicate the late diagenetic–early metamorphic grade, and if combined with measurements of the rank of associated coal or phytoclasts (Section 8.6.2) they can give an indication of the temperatures to which the sequence as a whole has been subjected. Carboniferous sediments of the South Wales Coalfield (Gill *et al.*, 1977) and Tertiary sandstones of the Alps (Stalder, 1979) have been treated in this manner.

3.7   Mudrocks and their depositional environments

Three major groups of mudrock can be distinguished: (i) those formed through contemporaneous processes of weathering and soil formation upon pre-existing

89

rocks and sediments, (ii) those formed through the normal sedimentary processes of erosion, transporation and deposition and (iii) those formed through in-situ weathering and/or later alteration of volcaniclastic deposits.

3.7.1 RESIDUAL MUDROCKS AND SOILS

Soils and weathering mantles developed on bedrock are in fact not commonly preserved in the geological record. They should be preserved at unconformities but in the majority of cases they have been eroded before deposition of the overlying beds. Weathering profiles on igneous and metamorphic rocks are occasionally preserved and examples include the weathered Lewisian gneiss beneath the Torridonian Sandstone in N.W. Scotland (Williams, 1968) and a pre-Eocene kaolinitic soil upon granodiorite in Mexico (Abbott *et al.*, 1976).

Of the soils developed within sediments, calcrete (or caliche) and seatearth are common in the geological record. Calcretes vary from scattered to densely-packed nodules of $CaCO_3$ (Fig. 4.54), and are typical of semi-arid climatic areas where evaporation exceeds precipitation (Section 4.10.1). They occur in many river floodplain sediments and clay minerals formed in these soils include sepiolite and palygorskite. In deltaic sequences, such as occur in the Upper Carboniferous of Britain (Wilson, 1965) and the Mississippian of eastern North America, clayey soils known as seatearths or underclays occur beneath coal seams. These formed under a humid tropical climate and are typically massive with rootlets and siderite nodules. If subjected to excessive leaching they may be rich in kaolinite (Staub & Cohen, 1978). Some of these seatearths, known as fireclays, have refractory qualities. Soils which form within more sandy sediments may be leached to give a ganister (Fig. 2.40; Sections 2.7.1; 8.7). Some kaolinite-rich beds occurring within or above coal seams and referred to as tonsteins, have formed form the alteration of volcanic ash (Section 10.5).

3.7.2 DETRITAL MUDROCKS

The majority of clay minerals and silt-grade quartz in mudrocks are derived from the erosion of contemporary land masses. These fine terrigenous clastic particles are largely transported in suspension by water, with deposition taking place in quiet, low energy environments. Rivers in particular transport vast quantities of silt and clay in suspension for deposition in floodplains, lakes, deltaic environments and nearshore and offshore marine environments. Wind is another major agent of transporation, carrying dust up to thousands of kilometres from source areas which are typically deserts and glacial outwash plains. The dust may be deposited on land as loess (aeolian silt) or carried into the oceans where it contributes towards hemipelagic sedimentation. On continental shelves and slopes mud can be re-suspended by storms to form clay- and silt-rich clouds referred to as nepheloid layers which are thought to be an important mechanism for transporting fine sediment to ocean floors and abyssal plains. Slumping of slope and outer shelf muds and the generation of clay-laden, low density turbidity currents are also important in supplying ocean basins with fine sediment. On alluvial fans, in association with volcanoes (Section 10.3), in glacial–proglacial regimes and on submarine fans mud can be transported in viscous, sediment-rich, water-poor flows known as mudflows. Glaciers themselves can transport fine terrigenous clastic material for deposition in tills.

Apart from colour and sedimentary structures, the differentiation of one detrital mudrock from another may depend on detailed mineralogical and geochemical studies. Where there are significant variations in the clay mineral assemblage through a mudrock sequence, these will usually indicate some major event in the source area. As an example, the incoming of chlorite into illite-dominated shales of the Lower Palaeozoic in southern Norway is interpreted by Bjørlykke (1974) as reflecting volcanic activity of distant island arcs. One avenue of mudrock geochemistry which has been explored is the use of the boron content of illite as a palaeosalinity indicator (higher values in marine than freshwater muds). Although the technique has been much criticized, it can give meaningful results (Bohor & Gluskoter, 1973). For many mudrocks, interpretations of depositional environment may rely heavily on the fauna within the mudrocks and on the characteristics of associated and interbedded sediments.

Marine mudrocks

The typical mudrocks of the marine open shelf are various shades of grey and rich in fossils. The fossils are both epifaunal (living on the sediment surface) and infaunal (living within the sediment) with some pelagic forms (free-swimming and free-floating species). Bioturbation is common. Thin, sharp-based and graded beds of sandstone and limestone within the mudrocks may represent storm deposits. Mudrocks of the marine shelf are mainly deposited below wave-base, at depths in excess of 20–50 m. They are a major component of all shelf, platform and epeiric-sea sequences of the geological record and so frequently pass laterally (shorewards) or vertically into limestones or sandstones of the shallower water, near-shore zone.

Muds and mudrocks deposited in deeper water largely from suspension, are termed hemipelagic. Many of these are grey in colour although red, brown, green and black varieties are not uncommon. They are characterized by a fauna which is dominantly pelagic, such as diatoms, planktonic foraminifera and coccolithophoridae from the Mesozoic to the present day; radiolaria from the Palaeozoic; cephalopods in the later Palaeozoic and Mesozoic; and graptolites in the Lower Palaeozoic. Hemipelagic mudrocks frequently contain interbedded siliciclastic and carbonate turbidites (Figs. 2.22, 4.61). They may also grade laterally or vertically into pelagic limestones (Section 4.10.2) or radiolarites (Section 9.3), which form in areas or at times of minimal clay and silt sedimentation. Hemipelagic sediments accumulating below the carbonate compensation depth (Section 4.10.2) are red and brown clays. These cover the abyssal plains of the central Pacific and occur in the Atlantic and Indian oceans. They consist of detrital clay and silt, clay minerals and zeolites derived from alteration of volcanic ash (Section 10.5), and radiolarians, diatoms and sponge spicules. Examples of hemipelagic mudrocks are to be found in all basinal sequences throughout the geological record. The graptolitic shales of the Lower Palaeozoic of western Europe and the Cretaceous–Tertiary foraminiferal marls of the Alpine–Tethyan region are good examples.

Mudrocks of paralic (shoreline) environments such as lagoons, tidal flats, deltaic interdistributary bays and coastal swamps can be identified by their fauna and flora, sedimentary structures (rippled lenses and desiccation cracks for example) and associated coarser clastic deposits such as channel, beach or barrier sandstones. Restricted fossil assemblages may suggest brackish water or hypersaline conditions and the presence of rootlets may indicate emergence. These mudrocks are generally a dark grey colour, a function of the high organic content.

Non-marine mudrocks

The mudrocks of river floodplains are best identified by their association with fluvial channel sandstones. In many cases the mudrocks are overbank deposits and constitute the upper part of fining upward sequences generated through lateral migration of the river channel or by channel abandonment (Section 2.10.1). Floodplain silts and clays deposited under a semi-arid climate are invariably red and frequently contain calcareous nodules of pedogenic origin (calcretes or caliches, Section 4.10.1).

Mudrocks deposited in lakes vary considerably depending on the chemistry of the lake waters, organic productivity and climate. In the majority of cases the clay minerals are detrital, but clay minerals and the related zeolites can be precipitated on the lake floor or within the surficial sediments. Clays which may be neoformed in this way include sepiolite, palygorskite and corrensite, chiefly occurring in alkaline and hypersaline lacustrine sediments. Lacustrine sequences with such clays are the Eocene Green River Formation of western U.S.A. (Picard & High, 1972) and the Triassic Keuper Marl of western Europe (Jeans, 1978).

One of the characteristic features of lacustrine sediments is a rhythmic lamination on the scale of millimetres (Figs. 3.1b,c). In non-glacial lakes, it frequently results from a seasonal clastic influx, coupled with phytoplankton growth. Some oil shales possess a lamination of this type (Section 8.8). Glacial lake sediments are typically varved, that is they consist of a rhythmic alternation of coarse and fine laminae (Fig. 3.1a). The coarse laminae, of silt to fine sand grade, are deposited from low density suspension currents during spring melting, and the fine laminae, of clay-grade material, are deposited from suspension during summer and winter. The mechanism of varve formation has been recently considered by Sturm & Matter (1978) who concluded that thermal stratification of the lake is of critical importance. Organic productivity in lakes can supply much sediment. Photosynthesis by planktonic algae can cause precipitation of $CaCO_3$, giving rise to marl, or if there is a paucity of clay, then limestone. The accumulation of the organic matter itself can lead to the formation of bituminous mudrocks, notably oil shales (see below and Section 8.8). Diatoms inhabit many freshwater lakes and the accumulation of their tests produces siliceous muds (diatomaceous earths) and diatomites (Section 9.5).

92

One particularly important group of mudrocks are those rich in organic matter; these are black shales and carbonaceous and bituminous mudrocks, which typically contain 3–10% organic carbon. With an increasing organic content, organic-rich mudrocks pass into oil shales which yield a significant amount of oil on distillation (Section 8.8). In many depositional environments organic matter is decomposed and destroyed at the sediment surface but if the rate of organic productivity is high, then organic matter can be preserved. Much of this organic matter is sapropelic, supplied by phytoplankton (Section 8.2). The accumulation of organic matter is favoured if the circulation of water is restricted to some extent so that insufficient oxygen reaches the bottom sediments to decompose the organic material (Didyk *et al.,* 1978). Locations where this commonly takes place are lakes, fjords, restricted basins (such as the Black Sea), sediment-starved basins (such as the Gulf of California) and deep-ocean trenches (such as the Cariaco Trench). As a result of the poor circulation and restriction, the sea or lake floor may become oxygen-deficient or totally anoxic. Where there is an oxygen deficiency, organic matter will be preserved, but the surface sediments could still support a benthic epifauna, although of low diversity. Where there are anoxic conditions on the seafloor, there is invariably much H_2S in the water and benthic organisms are absent (Fig. 3.7). This is the case with the Black Sea and Cariaco Trench at the present

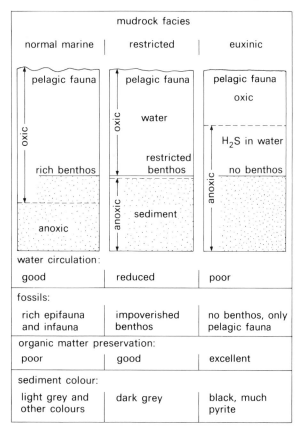

Fig. 3.7 The relationships between mudrock facies and oxicity–anoxicity, fauna and organic matter content.

time. Mudrocks deposited in an anoxic environment would contain only pelagic fossils. Such changes in fauna with increasing oxygen deficiency have been described from Jurassic marine mudrocks of N.E. England (Morris, 1979). Pyrite is common in marine organic-rich mudrocks and siderite in non-marine ones (Section 6.3).

There are many organic-rich mudrocks in the geological record; their importance lies in their potential as source rocks for petroleum, if buried to suitable depths and subjected to appropriate temperatures (Section 8.9). Examples include Jurassic and Cretaceous bituminous shales, with organic matter derived from radiolarians and Foraminifera, which have supplied much of the petroleum for the Middle East oilfields, and the organic-rich Kimmeridge Clay, the chief source rock for North Sea oil.

Thin horizons of Cretaceous black shales and organic-rich mudrocks have been cored from the Atlantic, Pacific and Indian Oceans and equivalent horizons can be found in sections on land. These organic-rich layers, which signify short-lived periods of poor circulation and oxygen deficiency in the world's oceans, are thought to result from increased organic productivity associated with transgressive events which produced widespread shallow seas (Schlanger & Jenkyns, 1976; Jenkyns, 1980). Similar black mud horizons occur in the Miocene–Pleistocene of the Mediterranean; the Pleistocene stagnations were caused by stratification of the water at times of glacial expansions (Kidd *et al.*, 1978).

One feature of organic-rich sediments is that they contain high concentrations of certain trace elements, in particular Cu, Pb, Zn, Mo, V, U and As. The trace elements are adsorbed on to the organic matter and also on to the clay minerals. It is likely that the source of these elements is seawater and that they are scavenged by the organisms and organic matter. An example is the organic-rich Permian Marl Slate of northeast England and the North Sea and its equivalent in Germany and Poland, the Kupferschiefer, which has a high metal sulphide content. It was deposited in a thermally-stratified sea with anoxic bottom waters.

Loess

Loess is a yellow to buff-coloured clastic deposit composed principally of silt-sized quartz grains, generally in the size range of 20–50 μm. A distinctive feature is the well-sorted nature of the silt, together with a dominantly angular shape to the grains. Loess is usually unstratified and unconsolidated, but it may contain shells of land snails and concretions formed around roots. Loess deposited during the late Pleistocene occurs over vast areas of central Europe, the Mississippi valley region of the United States, eastern South America and China. There are few unequivocal accounts of pre-Pleistocene wind-blown silt; one recently documented 'loessite' is from the Late Precambrian of North Norway and Svalbard (Edwards, 1979).

Loess is primarily regarded as an aeolian deposit, but once deposited it can be considerably modified by fluvial reworking and pedogenesis. Two types of loess are distinguished by many authors: loess of cold, periglacial regimes derived from deflation of glacial outwash plains (this accounts for most of the late Pleistocene occurrences and the Precambrian loessite) and loess derived from hot, arid, desert areas. Glacial grinding is considered the most effective mechanism for producing vast quantities of quartz silt; loess from desert weathering and deflation is probably of minor geological importance. For the literature on loess see Smalley (1976).

3.7.3 MUDROCKS OF VOLCANICLASTIC ORIGIN

Mudrocks formed from the alteration of volcaniclastic material are known as bentonites (also fuller's earth) if montmorillonite is the main clay mineral present and tonsteins if kaolinite is dominant. The volcaniclastic deposit may be subaerial or subaqueous, but because of the metastable nature of volcanic glass, devitrification soon takes place and clay minerals form. The identification of these mudrocks is based on the presence of glass shard pseudomorphs, euhedral crystals and geochemistry (Section 10.5).

Further Reading

Text-books with much relevant material are:

Grim, R.E. (1968) *Clay Mineralogy,* pp. 596. McGraw-Hill, New York.

Millot, G. (1970 *Geology of Clays,* pp. 429. Springer-Verlag, New York.

Velde, B. (1977) *Clays and clay minerals in natural and synthetic systems,* pp. 218. Elsevier, Amsterdam.

Rieke, H.H. & Chilingarian, G.V. (1977) *Compaction of argillaceous sediments,* pp. 424. Elsevier, Amsterdam.

Müller, G. (1967) Diagenesis in argillaceous sediments, In: *Diagenesis in sediments* (Ed. by G. Larsen and G.V. Chilingar), pp. 127–178. Elsevier, Amsterdam.

Potter, P.E., Maynard, J.B. & Pryor, W.A. (1980) *Sedimentology of Shale,* pp. 270. Springer-Verlag, Berlin.

4

Limestones

4.1   Introduction

Biological and biochemical processes are dominant in the formation of carbonate sediments; with a few notable exceptions inorganic precipitation of $CaCO_3$ from seawater can rarely be demonstrated. Once deposited, the chemical and physical processes of diagenesis can considerably modify the carbonate sediment. Limestones occur throughout the world in every geological period from the Cambrian onwards, and reflect the changing fortunes, through evolution and extinction of invertebrates with carbonate skeletons. In the Precambrian, the limestones are often dolomitic, and many contain algal stromatolites, produced largely by the blue-green algae.

The economic importance of limestones today lies chiefly in their reservoir properties since many of the world's major petroleum reserves are contained within carbonate rocks (Chilingarian *et al.*, 1972). Limestones are also hosts to epigenetic lead and zinc sulphide deposits of the Mississippi Valley type (Monseur & Pel, 1973), and they have a wide variety of chemical and industrial uses.

As a result of recent geological events (in particular the Pleistocene glaciation), shallow marine carbonate sediments are not so widely developed. In the past shallow epeiric seas periodically covered vast continental areas so that limestones were deposited over thousands of square kilometres. Organisms with carbonate skeletons occur throughout the world's seas and oceans so that carbonate sediments can develop anywhere. However, there are several factors, of which the most important are temperature, salinity, water depth and siliciclastic input, that control carbonate deposition (Lees, 1975). Many carbonate skeletal organisms, such as the reef-building corals and many calcareous green algae, require warm waters in which to flourish. The majority of carbonate sediments therefore occur in the tropical–subtropical belt, some 30° north and south of the equator, and most limestones of the Phanerozoic formed in low latitudes. Carbonate sands do occur in higher latitudes, such as along the western coast of Ireland and Norway where calcareous red algae (especially *Lithothamnion*) and mollusca dominate the sediments, and also off south Australia. However, very few ancient limestone sequences were deposited in temperate latitudes; one recently documented exception being in the Oligocene of New Zealand (Nelson, 1978). Many carbonate skeletal organisms are affected by salinity and water depth and occur preferentially in the shallow agitated part of the photic zone in seawater of normal salinity. Carbonate oozes are present on the deep ocean floors and these are composed principally of pelagic organisms which live in the photic zone. High rates of carbonate dissolution at depths of several kilometres result in few carbonate sediments below this. One of the overriding controls of carbonate deposition is a lack of terrigenous detritus. The influx of much siliciclastic material will inhibit the formation of limestones.

For the petrographic study of limestones, acetate peels are commonly used. Limestone surfaces are polished and etched with dilute acid and then covered in acetone. A piece of acetate sheet is then rolled on, taking care not to trap air bubbles. After 10 minutes or so, the acetate is peeled off and, between glass, is ready for the microscope. All microscopic details are faithfully replicated, but of course polarizers cannot be crossed.

4.2   Mineralogy of carbonate sediments

In Recent and subrecent sediments, two calcium carbonate minerals predominate: aragonite (orthorhombic) and calcite (trigonal). Two types of calcite are recognized depending on the magnesium content: low magnesium calcite with less than $4 \, mol \, \%$ $MgCO_3$ and high magnesium calcite with greater than $4 \, mol \, \%$, but typically ranging between 11 and $19 \, mol \, \%$ $MgCO_3$. By comparison, aragonite normally has a very low Mg content (less than 5000 ppm) but it may contain up to 10000 ppm (1%) strontium substituting for calcium. The mineralogy of a modern carbonate sediment depends largely on the skeletal and non-skeletal grains present. Carbonate skeletons of organisms have a specific mineralogy or mixture of mineralogies (Table 4.1) although the actual magnesium content of the calcites may vary, being partly dependent on water temperature.

Since aragonite is metastable and in time high Mg calcite loses its Mg, all carbonate sediments are eventually converted to low Mg calcite. Dolomite, $Ca \, Mg \, (CO_3)_2$, may develop within carbonate sediments and rocks (Section 4.8). Non-carbonate minerals in limestones include terrigenous quartz and clay, and pyrite, hematite, chert and phosphate of diagenetic origin. Evaporite minerals, in particular gypsum–anhydrite, may be associated with limestone sequences (Chapter 5).

4.3   Components of limestones

Limestones are very varied in composition but broadly the components can be divided into four groups: i non-skeletal grains, ii skeletal components, iii micrite and iv cement. The common cement, sparite, and others, are discussed in the diagenesis section, 4.7.

4.3.1   NON-SKELETAL GRAINS (Fig. 4.1)

Ooids

Ooids are spherical–subspherical grains consisting of one or more regular concentric lamellae around a nucleus, often a carbonate particle or quartz grain (Figs. 4.2, 4.3). Sediment composed of ooids is referred to as an oolite. Although the term ooid has been restricted to grains less than 2 mm in

Table 4.1 The mineralogy of carbonate skeletons (× = dominant mineralogy, (×) = less common). During diagenesis, these mineralogies may be altered or replaced; in particular, aragonite is metastable and is invariably replaced by calcite, and in time, high Mg calcite loses its Mg.

Mineralogy / Organism	aragonite	low Mg calcite	high Mg calcite	aragonite + calcite
Mollusca:				
Bivalves	×	×		×
Gastropods	×			×
Pteropods	×			
Cephalopods	×		(×)	
Brachiopods		×	(×)	
Corals:				
Scleractinian	×			
Rugose + tabulate		×	×	
Sponges	×	×	×	
Bryozoans	×		×	×
Echinoderms			×	
Ostracods		×	×	
Foraminifera:				
benthonic	(×)		×	
Pelagic		×		
Algae:				
Coccolithophoridae		×		
Rhodophyta	×		×	
Chlorophyta	×			
Charophyta		×		

diameter, with the term pisolite for similar grains of a larger diameter, the majority of ooids range from 0.2–0.5 mm in diameter. Ooids typically form in agitated waters where they are frequently moved as sand waves, dunes and ripples by tidal and storm currents, and wave action. On the Bahama platform, the ooids form shoals close to the edge of the platform (Ball, 1967; Hine, 1977); in the Arabian Gulf the ooids form in tidal deltas at the mouth of tidal inlets between barrier islands (Loreau & Purser, 1973). Depths of water where ooids occur are usually less than 5 m, but they may reach 10–15 m.

If only one lamella is developed around a nucleus, then the term superficial ooid or coated grain is applied (Fig. 4.1). Composite ooids consist of several small ooids enveloped by concentric lamellae. Practically all marine ooids forming today, such as in the Bahamas and Arabian Gulf, are composed of aragonite and they have a high surface polish. High Mg calcite–aragonite ooids have been recorded from Baffin Bay, Texas, and early Holocene ooids of high Mg calcite occur off the Great Barrier Reef of Queensland (Marshall & Davies, 1975) and on the Amazon Shelf (Milliman & Barretto, 1975). All ooids in ancient limestones are composed of low Mg calcite, unless dolomitized or silicified. The characteristic microstructure of modern marine ooids is a

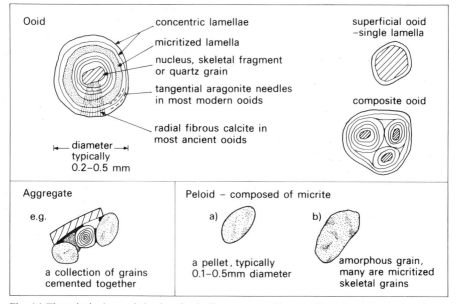

Fig. 4.1 The principal non-skeletal grains in limestones: ooids, peloids and aggregates.

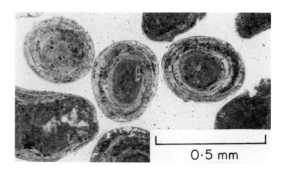

Fig. 4.2 Modern ooids composed of aragonite from the Bahamas. Peloids have formed the nuclei. Plane polarized light.

0·5 mm

tangential orientation of acicular aragonite crystals or needles, 2 μm in length. Lamellae of microcrystalline aragonite and of randomly-oriented aragonite needles also occur. The subrecent high Mg calcite ooids have a radial fabric. Ooids contain organic matter, located chiefly between lamellae and in microcrystalline layers.

Ooids also form in lagoons, lakes and rivers and on tidal flats, in fresh and hypersaline water. Examples include Baffin Bay, Texas (Land *et al.*, 1979), protected lagoons of the Arabian Gulf (Loreau & Purser, 1973), and Great Salt Lake, Utah (Sandberg, 1975; Halley, 1977). The ooids are often dull, some with a cerebroid (bumpy) surface, and they may contain inclusions of clay. A characteristic feature is a radial fabric of aragonite crystals. Ooids from lagoonal and hypersaline settings are commonly broken, because of the primary radial fabric (Fig. 4.4), and this contrasts with normal marine ooids which are rarely fractured.

The calcitic ooids of ancient limestones typically have a radial fabric consisting of large, wedge-shaped fibrous crystals (Figs. 4.3, 4.4). The

99

Fig. 4.3 Ooids formed around peloids and skeletal grains (bivalve fragments and echinoid spine, upper right). Ooids are composed of calcite; some have a radial, as well as concentric structure while others are micritic. Also gastropod in transverse section, middle right. Plane polarized light. Osmington Oolite, Jurassic; Dorset, England.

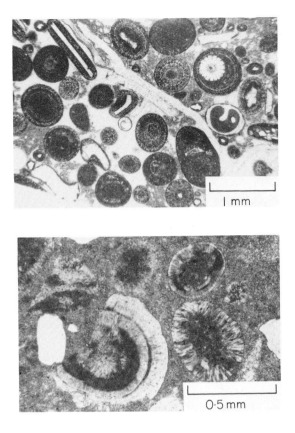

Fig. 4.4 Broken ooid and ooids with strong radial fabric. Oomicrite of lagoonal origin, Lower Carboniferous; Gloucestershire, England. Courtesy of T.P. Burchette.

concentric pattern of growth lines is still present. With crossed polars, a pseudo-uniaxial cross is seen. Other ancient ooids are micritic with a poorly-defined concentric structure (Fig. 4.3). As with modern ooids, such loss of structure is attributable to micritization by endolithic algae (Section 4.3.3).

Origin of ooids

There has been much discussion on the origin of ooids; current ideas invoke biochemical or inorganic processes, an algal origin being largely discarded now. Although a precise mechanism of inorganic precipitation has not been demonstrated, seawater in shallow tropical areas is supersaturated with respect to $CaCO_3$, so that this, together with water agitation and elevated temperature, might be sufficient to bring about carbonate precipitation on nuclei. A biochemical origin hinges on the organic mucilage which coats and permeates the ooids. One view is that bacterial activity within the organic matter creates a microenvironment conducive to carbonate precipitation. The identification of a proteinaceous matrix in ooids by Mitterer (1971) suggests another biochemical process, since in organisms it is considered that amino acids induce calcification. Laboratory synthesis of ooids by Davies *et al.,* (1978) has suggested that organic compounds in the water are instrumental in the formation of quiet-water ooids with their radial fabric, but that ooids formed in agitated conditions are precipitated inorganically.

100

With ancient ooids, discussion is centred on two points: (i) the original mineralogy—whether they were precipitated as calcite, their present mineralogy, or as aragonite, like most modern marine ooids, and (ii) the radial structure—whether this is a primary feature or replacement of an original tangential arrangement of crystals. The dominance of aragonite ooids with tangential fabric in modern marine environments led Shearman *et al.* (1970) to propose a diagenetic model for ancient calcite ooids whereby aragonite dissolution was followed by radial growth of calcite. It was suggested that the concentric lamellar structure of the ooid was retained by the organic matter, which acted as a framework or template for calcite precipitation. However, as mentioned above, primary radial fabrics do occur in aragonitic ooids, although from quiet-water environments, and they are present in subrecent high Mg calcite ooids. Sandberg (1975), on the basis of Great Salt Lake ooids, and Wilkinson & Landing (1978) from a study of Jurassic ooids, have argued for a primary origin of the radial fabric in ancient ooids and an original calcite (probably high Mg) mineralogy.

Peloids

Peloids are spherical, cylindrical or angular grains, composed of microcrystalline carbonate, but with no internal structure (Figs. 4.1, 4.5). As with ooids, the size of peloids may reach several mm, but the majority are in the range of 0.1–0.5 mm in diameter. Most peloids are of faecal origin and so can

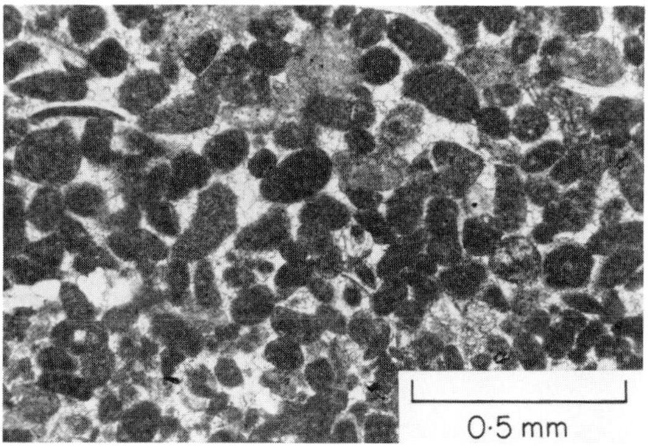

Fig. 4.5 Peloids, many are probably faecal pellets, in a sparite cement. Plane polarized light. Purbeck limestone, Jurassic; Dorset, England.

0·5 mm

be referred to as pellets. Organisms such as gastropods, crustaceans, and polychaetes produce pellets in vast quantities. Faecal pellets have a regular shape and they are rich in organic matter. They are most common in the sediments of protected environments such as lagoons and tidal flats. Pellets are very common in limestones and many micritic limestones, seemingly without sand-sized grains, may actually be pelleted. The definition of pellets is often lost as a result of diagenetic processes and the limestones may show a flocculent or clotted texture, referred to as *structure grumeleuse*.

Amorphous grains are irregularly-shaped peloids formed by the algal micritization of skeletal fragments (Section 4.3.3). They are an important component of modern Bahamian carbonate sediments.

101

Aggregates consist of several carbonate particles cemented together by a microcrystalline cement or bound by organic matter (Fig. 4.1). Such grains in the Bahamas are known as grapestones and form in relatively protected shallow subtidal areas, often beneath a thin surficial algal mat.

Intraclasts are fragments of lithified or partly lithified sediment. A common type of intraclast in carbonate sediments is a micritic flake or chip, derived from desiccated tidal flat muds (Fig. 4.6). An abundance of these flakes produces edgewise conglomerates or flakestones.

Fig. 4.6 Intraclasts (flakes) being generated by desiccation (in central part) and forming the sediment in the upper part. Late Precambrian; Finnmark, Norway.

4.3.2 SKELETAL COMPONENTS (EXCLUDING ALGAE)

The skeletal components of a limestone are a reflection of the distribution of carbonate-secreting invertebrates through time and space (Fig. 4.7). Environmental factors, such as depth, temperature, salinity, substrate and turbulence, controlled the distribution and development of the organisms in the various carbonate subenvironments at a point in time. Throughout the Phanerozoic, various groups expanded and evolved to occupy the niches left by others on the decline or becoming extinct.

The main skeletal contributors to limestones are discussed in the following sections, with comments on their recognition. Detailed accounts of skeletal structure and thin-section appearance are given in Majewske (1969), Horowitz & Potter (1971), Milliman (1974), Bathurst (1975), Flügel (1978) and Scholle (1978). For the identification of skeletal particles in thin section the points to note are:

(1) shape (and size), bearing in mind that under the microscope one is only getting a two-dimensional view; look for other sections of the same fossil to determine the three-dimensional shape.

(2) internal microstructure, which may be modified or obliterated by diagenesis.

(3) mineralogy; although in a limestone everything will be calcite, unless dolomitized or silicified, fabric evidence can be used to decide if a skeletal particle was originally aragonitic. Staining a thin section for ferroan calcite and dolomite (by using alizarin red S and potassium ferricyanide) may give

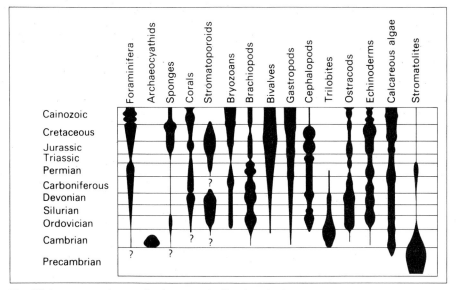

Fig. 4.7 Age range and generalized taxonomic diversity of principal carbonate-secreting organisms (modified from Horowitz & Potter (1971)).

additional information; for example skeletal components originally of high Mg calcite may be preferentially replaced by ferroan calcite. For details of the staining technique see Dickson, 1966 or Hutchison, 1974.

(4) other features likely to be diagnostic, such as presence of spines or pores.

Mollusca

Bivalves, gastropods and cephalopods occur in limestones from the Lower Palaeozoic onwards. The *bivalves* are a very large group with species occupying most marine, brackish and fresh water environments. Bivalves have been important contributors to marine carbonate sediments, particularly since the Tertiary following the decline of the brachiopods. The modes of life are very varied too, including infaunal (living within the sediment), epifaunal (attached to a hard substrate), vagile (crawlers), nektonic (free-swimming) and planktonic (free-floating). Certain bivalves, such as oysters, may form reef-like structures. During the Cretaceous, masses of aberrant, coral-like bivalves called rudists, formed reefs in Mexico, southern Europe and the Middle East for example. Fresh and brackish water limestones may be largely composed of bivalves; examples occur in the Upper Carboniferous, Upper Triassic (Rhaetic) and Upper Jurassic (Purbeck) of western Europe.

The majority of bivalve shells are composed of aragonite; some are of mixed mineralogy; others, such as the oysters, are calcitic. Bivalve shells consist of several layers of specific internal microstructure, composed of micron-sized crystallites (Bathurst, 1975). As an example, a common shell structure is an inner nacreous layer consisting of sheets of aragonite tablets, and an outer prismatic layer of aragonite (or calcite) prisms. If composed of aragonite, the internal structure of the bivalve may not be preserved since the metastable aragonite will have been replaced by calcite during diagenesis

(Section 4.7). For this reason many bivalve fragments in limestones are composed of clear, coarse drusy sparite (Figs. 4.8, 4.9). Calcite bivalves, such as oysters, will invariably retain their original structure. Bivalve fragments in thin section will be seen as elongate, rectangular and curved grains, typically disarticulated.

Gastropods are ubiquitous throughout shallow marine environments. They also occur in vast numbers, but low species diversity, in hypersaline and brackish waters, such as on tidal flats and in estuaries, since they are able to tolerate fluctuations and extremes of salinity. Most gastropods are benthic, vagile creatures. The encrusting vermiform gastropods, often confused with

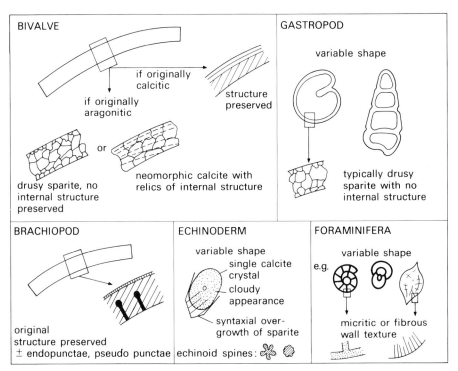

Fig. 4.8 Typical thin section appearance of bivalve, gastropod, brachiopod, echinoderm and foraminiferal skeletal grains in limestone.

Fig. 4.9 Bivalve fragments (the elongate grains) consisting of drusy sparite (calcite) since the original aragonite dissolved leaving a void. The bivalves possess a micrite envelope which in several cases has fractured as a result of compaction. Also present: ooids, an aggregate grain containing quartz silt; some micritic sediment, and a sparite cement. Plane polarized light. Osmington Oolite, Jurassic; Dorset, England.

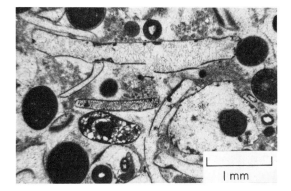

serpulids, form reef-like structures in the tropics. The small, conical pteropods are important in Cainozoic pelagic sediments.

The majority of gastropods have shells of aragonite with similar internal microstructures to bivalves. The internal microstructure of fossil gastropods is also rarely seen since the original aragonite is mostly replaced by calcite. Gastropod fragments can easily be recognized under the microscope by their shape, although this is very much dependent on the plane of section (Figs. 4.8, 4.10). Gastropods may resemble certain Foraminifera, but the latter are usually much smaller and composed of dark micritic calcite.

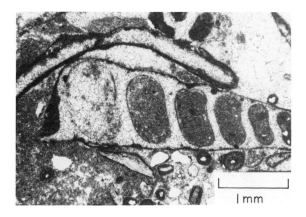

Fig. 4.10 Gastropod, in long section, preserved as drusy sparite since the shell was originally composed of aragonite and this dissolved leaving a void. Most chambers are filled with micritic sediment. Also present: a large bivalve fragment with micrite envelope. Plane polarized light. Osmington Oolite, Jurassic; Dorset, England.

Of the *Cephalopods*, nautiloids and ammonoids are relatively common in limestones of the Palaeozoic and Mesozoic and belemnoids occur in Mesozoic limestones. They were wholly-marine animals with a dominantly nektonic or nekto-planktonic mode of life, as with the modern *Nautilus,* octopus and cuttlefish. The cephalopods are more common in pelagic, often relatively deep water deposits. Examples include the Ordovician–Silurian *Orthoceras* limestones of Sweden, the Devonian Cephalopodenkalk and Griotte of western Europe and the Jurassic Ammonitico Rosso of the Alps. Nautiloid and ammonoid shells were originally aragonitic and so in limestones are typically composed of sparite with little internal structure. The shape, often large size, presence of septa and a siphuncle are the features to note.

Brachiopods

Brachiopods are particularly common in Palaeozoic and Mesozoic limestones of shallow marine origin. These were largely benthic sessile organisms; a few species were infaunal. Only in rare cases such as in the Permian of west Texas, did the brachiopods contribute to reef development. At the present time brachiopods are a less prominent group of marine invertebrates.

Although in section, brachiopod shells are similar to those of bivalves in shape and size, articulate brachiopods were composed of low or high Mg calcite so that the internal structure is invariably well preserved. The common structure is a very thin, outer layer of calcite fibres oriented normal to the shell surface, and a much thicker inner layer of oblique fibres (Figs. 4.8, 4.11). In the punctate brachiopods, such as the terebratulids, fine tubes (endopunctae)

105

Fig. 4.11 Brachiopod shell (impunctate) showing preservation of internal structure, consisting of obliquely arranged fibres. Rhynchonellids and spiriferids have impunctate shells. Plane polarized light. Lower Carboniferous, Clwyd, Wales.

perpendicular to the shell surface perforate the inner layer and are filled with sparite or micrite. With pseudopunctae, as occur in the strophomenid group, the calcite fibres of the shell wall are modified adjacent to the punctae (Fig. 4.12). Inarticulate brachiopods, mostly composed of chitin or chitinophosphate, are rare in limestones.

Cnidaria (*especially corals*)

The Cnidaria include the Anthozoa (corals) of which two ecological groups exist today: hermatypic corals which contain symbiotic dinoflagellate algae (zooxanthellae) in their polyps and ahermatypic corals without such algae. Because of the algae, hermatypic corals require shallow, warm and clear seawater. They are the reef-forming corals at the present time, being mainly responsible for the reef framework, which is often reinforced by red algae. Ahermatypic corals can occur at much greater depths and tolerate much colder waters. They locally form buildups and barriers. The rugose and tabulate corals were important in Silurian and Devonian reefs, and many Triassic reefs contain scleractinian corals. Some of the latter may well have been ahermatypic. Corals, both solitary and colonial, and coral debris occur in many non-reefal limestones.

The Palaeozoic rugose and tabulate corals were composed of calcite, often of high Mg calcite, so that preservation is invariably perfect (Fig. 4.13). scleractinian corals. Some of the latter may well have been ahermatypic.

Fig. 4.12 Brachiopod shell with pseudopunctae recognized by the irregularities in the shell structure adjacent to the coarse calcite of the punctae. Plane polarized light. Productid brachiopod, Lower Carboniferous; Clwyd, Wales.

106

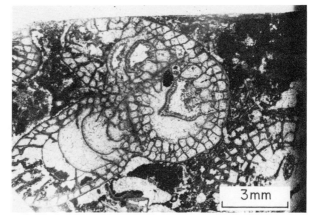

Fig. 4.13 Rugose coral (*Lithostrotion* sp.) in thin section showing internal plates (septa, tabulae and disseppiments). The coral is infilled with drusy sparite cement. Plane polarized light. Lower Carboniferous, S.E. Scotland.

skeletons and so are often poorly preserved in limestones. Identification of coral is based on such internal features as septa and where present other internal plates in the Rugosa and Scleractinia, and tabulae in the Tabulata. Corallite form and colonial organization are also important. The microstructure of Palaeozoic calcitic and later aragonitic corals is very similar, chiefly consisting of fibres in spherulitic or parallel arrangements, which form linear structures called trabeculae, or sheets.

Echinodermata

Echinoderms are wholly marine organisms which include the echinoids (sea urchins) and crinoids (sea lilies). In modern seas, echinoids inhabit reef and associated environments, sometimes in great numbers, but crinoids are restricted to deeper waters and are insignificant as producers of carbonate sediment. In the Palaeozoic and Mesozoic, fragments of echinoderms, especially the crinoids, are a major constituent of bioclastic platform carbonates. Many deep water limestone turbidities are composed of crinoidal debris, derived from shallow platforms.

Echinoid and crinoid skeletons are wholly calcitic; modern forms generally have a high Mg content. Echinoderm fragments are easily identified since they are composed of large, single calcite crystals, individual grains thus showing unit extinction. In many cases, a sparite cement crystal has grown syntaxially around the echinoderm fragment (Figs 4.14, 4.44). Echinoderm grains have a dusty appearance, especially relative to a sparite cement overgrowth, and they may show a porous structure infilled with micrite or sparite (Fig. 4.14). Echinoid-spines are occasionally encountered in thin section and are distinguished by their radial stellate shape (Fig. 4.3).

Bryozoa

Although these small colonial marine organisms are only locally significant suppliers of carbonate sediment at the present time, they have in the past contributed to the formation of reef and other limestones, particularly in the

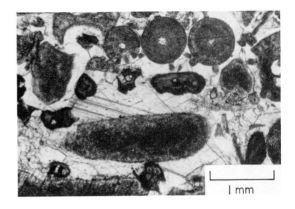

Fig. 4.14 Echinoderm fragments: rounded to elongate in section, with syntaxial cement overgrowths (the latter identified here by the continuous twin planes from skeletal fragment to calcite cement). The darker grains with large 'holes' (zooecia) are fenestellid bryozoan fragments. Plane polarized light. Lower Carboniferous, central Scotland.

| mm

Palaeozoic. Examples include the Mississippian mud mounds of southwestern U.S.A. and Europe, and the Permian reefs of Texas and elsewhere.

Modern bryozoan skeletons are composed of either aragonite or calcite (often high Mg calcite) or a mixture of both. There are many types of bryozoans but the fenestrate variety, including the fenestellids, are most frequently seen in sections of Palaeozoic limestones. They appear as strings of cells (zooecia) around 100 μm in diameter, infilled with sediment or sparite, and joined by dark laminated calcite of the stem (Fig. 4.15).

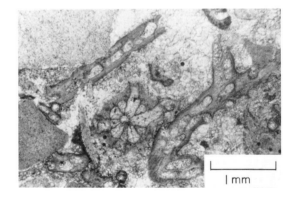

Fig. 4.15 Fenestellid bryozoans in thin section, recognized by the preservation of the wall structure and presence of cavities (zooecia). Plane polarized light. Lower Carboniferous, central Scotland.

| mm

Foraminifera

Foraminifera are dominantly marine Protozoa, largely of microscopic size. Planktonic Foraminifera dominate some pelagic deposits, such as in the *Globigerina* oozes of ocean floors and some Cretaceous and Tertiary chalks and marls. Benthic Foraminifera are common in warm shallow seas, living within and on the sediment, and encrusting hard substrates.

Foraminifera are composed of low or high Mg calcite, rarely of aragonite. Foraminifera are very diverse in shape but in section many common forms are circular to subcircular with chambers (Figs. 4.8, 4.16). The test wall is microgranular in many thin-walled Foraminifera such as the endothyracids and often fibrous in larger, thicker species, such as the rotaliids.

108

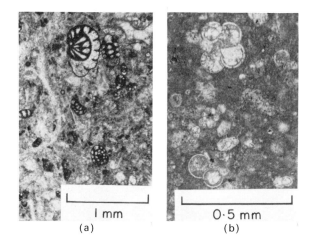

Fig. 4.16 Foraminifera. 4.16a Endothyracid Foraminifera from the Lower Carboniferous of Clwyd, Wales, 4.16b *Globigerina* Foraminifera from the Chalk, Upper Cretaceous; Lincolnshire England.

(a) 1 mm   (b) 0·5 mm

Other carbonate-forming organisms

There are many other organisms which have calcareous skeletons but contributed in a minor way to limestone formation, or were important for only short periods of geological time.

Sponges: Spicules of sponges, which may be composed of silica or calcite, occur sporadically in sediments from the Cambrian. The importance of spicules is as a source of silica for the formation of chert nodules and silicification of limestones (Section 4.8.6). At times, sponges provided the framework for reefs and mounds; examples include lithistid sponges in the Ordovician, calcisponges in the Permian of Texas and Triassic of the Alps, and silicisponges (now calcitized) in the Jurassic of southern Germany. Sclerosponges are the dominant reef-forming organism in some modern Caribbean reefs. *Stromatoporoids,* once considered as hydrozoans, are now classified as a subphylum of the Porifera. Stromatoporoids were marine colonial organisms which had a wide variety of growth forms, ranging from spherical to laminar, depending on species and environmental factors. Stromatoporoids were a major reef organism in the Silurian and Devonian, commonly in association with rugose and tabulate corals. *Archaeocyathids,* somewhat sponge-like but of uncertain affinities, formed reefs in the Cambrian of North America, Morocco and South Australia.

Arthropods: Of this group, the ostracods (Cambrian to Recent) are locally significant in Tertiary limestones and some others. They live at shallow depths in marine, brackish or fresh water environments. Ostracods have small (around 1 mm in length), thin bivalved shells, smooth or ornamented, composed of calcite with a radial-fibrous structure (one is present in Fig. 4.44). Trilobites (Cambrian–Permian) occur locally in Palaeozoic shelf limestones, but never in rock-forming quantities.

Calcispheres: These are spherical objects, up to 0.5 mm in diameter, composed of calcite (usually sparite), often with a micritic wall. They are ascribed to the algae, although an affinity with Foraminifera has been suggested. They occur in many Palaeozoic limestones, particularly fine-grained micrites of back reef or lagoonal origin.

109

Algae make a major contribution to limestones by providing skeletal carbonate particles, trapping grains to form laminated sediments and attacking particles and substrates through their boring activities. Many of the Precambrian limestones were at least in part produced by algae, and algal limestones are widely distributed throughout the Phanerozoic. Four groups are important: red algae (rhodophyta), green algae (chlorophyta), blue-green algae (cyanophyta) and yellow-green algae (chrysophyta). Three relevant texts are Johnson (1961), Walter (1976) and Flügel (1977).

Rhodophyta

Calcareous algae of the Rhodophyta, such as the Corallinaceae (Carboniferous–Recent) and Solenoporaceae (Cambrian to Miocene) have skeletons composed of cryptocrystalline calcite which is precipitated within and between cell walls. In section, a regular cellular structure is present (Fig. 4.17a).

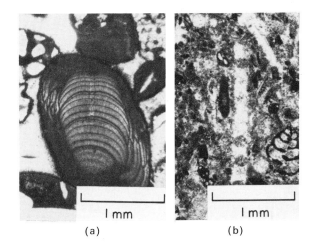

Fig. 4.17a Calcareous alga (*Lithothamnion* sp.), showing the fine cellular structure. Beachrock, Great Barrier Reef, Australia.
Fig. 4.17b Dasyclad alga (*Coelosporella* sp.): elongate grain preserved in sparite (therefore originally aragonitic), with circular 'holes' (utricles). Also present is a textularid Foraminifera. Lower Carboniferous, Clwyd, Wales. Courtesy of D.I. Gray.

(a)   (b)

Modern coralline algae have high Mg contents in the calcite, which is related to water temperature (higher values for a given species in warmer waters). Many of the coralline algae encrust substrates and if this is a pebble or shell then nodules referred to as rhodoliths develop. Encrustations may be massive and rounded, or delicately branched, depending on ecological factors (Adey & MacIntyre, 1973). One of the most important roles of these red algae is in binding and cementing the substrate, particularly in modern reefs. In temperate and arctic carbonate sands, the red alga *Lithothamnion* is a major contributor. In the Palaeozoic and Mesozoic, red algae of the Solenoporaceae are locally abundant and in some cases participate in reef formation.

Chlorophyta

Three algal groups are important: the Codiaceae, Dasycladaceae (both Cambrian–Recent) and Characeae (Silurian–Recent). The Characeae are only partly calcified (low Mg calcite), so that only stalks and reproductive capsules are found in limestones. Modern (and fossil) forms are restricted to fresh or brackish water.

With the dasyclad algae, calcification is also incomplete and involves the precipitation of an aragonitic crust around the stem and branches of the plant. Dasyclads are marine algae which tend to occur in shallow protected lagoonal areas of the tropics. Under the microscope, circular, ovoid or elongate shapes are seen, representing sections through the stems or branches (Fig. 4.17b).

The Codiacean algae include *Halimeda* and *Penicillus,* two common genera of Caribbean and Pacific reefs and lagoons. *Halimeda* is a segmented plant which on death and disintegration generates coarse, sand-sized particles. *Penicillus,* the shaving-brush alga, is less rigid, consisting of a bundle of filaments coated in needles of aragonite. Death of this and other algae provides fine-grained carbonate sediment: lime mud or micrite (Section 4.3.4).

Phylloid algae are a group of Late Palaeozoic algae which have a leaf or potato crisp-like shape. Some belong to the codiaceans, while others have more affinities with the red algae. Phylloid algae are one of the main components of shelf-margin mud mounds in the Upper Carboniferous–Lower Permian of S.W. U.S.A.

Chrysophyta (Coccoliths)

Coccolithophorids (Jurassic-Recent) are planktonic algae which have a low Mg calcite skeleton consisting of a spherical coccosphere (10–100 μm diameter) composed of numerous calcareous plates, called coccoliths. Because of their size these algae are studied with the scanning electron microscope. The coccoliths are chiefly disc-shaped, often with a radial arrangement of crystals (Fig. 4.18). Coccoliths are a significant component of modern deepwater carbonate oozes, particularly those of lower latitudes. They form the bulk of Tertiary and

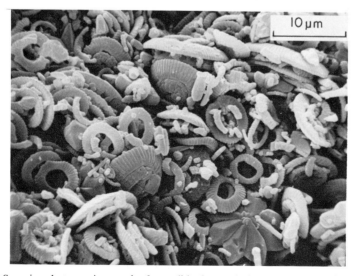

Fig. 4.18 Scanning electron micrograph of coccoliths from pelagic ooze, Shatsky Rise, northwest Pacific. Also present (lower right) is a larger discoaster, another type of nannoplanktonic alga. Courtesy of A. Matter and Deep Sea Drilling Project, Scripps I.O., California.

Cretaceous chalks and occur in red pelagic limestones of the Alpine Jurassic (see Fig. 4.27).

Cyanophyta, micrite envelopes and stromatolites

Although few blue-green algae are calcified, they have a profound effect on sediments through their boring activities and algal mat formation. A large proportion of skeletal fragments in modern and ancient carbonate sediments possess a dark *micrite envelope* around the grains (Figs. 4.9, 4.19; Bathurst,

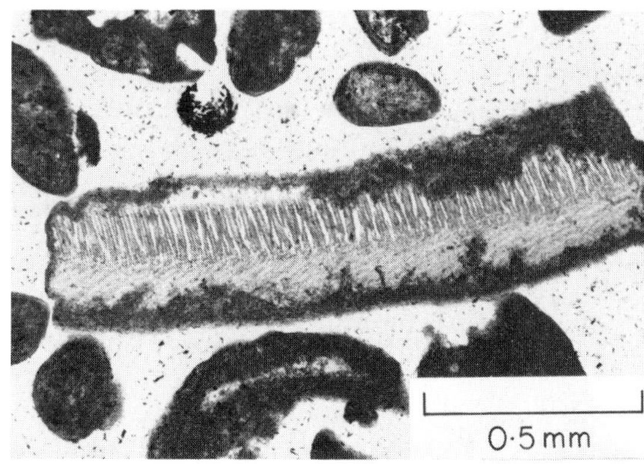

Fig. 4.19 Micrite envelope around bivalve fragment (the internal structure of the bivalve is well displayed, but probably this would not be preserved since the shell is composed of the metastable aragonite). Modern beach sand, Trucial Coast, Arabian Gulf.

1966, 1975). The envelope is produced by endolithic coccoid algae which bore into the skeletal debris. Following vacation of the bores (5–15 μm diameter), they are infilled with micrite. Repeated boring and infilling result in a dense micrite envelope. It should be emphasised that the envelope is the altered outer part of the skeletal grain; it is not a coating like that of an ooid around a nucleus. This process of grain degradation may eventually produce a totally micritized grain i.e. a peloid, (Section 4.3.1), devoid of the original skeletal structure. Many other organisms bore into skeletal grains and carbonate substrates: examples include clionid sponges, bivalves (such as *Lithophaga*) and polychaetes, all of which produce large bores and cavities, and fungi, which produce bores of 1–2 μm diameter. The micrite filling the algal bores, of aragonite or high Mg calcite in recent cases, may be precipitated physico-chemically, or biochemically through decomposition of the algae, possibly aided by bacteria. Micrite envelopes due to endolithic algae can be used as a depth criterion, indicating deposition within the photic zone (less than 100–200 m), but the algal, rather than fungal, origin must be demonstrated, and it must be borne in mind that grains can be transported to greater depths (Zeff & Perkins, 1979).

An important role of blue-green algae is in the formation of algal mats. In low latitude environments, such as shallow subtidal through to supratidal marine areas, and fresh to hypersaline lakes and marshes, organic mats of blue-green algae, together with bacteria, may cover the sediment surface (Fig. 4.20) or form columns and domes (Fig. 4.21). The algae are mainly filamentous varieties, common mat-forming genera being *Lyngbya, Microcoleus,*

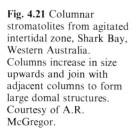

Fig. 4.20 Tidal flat algal mats from Abu Dhabi, Arabian Gulf. The algal mats (dark areas) are desiccated into small polygons and have a thin covering of recently-deposited carbonate sediment. Large polygonal desiccation cracks occur in the slightly higher, partly-lithified gypsiferous carbonate sediment (light area). Knife (circled) for scale. Courtesy of J.H. Powell.

Fig. 4.21 Columnar stromatolites from agitated intertidal zone, Shark Bay, Western Australia. Columns increase in size upwards and join with adjacent columns to form large domal structures. Courtesy of A.R. McGregor.

Schizothrix, Scytonema and *Oscillatorea,* although unicellular coccoid forms such as *Endophysalis* also occur. A mat usually has a specific algal community, which together with environmental factors produce a mat of particular morphology and structure. Areas where algal mats of various types are developing today include the Bahamas (Black, 1933; Monty, 1967; Gebelein, 1969; Monty & Hardie, 1976), the Arabian Gulf (Kendall & Skipwith, 1968) and Shark Bay, Western Australia (Davies, 1970a; Logan *et al.,* 1974; Hoffman, 1976; Playford & Cockbain, 1976).

The blue-green algae are mucilaginous and this, together with the filamentous nature, results in the trapping and binding of sedimentary particles to produce a laminated sediment or *stromatolite* (Figs. 4.22, 4.23). Stromatolites occur throughout the geological record but are particularly important in the Precambrian where they comprise thick limestone or dolomite sequences and have been used for stratigraphic correlation. The lamination in modern algal mats consists of couplets of dark organic-rich layers alternating with light carbonate-rich laminae. Laminae are usually less than several millimetres thick, but some carbonate laminae may reach a centimetre or more. In ancient stromatolites, the laminae are often alternations of dense micrite, sometimes dolomitized, and grains, such as pellets and skeletal debris. Algal laminated sediments frequently show small corrugations and irregularities in thickness, which serve to distinguish them from laminae deposited purely by

113

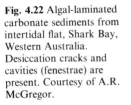

Fig. 4.22 Algal-laminated carbonate sediments from intertidal flat, Shark Bay, Western Australia. Desiccation cracks and cavities (fenestrae) are present. Courtesy of A.R. McGregor.

Fig. 4.23 Precambrian stromatolites consisting of spaced domes with some lateral linkage by planar laminations; part of a stromatolite bioherm. Late Precambrian; Finnmark, Norway.

physical processes. The laminae may also constitute larger-scale domes and columns. The alternating laminae reflect growth of the algal mat (organic layer) followed by sedimentation, and then trapping and binding of the sediment particles into the mat, as the algae grow through to form a new organic layer at the surface once more. One point of discussion is the time interval represented by the laminae in ancient stromatolites, since if periodic, its rhythmicity could be used to estimate variations in length of the month or year in the past (Pannella, 1976). Although a diurnal growth pattern has been demonstrated for some subtidal algal mats (Monty, 1967; Gebelein, 1969), in other cases, such as on tidal flats (e.g. Park, 1976) and in ephemeral lakes, algal mat growth is probably seasonal or related to periodic wettings, and sedimentation is erratic, being largely controlled by storm floodings.

Algal mats result in a range of laminated structures. The simplest are planar or slightly undulating and have been referred to as cryptalgal laminites (Aitken, 1967). Planar stromatolites typically develop on protected tidal flats and so may show desiccation polygons (Fig. 4.20) and contain laminoid fenestrae (elongate cavities) (Fig. 4.33 Section 4.6.3) and evaporite minerals or their pseudomorphs. Domed stromatolites, where laminations are continuous from one algal dome to the next, occur on the scale of centimetres to metres. Columnar stromatolites are individual structures, which may be several metres high (Fig. 4.21). Complex stromatolites, such as commonly occur in Precambrian strata, may be combinations of domes and columns, such as a large columnar structure with linked domes internally, or they may show branching of columns. One further type of algal-sediment structure is a nodule or oncolite with concentric laminations internally (Fig. 4.24).

In a notation for describing stromatolites (Fig. 4.25) Logan *et al.* (1964) referred to domed stromatolites as laterally-linked hemispheroids (LLH), columnar stromatolites as vertically-stacked hemispheroids (SH) and oncolites as spherical structures (SS). There is also a binomial classification, with form genera and species, particularly used for Precambrian examples. Common genera include *Conophyton, Collenia* and *Cryptozoon.*

The morphological variation of stromatolites largely depends on environmental factors such as water depth, tidal and wave energy, frequency of exposure and sedimentation rate. For example, large columnar structures in

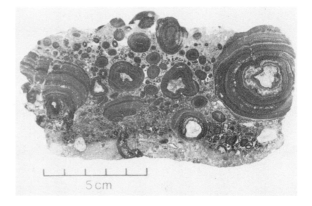

Fig. 4.24 Oncolites: spherical to irregular algal balls, some showing marked asymmetry through periods of stationary growth. Lower Carboniferous, eastern Scotland.

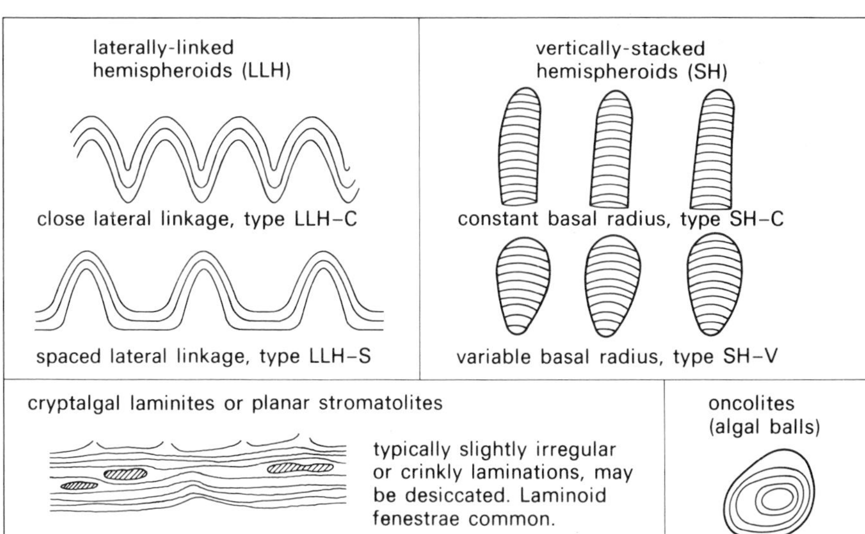

5 cm

laterally-linked hemispheroids (LLH)	vertically-stacked hemispheroids (SH)
close lateral linkage, type LLH–C	constant basal radius, type SH–C
spaced lateral linkage, type LLH–S	variable basal radius, type SH–V
cryptalgal laminites or planar stromatolites	oncolites (algal balls)
typically slightly irregular or crinkly laminations, may be desiccated. Laminoid fenestrae common.	

Fig. 4.25 Common types of stromatolite with terminology after Logan *et al.*, 1964.

115

Shark Bay are restricted to intertidal and subtidal areas in the vicinity of headlands, small columns and domes occur in less agitated bay waters, and low domes and planar algal mats dominate protected tidal flats (Hoffman, 1976). The microstructure of algal mats and stromatolites is also variable; it is thought largely to be a reflection of the algal community (Gebelein, 1974; Logan et al., 1974).

At the present time, algal mats are restricted in their occurrence, being well developed only in hypersaline tidal flat and freshwater environments. The dearth of algal mats in the normal marine environment, particularly in the subtidal, now and in the Phanerozoic, is largely due to grazing organisms, especially the gastropods. The absence of such metazoans in the Precambrian was one of the main factors in the widespread and diverse stromatolite development at that time, including many in subtidal and deeper water settings (Hoffman, 1974).

Marine algal mats are largely unlithified, whereas those of fresh and hypersaline waters may be cemented by algally-induced precipitation or physico-chemical precipitation of carbonate. Certain algae, common in the Palaeozoic are thought to be related to blue-greens, but have calcified filaments. The group name Porostromata has been applied to these algae which include *Girvanella, Ortonella, Garwoodia* and *Cayeuxia*. The algae consist of tubes or filaments with micritic walls, considered to be calcified sheaths, arranged in an irregular spaghetti-type or more-ordered, radial fashion (Fig. 4.26). The algae typically form nodules (oncolites) and also stromatolites. In some cases the algae are intimately associated with encrusting Foraminifera as in *Osagia* and *Sphaerocodium* nodules.

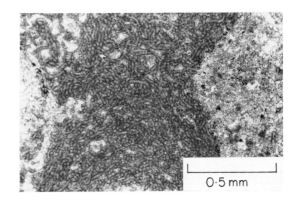

Fig. 4.26 *Girvanella* sp., a calcified filamentous alga common in Palaeozoic limestones. Plane polarized light. Ordovician, S.W. Scotland.

0·5 mm

4.3.4   MICRITE

Many limestones have a fine-grained usually dark matrix or are composed entirely of fine grained carbonate. This material is micrite (microcrystalline calcite), with a grain size generally less than 4 μm. Electron microscope studies (Fischer et al., 1967; Flügel et al., 1968; Steinen, 1978) show that the micrite is not homogeneous but has areas of finer or coarser crystals, and intercrystalline boundaries which may be planar, curved, irregular or sutured (Fig. 4.27). Micrites are susceptible to diagenetic alteration and may be replaced by coarser

Fig. 4.27 Transmission electron micrograph of a Jurassic micritic pelagic limestone showing variation in size and shape of micrite crystals, and two coccoliths consisting of elongate, radially-arranged crystals. Ammonitico Rosso, Jurassic, Austrian Alps.

mosaics of microspar (5–15 μm) through aggrading neomorphism (Section 4.7.3).

Carbonate muds are accumulating in many modern environments from tidal flats and shallow lagoons to the deep sea floor. There are many possible sources of lime mud, some important in a particular environment, others insignificant or hypothetical. Carbonate muds of the Bahaman Platform have been the subject of much research. They occur in the shallow subtidal, less-agitated central parts of the platform to the west of Andros Island, and in lagoons and gulfs, such as the Bight of Abaco. Fine carbonate sediments also occur on tidal flats and on the slopes and in the deep basins around the platform. The mud in subtidal areas consists predominantly of aragonite needles and laths a few microns in length; only some 20% of the sediment is recognizably biogenic. One of the early suggestions on the origin of the mud involved precipitation by denitrifying bacteria. Inorganic precipitation has been widely postulated (Cloud, 1962) as a result of seawater evaporation in areas remote from cold oceanic waters. The occasional 'whiting', a sudden milkiness of the sea due to suspended aragonite needles, may be the actual inorganic precipitation taking place, although stirring of bottom muds by shoals of fish can produce the same effect. The disintegration of calcareous green algae is now widely regarded as the main process of lime mud production (Stockman *et al.,* 1967; Neumann & Land, 1975). When algae such as *Penicillus* break down, a vast quantity of aragonite needles is released to the sediment. Measurements of growth rates and calculations of the mass of aragonite produced have shown that sufficient mud is produced to account for all the fine grained sediments. Indeed, it appears that there is an over-production so that algal disintegration in lagoons could be the source of mud for neighbouring tidal flats and deeper water off-platform areas (Fig. 4.28). Three other processes which produce lime mud, but in variable or limited quantities are: (i) bioerosion, where organisms such as boring sponges and algae attack carbonate grains and substrates; (ii) mechanical breakdown of skeletal grains through waves and currents; and (iii) biochemical precipitation through algal photosynthesis and decomposition,

117

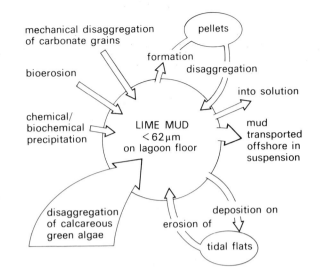

mechanical disaggregation
of carbonate grains

pellets

bioerosion

formation

disaggregation

into solution

chemical/
biochemical
precipitation

LIME MUD
<62μm
on lagoon floor

mud
transported
offshore in
suspension

disaggregation
of calcareous
green algae

erosion of

deposition on

tidal flats

Fig. 4.28 Lime mud budget for a lagoon in the Bahamas (based on Neumann & Land, 1975).

chiefly in algal mats of supratidal hypersaline and freshwater settings at the present time (Dalrymple, 1966; Monty & Hardie, 1976). Lagoonal lime muds and silts off British Honduras are largely formed from biological and mechanical breakdown of molluscan, foraminiferal, coral and algal skeletons (Matthews, 1966), although coccoliths are a major component of the finest fraction (Scholle & Kling, 1972). Carbonate mud, largely of skeletal origin, forms subtidal banks in Florida and Shark Bay (Turmel & Swanson, 1976; Davies, 1970b) where sea-grasses and algae trap and bind the sediment. Carbonate muds and oozes of the deep ocean floors are chiefly composed of coccoliths, with larger foraminiferal and pteropod grains.

One locality where inorganic precipitation is probably taking place is in lagoons along the Trucial Coast of the Arabian Gulf. Aragonite needle muds contain high strontium values (9400 ppm)—close to the theoretical for direct precipitation from that lagoonal water (Kinsman & Holland, 1969). There is also a paucity of calcareous algae in the region and the possibility of other aragonitic skeletons contributing is precluded by their low strontium values.

There is often little evidence in a limestone for the origin of the micrite. Nannofossils, in particular coccoliths, can be recognized with the electron microscope in some pelagic limestones (Fig. 4.27), but on the whole a biogenic origin is not obvious for the majority of ancient micrites. From the studies of modern lime muds, it is tempting to interpret ancient shallow marine micrites as the product of calcareous green algal disintegration. However, the possibility of inorganic precipitation in the past cannot be ignored and an origin through biological and mechanical breakdown of larger grains is always likely bearing in mind the masking effects of diagenesis on fine-grained limestones. In biosparites, grainstones and other coarse limestones, micrite could well be a cement, rather than a matrix (Section 4.7.2). In addition, fine carbonate sediment up to silt grade, may filter into a porous limestone soon after deposition or during early diagenesis. Such geopetal or internal sediment can be recognized by its cavity-filling nature.

4.4　Classification of limestones

Three classification systems are commonly used, each with a different emphasis.
(1) A very simple but often useful scheme divides limestones on the basis of grain size into calcirudite (most grains >2 mm), calcarenite (most grains between 2 mm and 62 μm) and calcilutite (most grains less than 62 μm).
(2) The classification scheme of Folk (1959, 1962) (Fig 4.29), based mainly on composition, distinguishes three components: (i) the allochems (particles or grains), (ii) matrix, chiefly micrite and (iii) cement, chiefly drusy sparite. An

Principal allochems in limestone	Limestone types	
	cemented by sparite	with a micritic matrix
skeletal grains (bioclasts)	biosparite	 biomicrite 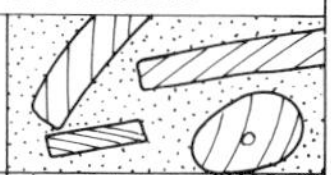
ooids	oosparite	 oomicrite 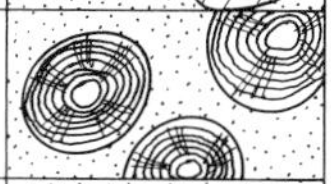
peloids	pelsparite	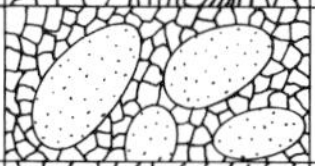 pelmicrite 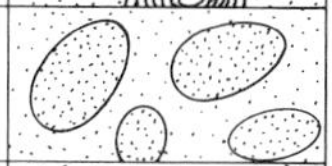
intraclasts	intrasparite	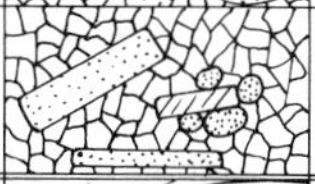 intramicrite 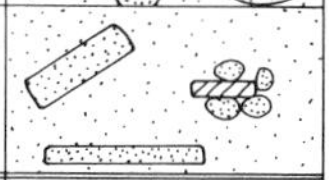
limestone formed in situ	biolithite	 fenestral limestone –dismicrite 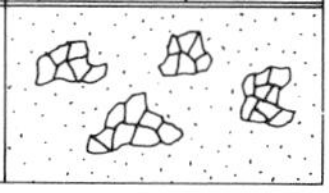

Fig. 4.29 Classification of limestones based on composition (after Folk).

abbreviation for the allochems (*bio*-skeletal grains, *oo*-ooids, *pel*-peloids, *intra*-intraclasts) is used as a prefix to *micrite* or *sparite* whichever is dominant. Terms can be combined if two types of allochem dominate, as in biopelsparite or bio-oosparite. Terms can be modified to give an indication of coarse grain size, as in biosparrudite or intramicrudite. Other categories of Folk are *biolithite*, referring to a limestone formed *in situ*, such as a stromatolite or reef-rock; and *dismicrite*, referring to a micrite with cavities (often spar-filled), such as a birdseye limestone (Section 4.6.3).
(3) The classification of Dunham (1962) (Fig. 4.30) divides limestones on the basis of texture into: *grainstone*, grains without matrix (such as a bio- or oo-sparite); *packstone*, grains in contact, with matrix (such as a biomicrite); *wackestone*, coarse grains floating in a matrix (could also be a biomicrite); and a *mudstone*, micrite with few grains. Several additional terms were added by Embry & Klovan (1972) to give an indication of coarse grain size (*floatstone* and *rudstone*), and of the type of organic binding in the *boundstones* during deposition (*bafflestone, bindstone* and *framestone*). The terms can be qualified to give information on composition, e.g. oolitic grainstone, pellet mudstone or crinoidal rudstone.
As a result of diagenetic modifications to limestones, care must often be exercised in naming the rock. For example: homogeneous-looking micrites may

119

Original components not organically bound during deposition						Original components organically bound during deposition		
of the allochems, less than 10% > 2 mm diameter				of the allochems more than 10% > 2 mm		boundstone		
contains carbonate mud (particles less than 0.03 mm diameter)			mud absent	matrix supported	grain supported	organisms acted as baffles	organisms encrusting and binding	organisms building a rigid framework
mud-supported		grain supported	grain supported					
less than 10% grains	more than 10% grains							
mudstone	wackestone	packstone	grainstone	floatstone	rudstone	bafflestone	bindstone	framestone

Fig. 4.30 Classification of limestones based on depositional texture (after Dunham, with modifications of Embry & Klovan).

be pelmicrites, and micrite in a bioclastic, grain-supported rock could be cement, compacted pellets (i.e. grains), primary sediment (i.e. matrix) or internal sediment (infiltrated geopetal sediment). The second example is basically separating grainstones from packstones-wackestones.

Of the many different types of limestone, five varieties are more common than others. These are the grainstones: oosparites and biosparites, the wackestones: biomicrites and pelmicrites, and the biolithites-boundstones. The depositional environments and facies of limestones are considered in section 4.10, where the typical limestone rock-types of each environment are noted. In essence, oosparites are of similar aspect in all their occurrences from the Precambrian to Recent. Oosparites typically possess cross bedding and most have formed in turbulent shallow-water environments. Biosparites also show current structures and are characteristic of shallow agitated settings. Their skeletal composition varies enormously, reflecting the organisms which were living in the area at that time of the Phanerozoic (Section 4.3.2). Biomicrites and pelmicrites are typical of less agitated environments such as lagoons, tidal flats and deeper water areas of shelves, platforms and basins. Biolithites or boundstones, characterized by a massive appearance, are very diverse in composition, depending on which skeletal organisms have contributed to the carbonate build-up.

4.5   Limestone grain size and texture

For the most part carbonate sediments are formed *in situ*. Although some may be transported from shelf to basin by turbidity currents or slumps, the majority accumulate where the component grains were formed or have been subjected to only limited transport by wave and tidal currents. The skeletal grains of carbonate sediments vary greatly in size and shape. Interpretations of limestone deposition are thus to a large extent based on the types of grain present since these will often provide concise information on the depth, salinity, degree of agitation, etc. This is not to say that the grain size and degree of sorting and rounding are unimportant. Although the grain size will largely be a reflection of the size of the carbonate skeletons of the organisms living in the area and of the many biological factors involved in their breakdown, the physical factors of waves and currents will also contribute, and in cases dominate. A measure of the grain size then will often give useful additional information, reflecting the energy level of the environment, or energy gradient of the area. Where one is dealing with grainstones or calcarenites, the grain size parameters discussed in Section 2.2.1 can be applied. It must be borne in mind that carbonate particles are hydrodynamically different from quartz grains; apart from complications arising out of shape, carbonate grains commonly have a lower density because of pores and contained organic matter. The degree of sorting or rounding of skeletal grains can be useful in certain bioclastic rocks, such as those of shelf areas where changes in these features could indicate proximity to a shoreline or zone of higher wave and tidal current activity. Some limestones of course, such as oosparites and pelsparites, are very well sorted and rounded anyway.

Measurements of carbonate grain size in ancient limestones are made by point counting slides or acetate peels under the microscope. For loose carbonate sands, the use of a settling chamber-sedimentation balance is recommended since this gives a better indication of the hydraulic behaviour than the grain size distribution obtained by sieving (Braithwaite, 1973).

In a general way the amount of micrite or lime mud in a limestone reflects the degree of agitation; lime muds tend to be deposited in quiet lagoonal or outer shelf areas. Increasing agitation leads to a decrease in the micrite content and increase in grain-support fabric and sparite content; sorting and rounding of grains then develops in the grainstone/biosparite. Interpretations must be made with care though since lime muds can accumulate in higher energy environments, trapped and stabilized by sea-grass or a surficial algal mat which leaves no record in the sediment, and micrite can be precipitated as a cement during early diagenesis (Section 4.7.2).

4.6   Sedimentary structures of limestones

Limestones contain many of the sedimentary structures common to sandstones, and some restricted to carbonate sediments.

4.6.1   BEDDING PLANES, HARDGROUNDS AND PALAEOKARSTIC SURFACES

As in siliciclastic sediments, bedding planes represent a change in the conditions of sedimentation. The changes are often subtle or short-lived. With limestones it is not uncommon to find that bedding planes have been affected by pressure solution due to overburden (Section 4.7.4). Through this, originally gradational bed boundaries may become sharp. Many limestones, particularly those interbedded with mudrocks, are nodular and have hummocky, irregular surfaces. Such nodular limestones may pass laterally into discrete nodules (Section 3.2). In many cases an early diagenetic segregation of carbonate, a sort of unmixing process, has contributed to the nodularity; burrowing may also give rise to a nodular appearance. Later pressure solution frequently accentuates the nodular structure. Nodular limestones occur in the Lower Lias of Britain (Hallam, 1964), Cretaceous chalk (Kennedy & Garrison, 1975a) and deeper water sequences of the Mesozoic and Devonian of Europe (Jenkyns, 1974).

A particular type of bedding plane is *a hardground surface*. Hardgrounds are horizons of synsedimentary cementation, taking place at or just below the sediment surface. Where a hardground surface formed the seafloor, it was often encrusted by oysters, Foraminifera and crinoids, and bored by polychaete annelids, certain bivalves and sponges. Hardground surfaces frequently cut across fossils and sedimentary structures in the limestone. Two types of hardground surface can be recognized: a smooth, planar surface formed by abrasion (Fig. 4.31), and an irregular, angular surface formed by solution (a corrosional hardground surface) The first type is more common in shallow subtidal sediments where waves and currents are able to move oolitic and skeletal sands across a hardground to produce a planar erosional surface; corrosional hardground surfaces are more common in pelagic sediments of deeper water where periods of non-sedimentation allow seafloor cementation and solution. The identification of a hardground is important since it

Fig. 4.31 Planar hardground surface encrusted with *Ostrea* (oysters) and penetrated by annelid and bivalve borings (seen on cut face). These features demonstrate synsedimentary seafloor cementation of the sediment (see also Fig. 4.46). Inferior Oolite, Jurassic; Gloucestershire, England.

5 cm

demonstrates synsedimentary submarine cementation. Hardgrounds occur throughout the Phanerozoic and have been described by Purser (1969), Bromley (1975), Kennedy & Garrison (1975a), Fürsich (1979) and others. Hardgrounds are forming at the present day off Qatar in the Arabian Gulf (Shinn, 1969) and on Eleuthera Bank, Bahamas (Dravis, 1979) (also Section 4.7.1).

Another particular type of bedding discontinuity peculiar to limestones is a palaeokarstic surface. When carbonate sediments become emergent, then solution through contact with fresh meteoric water produces an irregular, pot-holed surface. This solution commonly takes place beneath a thin soil cover, and the soil itself may be preserved as a discontinuous clay seam or bed immediately above the solution surface. The term karst is applied to subaerial limestone solution features, typical of more humid climatic areas, so that palaeokarstic surface refers to a subaerial solution horizon in a limestone. Laminated crusts form upon and within uplifted carbonate sediments, as a type of caliche (Section 4.10.1). They have been described from Florida and Barbados by Multer & Hoffmeister (1968) and James (1972). In ancient limestones laminated crusts could be mistaken for stromatolites, but their association with palaeokarstic surfaces and palaeosols indicates a subaerial, pedogenic origin. Palaeokarstic horizons occur in the Carboniferous of Britain (Walkden, 1974) and Kentucky (Walls *et al.*, 1975).

4.6.2 CURRENT STRUCTURES

All the current structures of siliciclastic rocks occur in limestones: wave and current ripples, cross laminations, cross bedding on all scales and planar bedding. Large channel structures to small scours, sole structures, graded bedding, convolutions and dewatering structures, can all be found. For details of these sedimentary structures see Section 2.3. The lack of clay in many limestones, together with the effects of surface weathering, may make internal structures difficult to discern. Careful observations perhaps aided by staining of cut blocks will often bring them to light.

The same importance is attached to current structures in limestones as in sandstones. They are essential to environmental interpretation and facies analysis, giving valuable information on depositional process, palaeocurrents, depth and water turbulence. Many elongate fossils in limestones show a preferred orientation which can give an indication of palaeocurrent direction.

4.6.3 CAVITY STRUCTURES

Many cavities in limestones are partly filled with internal sediment, occupying the lower part of the cavity, with the space above occupied by sparite cement. Such cavity fills are known as *geopetal structures* and are a most useful way-up indicator (sparite at the top of course). Geopetal structures also record the horizontal at the time of sedimentation (acting as a spirit-level) and in some cases show that there was an original depositional dip (as in fore-reef limestones for example). *Umbrella structures* are simple cavities occurring beneath convex-up bivalve and brachiopod shells and other skeletal fragments. *Intra-skeletal cavities* occur in enclosed or chambered fossils, such as gastropods, Foraminifera and ammonoids. *Growth cavities* are formed beneath the skeletons of frame-building organisms, corals and stromatoporoids for example where they build out above the sediment or enclose space within the skeletons.

Fenestral cavities or 'birdseyes' are small cavities which occur particularly in pelleted micritic sediments of intertidal-supratidal environments. The majority are spar-filled only, but some may be sediment-filled. Three main types can be distinguished: (i) irregular fenestrae, the typical 'birdseyes' (Fig. 4.32), several

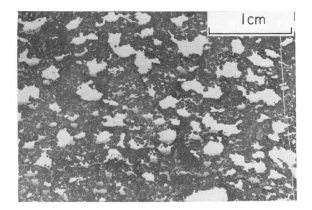

Fig. 4.32 Birdseyes (fenestrae) in pelsparite. Fenestrae are filled with sparry calcite although some contain internal sediment. Lower Carboniferous, Clwyd, Wales. Courtesy of D.I. Gray.

millimetres across, equidimensional to irregular in shape; (ii) laminoid fenestrae, several millimetres high and several centimetres long, parallel to bedding (Figs. 4.33, 4.22), and (iii) tubular fenestrae, cylindrical, vertical to subvertical in arrangement, several millimetres in diameter. Irregular fenestrae in abundance form the so-called birdseye limestones. They are ascribed to gas entrapment in the sediment and desiccation and so are a characteristic intertidal facies indicator (Shinn, 1968; Shinn *et al.*, 1969). Laminoid fenestrae develop in laminated sediments, particularly cryptalgal laminites, from the decay of organic matter and desiccation (Logan, 1974; Hardie, 1977). Tubular fenestrae are mainly formed by burrowing organisms but plant rootlets produce similar tubes. There are many descriptions of fenestral limestones, examples include the 'Loferites' of the Alpine Triassic (Fischer, 1964) and the Ordovician of Virginia (Grover & Read, 1978).

124

Fig. 4.33 Laminoid fenestrae and birdseyes infilled with calcite in cryptalgal and peloidal dolomite. The dolomite is probably penecontemporaneous in origin. Loferite, Triassic; Austrian Alps.

One further particular type of cavity is *Stromatactis* (Fig. 4.34). This is common in carbonate mud mounds of the Palaeozoic such as the Waulsortian reefs of the European Carboniferous and Devonian (e.g. Bathurst, 1959; Lees, 1964), the Devonian Tully Limestone of New York (Heckel, 1972a) and Ordovician mud mounds of Nevada (Ross *et al.*, 1975). Stromatactis cavities have an irregular unsupported roof and a flat floor, formed by internal sediment. The cement is invariably a first generation of fibrous calcite, followed by drusy sparite. Most Stromatactis cavities are a few cm long, but they may reach 10's of cm. The origin of the cavities has led to much discussion and speculation, earlier ideas being the decay of soft-bodied organisms and burrowing. Two current hypotheses are (i) sediment collapse (dewatering) i.e. an unmixing of water from the lime mud (Heckel, 1972a) and (ii) local seafloor cementation producing crusts, with cavities below.

Two types of cavity which form in partly lithified or cemented limestone are sheet cracks and neptunian dykes. *Sheet cracks* are cavities generally running parallel to the bedding which have planar walls, although some may have irregular roofs. *Neptunian dykes* cut across the bedding and may penetrate down many metres from a particular bedding plane. Both sheet cracks and neptunian dykes are filled with internal sediment, often with fossils a little younger than the adjacent limestone, if the cavities opened on to the seafloor. Spectacular examples of both types, in pelagic limestones and penetrating down

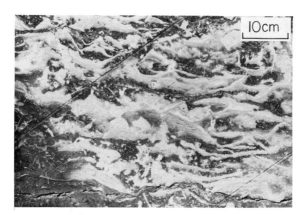

Fig. 4.34 *Stromatactis* cavities infilled with fibrous calcite and drusy calcite. Mudmound, Upper Devonian; Ardennes, Belgium.

125

into underlying platform carbonates, occur in the Triassic and Jurassic of the Alps (Wendt, 1971). Sheet cracks and neptunian dykes form through small tectonic movements during sedimentation and/or some slight downslope movement of sediment, causing fracturing of the lithified or partly lithified limestone mass.

Veins infilled with calcite, of various types, are common in limestone (Misik, 1971). Although they can be early structures, like neptunian dykes, the majority form through later fracturing.

4.7 Carbonate diagenesis

The onset of diagenesis is often difficult to define in a carbonate sediment. Sedimentational processes, such as algal micritization, may be contemporaneous with diagenetic processes such as cementation. Diagenesis in limestones is often referred to as isochemical, when there is no major change in chemistry of the sediment in its transformation into a rock, and allochemical when chemical changes are involved. The latter are chiefly brought about by dolomitization and silicification (Section 4.8). The principal diagenetic processes (isochemical) taking place are cementation (Sections 4.7.1, 4.7.2) and neomorphism (Section 4.7.3); their effects are summarized in Fig. 4.35. The formation of stylolites is also a common diagenetic phenomenon (Section 4.7.4).

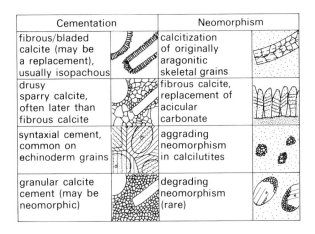

Cementation		Neomorphism	
fibrous/bladed calcite (may be a replacement), usually isopachous		calcitization of originally aragonitic skeletal grains	
drusy sparry calcite, often later than fibrous calcite		fibrous calcite, replacement of acicular carbonate	
syntaxial cement, common on echinoderm grains		aggrading neomorphism in calcilutites	
granular calcite cement (may be neomorphic)		degrading neomorphism (rare)	

Fig. 4.35 The common diagenetic fabrics produced by cementation and neomorphism.

4.7.1   CEMENTATION OF RECENT-SUBRECENT CARBONATE
        SEDIMENTS

Until the 1950's, it was tacitly assumed that carbonate sediments could only be converted into limestones through burial. It was then realized that Recent and Pleistocene carbonate sediments were being cemented through contact with fresh, meteoric waters and so cementation through uplift became the accepted process. Discoveries in the 1960's of cemented sediments on the seafloor in shallow and deep waters showed that neither burial nor uplift was necessary, and that in fact syn-sedimentary cementation was common. More recently cores through pelagic carbonates from the ocean floors have demonstrated cementation with increasing depth of burial. In ancient limestones there is often

126

evidence that cementation has taken place in several different diagenetic environments and over a lengthy period of time. To appreciate the problems of interpreting cements in ancient limestones it is necessary first to consider modern carbonate cements, which can be precipitated in: intertidal, shallow subtidal, deeper water, meteoric and burial environments (Fig. 4.36). Descriptions and reviews of carbonate cements are given by Bricker (1971), Milliman (1974), Bathurst (1975), Chilingarian *et al.* (1979) and Longman (1980).

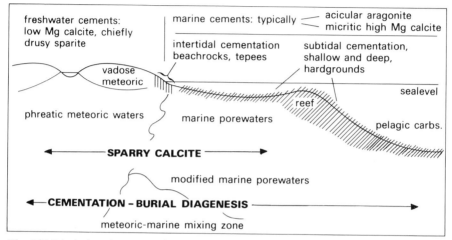

Fig. 4.36 Principal environments of cementation of carbonate sediments.

Intertidal Cementation. Cementation in the intertidal zone produces cemented beach sands known as beachrock. Beachrocks are most common in the tropics and subtropics but they do occur along temperate shorelines. They are composed of the same sediment as forms the surrounding loose beach sand; this is usually calcareous but it may have a substantial or even dominant siliciclastic component. Beachrocks can form quickly as is evidenced by the inclusion of man-made objects such as beer cans and World War II debris. In many cases, beachrock at the surface is being bored and eroded so it is thought likely that beachrock formation takes place a few tens of centimetres below the surface of the beach.

The cements in beachrocks are aragonite and/or high Mg calcite. Aragonite typically occurs as fringes from 10–200 μm thick of acicular crystals, oriented normal to the grain surfaces (Figs. 4.37, 4.38). High Mg calcite is a dark cryptocrystalline or micritic cement, often vaguely pelleted, which forms a coating around grains (Fig. 4.37) or completely fills pores. In many cases the cement fringes are isopachous, i.e. of equal thickness (Fig. 4.39), indicating marine phreatic (below the water table) precipitation where pores were constantly water-filled. Asymmetric cement fringes, thicker on the underside of grains, and meniscus cements, concentrated at grain contacts (Fig. 4.39), are recorded from some beachrocks and indicate precipitation in the marine vadose zone (Taylor & Illing, 1969).

Three processes of beachrock formation have been discussed: (i) a purely physico-chemical precipitation through evaporation of seawater and CO_2 loss,

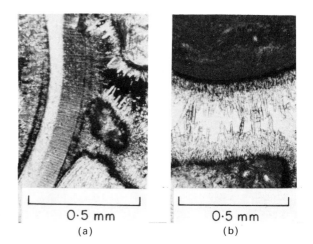

Fig. 4.37 Beachrock with cements of acicular aragonite and micritic high Mg calcite; the latter is the dark coating around grains. In 4.37a the aragonite cement has grown syntaxially on the bivalve grain as can be seen from the common extinction. Crossed polars. 4.37b plane polarized light. Great Barrier Reef, Australia. Sample courtesy of T.P. Scoffin.

0·5 mm
(a)

0·5 mm
(b)

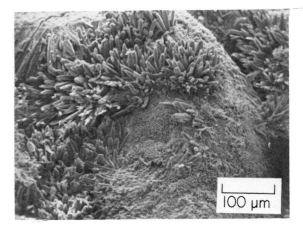

Fig. 4.38 Scanning electron micrograph of beachrock cement shown in Fig. 4.37. Where the coating of acicular aragonite crystals has come off the grain, the high Mg calcite micrite cement is visible.

100 μm

Fig. 4.39 The geometry of first generation cements: isopachous cement, indicative of precipitation in phreatic zones where all pores are filled with water (typical feature of low intertidal and subtidal cements); and gravity (stalactitic) and meniscus cements, indicative of vadose zone precipitation, as occurs in high intertidal, supratidal and shallow-subsurface continental situations.

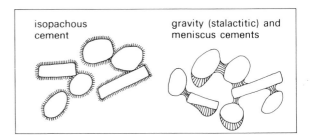

isopachous cement

gravity (stalactitic) and meniscus cements

particularly at times of low tide; (ii) a biochemical precipitation through decomposition of organic matter, chiefly algae, and (iii) precipitation from fresh ground water or seawater mixing with freshwater. The majority of workers have favoured (i), Hanor (1978) for example, but some beachrocks appear to have formed in a seawater-meteoric water mixing zone, and there is an indication of a biological or biochemical origin for some micritic cements.

Cemented surface crusts of intertidal-supratidal sediment occur in some carbonate areas. Along the Trucial Coast, Arabian Gulf, aragonite-cemented crusts are commonly brecciated or polygonally-cracked, and expansion of the surface crust as a result of the cementation may lead to pseudo-anticlines or tepee structures being formed (Assereto & Kendall, 1977). Supratidal crusts in the Bahamas are cemented by dolomite (Section 4.8.1).

Shallow subtidal cementation

Cements of acicular aragonite and micritic high Mg calcite occur within chambers and hollows of many skeletal grains on the shallow seafloor (e.g. Alexandersson, 1972). This intragranular cementation is often well developed in gastropod and foraminiferal grains, particularly in turbulent areas and can be physico-chemical or biochemical. Seafloor precipitation of cryptocrystalline aragonite produces grapestones and other aggregates in the Bahamas and elsewhere.

Shallow subtidal cementation of loose carbonate sand to produce surface crusts and lithified layers is rare but is taking place in a few metres of water off the Qatar Peninsula, Arabian Gulf (Shinn, 1969) and on Eleuthera Bank, Bahamas (Dravis, 1979). Off Qatar, these modern hardgrounds are being bored and encrusted by organisms, and polygonal cracks and tepee structures have formed through expansion of the cemented layer. The cements are mainly acicular aragonite, with some micritic high Mg calcite, and their precipitation is probably due to turbulent bottom conditions and the pumping of $CaCo_3$-supersaturated seawater through the sediments.

Cementation within modern reefs is now well documented (e.g. Schroeder, 1972; James *et al.*, 1976; MacIntyre, 1977). In general, high Mg calcite cements with micritic peloidal and bladed-fibrous habits, dominate over acicular aragonite. The cements are precipitated within primary cavities, between and within framework and other skeletons, and within secondary cavities formed by boring organisms. Internal sedimentation is often contemporaneous with cementation.

Aragonite cements of intertidal-shallow subtidal sediments often have a high strontium content, up to 10000 ppm, and the high Mg calcite cements are typically between 14 and 19 mol % $MgCO_3$. The controls over which cement is precipitated are not known, although in some cases there is a clear substrate influence. Cement crystals may preferentially nucleate upon a substrate of the same mineralogy and have a syntaxial relationship (Fig. 4.37a).

Deeper water cementation

Cemented carbonate sediments have been recovered from the ocean floors at depths down to 3500 m, mostly from areas of negligible sedimentation, such as seamounts, banks and plateaus (Fischer & Garrison, 1967). The limestones consist chiefly of planktonic Foraminifera, mollusca and coccoliths, and some benthic forms, cemented by a micritic calcite. The limestones are often bored and may be impregnated with phosphate and ferromanganese oxides. In areas of cold, deep water such as off the Bahama Platform (700–2000 m deep), the cement is a micritic low Mg calcite. Originally aragonitic skeletal grains in the sediment have been leached and high Mg calcite grains have lost their Mg (Schlager & James, 1978). In the Mediterranean and Red Sea, where there are and/or have been warmer bottom waters, thin surface crusts and nodules of lithified pelagic sediment are cemented by micritic high Mg calcite (Milliman & Müller, 1973; Müller & Fabricius, 1974). Acicular and micritic aragonite cements pteropod layers on the floors of the Red Sea (Milliman et al., 1969). Carbonate cement precipitation in these deeper water environments is mainly due to the very slow sedimentation rates, which allow interaction between sediment and seawater. The $CaCO_3$ is derived from seawater and the dissolution of less stable grains in the sediment, and the type of cement precipitated and its Mg content are determined by the water temperature (Schlager & James, 1978).

Meteoric cementation and diagenesis

The importance attached to meteoric diagenesis by earlier workers derives from the widespread occurrence of cemented Pleistocene limestones in the Caribbean and elsewhere. Studies of such limestones from Bermuda by Land et al. (1967) suggested five stages of freshwater cementation and diagenesis. Stage I is the initial sediment, consisting mainly of skeletal grains of aragonite and high Mg calcite. Stage II involves the precipitation of low Mg calcite on the surfaces of grains. The cement may be (i) isopachous (a uniform fringe) if precipitated in the phreatic zone, below the water table where all pores are completely filled with water or (ii) asymmetric, thicker on the underside of grains (a dripstone effect) or located at grain contacts (a meniscus effect) if precipitated in the vadose zone (Figs. 4.39, 4.40). Syntaxial overgrowths on echinoderm grains also develop at this early stage. Stage III involves the loss of Mg from the high Mg calcite, leaving a sediment of low Mg calcite and aragonite (Fig. 4.41). Stage IV is the main diagenetic event of aragonite dissolution and reprecipitation of the $CaCO_3$ as drusy sparry calcite. Aragonite skeletal grains are dissolved and the voids left are then infilled with calcite, the shape of the grains being maintained by the cement fringe of stage II or a micrite envelope (Section 4.3.3). The main feature of this dissolution-reprecipitation process is a loss of internal structure in the aragonitic skeletal grains. Although reprecipitation of calcite often

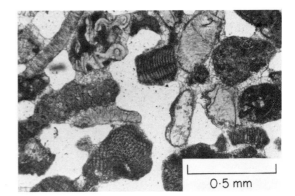

Fig. 4.40 Calcite cement at grain contacts, the first phase of meteoric cementation. Grains present here include calcareous algae, bivalve and echinoderm fragments. Pleistocene; Bermuda. Sample courtesy of H.C. Jenkyns.

0·5 mm

130

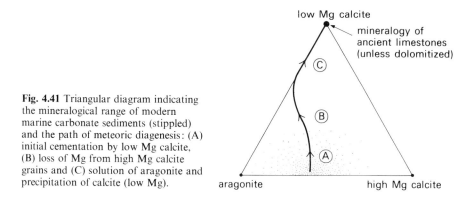

low Mg calcite

mineralogy of
ancient limestones
(unless dolomitized)

Fig. 4.41 Triangular diagram indicating
the mineralogical range of modern
marine carbonate sediments (stippled)
and the path of meteoric diagenesis: (A)
initial cementation by low Mg calcite,
(B) loss of Mg from high Mg calcite
grains and (C) solution of aragonite and
precipitation of calcite (low Mg).

aragonite   high Mg calcite

follows soon after aragonite dissolution, there may be a longer time gap, allowing some compaction of the skeletal voids before calcite precipitation and producing broken and fractured cement fringes and micrite envelopes (seen in Fig. 4.42). Some aragonite grains are replaced *in situ*, i.e. calcitized (Section 4.7.3), so that the internal structure is preserved to some extent. Stage V involves further precipitation of low Mg calcite to fill remaining voids. In most Pleistocene limestones, this final stage rarely produces a fully-cemented limestone, porosity may still be up to 20%. A further phase of cementation is thus required to reduce the porosity to less than 5%, the typical value of ancient limestones.

More recent work on meteoric diagenesis in Pleistocene limestones distinguishes between the phreatic and vadose zones (Steinen & Matthews, 1973; Pingitore, 1976). Carbonate sediments in the vadose zone have often suffered less diagenetic modification; aragonite may be preserved longer and cementation by low Mg calcite be less extensive; carbonate sediments in the phreatic zone on the other hand are entirely composed of low Mg calcite, have more cement and show a greater development of mouldic and dissolution porosity (Section 4.9). The $CaCO_3$ for meteoric cementation is thought to be partly derived from the 8% volume increase on aragonite solution, and from more complete dissolution of grains at certain, probably higher, near-surface, horizons.

Calcite cementation takes place in many continental sediments where pore waters are enriched in $CaCO_3$. Wadi gravels, scree sediments, and fluviatile and aeolian sandstones may all be cemented in this manner. Extensions of this meteoric calcite precipitation are the calcareous soils of semi-arid regions—calcretes (Section 4.10.1), and speleothems (stalactites and stalagmites), tufas and travertines of limestone districts.

Burial cementation

Cementation of carbonate sediments through burial diagenesis has been well demonstrated through studies of cores of pelagic ooze collected during deep-sea drilling. With increasing depth in a sediment core, there is a loss of porosity and an increase in the degree of lithification, from an ooze to chalk and then limestone at depths in excess of 500–600 m, in sediments of Lower Tertiary age

131

(Schlanger & Douglas, 1974; Matter, 1974). Cementation largely results from dissolution of the less stable, micron-sized calcite crystals of pelagic skeletal fragments and reprecipitation of the calcite upon larger crystals of skeletal origin. In spite of the cementation, porosities of the pelagic limestones are still in the region of 40%.

Because of fluctuations of sealevel in the last million years, it cannot be demonstrated that shallow-marine carbonate sediments have been cemented during and through burial. From studies of ancient limestones, however, it has been postulated that at least some sparry calcite, the ubiquitous late cement, is precipitated during burial diagenesis (Section 4.7.2).

To summarize, the main features of modern shallow-marine cements are the mineralogy: aragonite and high Mg calcite, and the fabrics: acicular, fibrous and bladed, or micritic. Meteoric cements are drusy sparites of low Mg calcite, associated with dissolution and replacement of aragonite skeletal grains. The mineralogical and fabric differences between shallow-marine and meteoric cements are related to the Mg/Ca ratio (Folk, 1974). The high Mg/Ca ratio of seawater (5) is thought to prevent the precipitation of ordinary (low Mg) calcite, but to allow the metastable forms aragonite and high Mg calcite to precipitate. The acicular and micritic habits of marine cements are also attributed to the effects of magnesium and possibly other ions, which poison crystal growth. Freshwater environments have low Mg/Ca values (around 2) and so permit precipitation of low Mg calcite, in the characteristic equant sparite habit. As can be observed at present cementation in marine, meteoric and burial-pelagic diagenetic environments rarely produces a fully-lithified limestone, in most cases porosity still exceeds 20%.

4.7.2   CEMENTS IN ANCIENT LIMESTONES

In many limestones, two distinct cement generations can be distinguished, based on crystal morphology, distribution and geochemistry (Figs. 4.35, 4.42, 4.43). The cement which occupies the majority of the original pore space is a clear, equant calcite, referred to as *sparite,* drusy sparite or orthospar. Sparite possesses a number of features, which, taken together, allow its cement interpretation. These are: (i) its location between grains and skeletons, and within original cavities, (ii) its generally clear nature, with few inclusions, (iii) the presence of planar intercrystalline boundaries, (iv) a drusy fabric, i.e. an increase in crystal size away from the substrate or cavity wall and (v) crystals with a preferred orientation of optic axes normal to the substrate. The fabric characteristics of sparite (iv) and (v) are reflections of the preferred growth direction of calcite, parallel to the c-axis. Calcite cement may also take the form of large poikilotopic crystals, several millimetres to centimetres across (Fig. 4.42), and more equigranular (idiotopic) mosaics (terms explained in Table 4.2).

Where echinoderm grains, and others composed of a single calcite crystal, are present in the limestone, then the sparite cement precipitates syntaxially (in optical continuity) upon the grain to produce an overgrowth (Fig. 4.44). Preferential cement growth upon such single crystal grains may envelop adjacent small, polycrystalline grains.

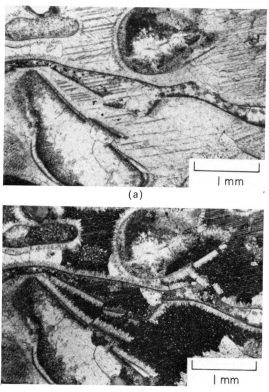

Fig. 4.42 Biosparite with two cement generations: an early fibrous cement which forms a fringe around grains and a later drusy calcite of large poikilotopic crystals. Compaction before precipitation of the later calcite has broken the early cement fringe and micrite envelope around the bivalve fragment. The bivalve was in a void stage at the time of compaction and is also infilled with the poikilotopic calcite. Compaction also broke the shell with its cement fringe below the bivalve fragment. 4.42a plane polarized light. 4.42b crossed polars. Middle Jurassic, Yorkshire, England.

(a)

1 mm

(b)

1 mm

In many limestones a cement generation was precipitated before the sparite. This early cement is typically a thin fringe of *fibrous calcite* around grains, which contrasts with the coarse blocky sparite (Figs. 4.42, 4.43). The fibrous fringe is usually less than 200 μm thick and consists of wedge-shaped crystals oriented normal to the substrate. The crystals may have unit or undulose extinction and they may be syntaxial with their substrate crystals. One particular type of fibrous calcite, referred to as radiaxial fibrous calcite (Fig. 4.45), possesses special fabrics: undulose extinction, convergent optic axes, subcrystals, curved twin planes and inclusions (Bathurst, 1959; Kendall &

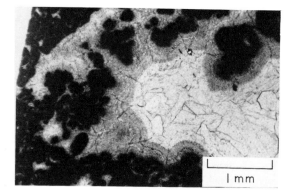

Fig. 4.43 Inclusion-rich fibrous cement fringe contrasting with clear drusy sparite infilling remainder of cavity. Plane polarized light. Fenestral limestone, Tertiary, Iraq.

1 mm

133

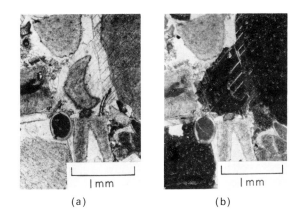

Fig. 4.44 Echinoderm fragments (dark grains with minute pores) with syntaxial overgrowths of clear sparite. Twin planes pass from echinoderm into overgrowth. Circular grain lower left is an ostracod (articulated) infilled with micrite. 4.44a plane polarized light. 4.44b crossed polars. Lower Carboniferous, central Scotland.

(a)   (b)

Tucker, 1973). This fibrous calcite occurs particularly in the larger cavities in limestones, such as occur in mud mounds (Stromatactis cavities, Section 4.6.3) and reef limestones.

Many fibrous calcite cements are isopachous (equal thickness), but occasionally asymmetric fringes occur, growing preferentially downwards into the cavity, and cements are concentrated at grain contacts. There may be clear evidence of a discontinuity between the fibrous calcite and sparite in the form of some internal sediment or a fracture of the fibrous calcite cement fringe resulting from compaction before sparite precipitation (Fig. 4.42).

One other type of early cement in grainstones is a *micrite,* although its distinction from matrix may not be easy. The distribution of the micrite, perhaps lining cavity walls or forming asymmetric fringes, together with a vague pelleted texture, may indicate a cement origin. The lithification of fine grained limestones is assumed to take place through the precipitation of a micritic cement, although this can rarely be demonstrated because of the fine grain size and likely neomorphism (Section 4.7.3; Steinen, 1978). Cementation of micritic limestones is thought to be an early diagenetic process in view of the almost total lack of compaction and full-bodied, 3-dimensional preservation of fossils which contrasts with fossils in mudrocks (Bathurst, 1970; although see Shinn *et al.* (1977) reporting experimental compaction of lime mud).

Further information on cements can be gained from the trace element

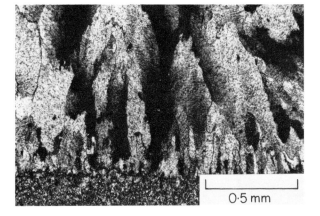

Fig. 4.45 Radiaxial fibrous calcite lining a cavity. Fibrous calcite appears dusty, has curved twin planes and undulose extinction. Crossed polars. Lower Carboniferous; Yorkshire, England.

0·5 mm

134

geochemistry, by the use of stains, cathodoluminescence and electron probe, and carbon and oxygen isotopes. Staining a limestone with potassium ferricyanide and alizarin red S will give an indication of the iron content of the calcite (Dickson, 1966). It is not uncommon to find that cement crystals are delicately zoned, implying changes in the chemistry of the pore-waters from which the calcite was precipitated. In many limestones one finds that the early cement generation is composed of non-ferroan calcite while the main cement phase, sparite, is a ferroan calcite (Oldershaw & Scoffin, 1967; Evamy, 1969). Cathodoluminescence can be used to distinguish manganese-rich and manganese-poor calcites, although a high iron content inhibits luminescence. By the use of stains and cathodoluminescence Meyers (1974; 1978), studying Mississippian limestones of New Mexico, has been able to correlate cement zones and generations over a wide area and so deduce a cement stratigraphy. Other geochemical differences between first and second generation cements have been found using strontium and magnesium (Benson & Matthews, 1971; Davies, 1977). Carbon and oxygen isotopes can also be used in the study of carbonate cements (Hudson, 1975; 1977; Allan & Matthews, 1977; Dickson & Coleman, 1980).

Interpretation of limestone cements

With cements in ancient limestones, discussion centres on the precipitational environment and original mineralogy and comparisons are made with modern carbonate cements.

For the first generation cements, the gross morphology may give an indication of precipitational milieu. Isopachous cement fringes will indicate phreatic conditions, marine or freshwater, whereas dripstone and meniscus cements will indicate the vadose zone, again either marine (high intertidal) or freshwater. The fibrous calcite (low Mg) cement fringes of ancient limestones are equivalent to acicular aragonite and bladed high Mg calcite cements of modern beachrocks and subtidal carbonates described earlier. In many cases isopachous fibrous cements in shallow marine limestones are interpreted as syn-sedimentary and submarine in origin. This is confirmed where hardground surfaces (Section 4.6.1) are present in the sequence and cements are cut by borings of organisms (Fig. 4.46; Purser, 1969). Some fibrous calcites may have been high Mg calcite cements which apart from losing their Mg, were little modified during diagenesis. Other fibrous calcites, such as the radiaxial variety, may be replacements of acicular aragonite or bladed high Mg calcite, through a calcitization, thin-film mechanism (Section 4.7.3). Evidence for a high Mg calcite precursor is provided by microdolomite inclusions (1–10 μm across) which give the fibrous calcite a cloudy appearance (Lohmann & Meyers, 1977), and by an enrichment in Mg relative to other cements in the same limestone (Davies, 1977). Fibrous calcite after aragonite can be identified from a higher Sr content, relative to other cements.

The trace elements iron and manganese in calcite cements are interpreted in terms of the Eh of the porewaters: low contents indicating oxidizing porewaters, and high values reflecting reducing porewaters. The non-ferroan character of fibrous calcites is consistent with a syn-sedimentary submarine or

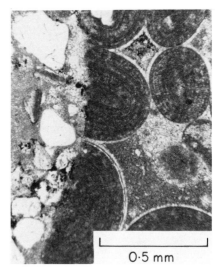

Fig. 4.46 Submarine cement from hardground; ooids and their isopachous fibrous cement fringe have been cut by an annelid boring (infilled with quartz grains and micrite) showing that the oolitic sediment had been cemented and that the fibrous fringe is a marine cement. Plane polarized light. Inferior Oolite, Jurassic; Gloucestershire, England.

0·5 mm

intertidal origin, while the ferroan nature of many sparry calcites is explained by precipitation in the meteoric phreatic zone where reducing conditions prevail (Evamy, 1969; Richter & Füchtbauer, 1978).

The precipitational environment of sparite, the main cement in all medium-coarse grained limestones, is still a problem. Drusy sparite is the typical cement of meteoric diagenesis, and this together with its ferroan nature has led to the belief that the main site of limestone cementation is the meteoric phreatic zone. An alternative, however, is the deep burial environment (Fig. 4.36). Evidence for sparite precipitation beneath tens to hundreds of metres of sediment could be the occurrence of broken and fractured fibrous cement fringes and micrite envelopes around formerly aragonitic skeletal grains, indicating some compaction due to overburden pressure after aragonite dissolution and before sparite precipitation (Fig. 4.42). The porewaters for deep burial cementation are likely to be chiefly connate (modified seawater buried with the sediments) although there may well be a degree of mixing with meteoric water. There are three possible sources for the $CaCO_3$: (i) the connate water itself, possibly together with meteoric water, (ii) pressure solution within the limestones or at deeper levels and (iii) solution of $CaCO_3$, mainly of skeletal aragonite, from calcareous shales interbedded with the limestones. For precipitation of calcite from trapped seawater the Mg/Ca ratio would have to be lowered. It has been suggested that adsorption of Mg on to clay minerals could have this effect (Folk, 1974). Pressure solution has been considered of major importance in cementation in view of the great quantities of $CaCO_3$ which have clearly been dissolved from pressure solution planes, and because of the vast quantities of $CaCO_3$ which are required from somewhere to produce the fully-cemented limestone sequences that are seen at the Earth's surface.

4.7.3   NEOMORPHISM

Some diagenetic processes involve changes in the mineralogy or fabric of the sediment. For these processes of replacement, often loosely referred to as recrystallization, Folk (1965) introduced the term neomorphism, to include all transformations between one mineral and itself or a polymorph. There are two aspects to neomorphism: the wet polymorphic transformation of aragonite to calcite and the wet recrystallization of calcite to calcite. Both processes are wet since they take place in the presence of water, through solution-reprecipitation; dry, solid state processes, such as the inversion of aragonite to calcite or recrystallization (*sensu stricto*) of calcite to calcite, are unlikely to occur in limestones, where diagenetic environments are always wet (see Bathurst, 1975 for discussion). Most neomorphism in limestones is of the aggrading type, leading to a coarser mosaic of crystals. Such neomorphism includes: (i) microspar-pseudospar formation in calcilutites; (ii) the calcitization of originally aragonitic skeletons and (iii) the replacement of acicular cements by fibrous calcite. Degrading neomorphism results in a finer mosaic of crystals.

A scheme for describing textures and fabrics of neomorphic limestones and dolomites, and other sediments such as evaporites which have been precipitated, crystallized or recrystallized, has been devised by Friedman (1965) and is given in Table 4.2.

Microspar-pseudospar formation in calcilutites. It is not uncommon to find that in fine grained limestones, the micritic matrix (less than 4 μm) has been locally or even totally replaced by microspar (crystal sizes between 4 and 10 μm) and pseudospar (10–50 μm) (Fig. 4.47). This *neomorphic spar* can be recognized by: (i) irregular or curved intercrystalline boundaries, often with embayments (contrasting with the plane intercrystalline boundaries of sparite cement, Section 4.7.2), (ii) very irregular crystal size distribution and patchy development of coarse mosaic, (iii) gradational boundaries to areas of neomorphic spar, and (iv) presence of skeletal grains floating in coarse spar.

Aggrading neomorphism involves the growth of certain crystals at the expense of others. It is likely that growth takes place in solution films and cavities between crystals, by syntaxial precipitation on pre-existing crystals. $CaCO_3$ will be derived from solution of submicron-sized crystals and inflowing

Table 4.2 Terms for describing textures and fabrics of crystal mosaics in sedimentary rocks (from Friedman, 1965).

For crystal shape:	anhedral – poor crystal shape subhedral – intermediate crystal shape euhedral – good crystal shape	
For equigranular mosaics:	zenotopic – majority of crystals anhedral hypidiotopic – majority of crystals subhedral idiotopic – majority of crystals euhedral	

For inequigranular mosaics:	porphyrotopic – where larger crystals (porphyrotopes)
	    enclosed in a finer-grained matrix
	poikilotopic – where larger crystals (poikilotopes)
	    enclose smaller crystals	

Size-scale:	micron-sized decimicron-sized centimicron-sized millimetre-sized centimetre-sized	0–10 μm 10–100 μm 100–1000 μm 1–10 mm 10–100 mm

137

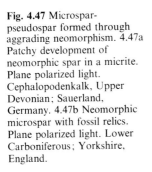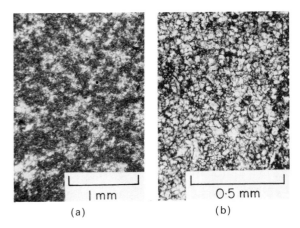

Fig. 4.47 Microspar-pseudospar formed through aggrading neomorphism. 4.47a Patchy development of neomorphic spar in a micrite. Plane polarized light. Cephalopodenkalk, Upper Devonian; Sauerland, Germany. 4.47b Neomorphic microspar with fossil relics. Plane polarized light. Lower Carboniferous; Yorkshire, England.

1 mm	0·5 mm
(a)	(b)

pore waters. Two factors affecting microspar formation are the clay content and Mg ions attached to the micrite. Bausch (1968) showed that aggrading neomorphism was restricted to Jurassic limestone with less than 2% clay. Folk (1974) and Longman (1977) have suggested that microspar formation is inhibited by a 'cage' of Mg ions, expelled from high Mg calcite and then located around micrite crystals. It is thought that if the Mg ions are removed, by flushing with meteoric water or adsorption on to clays, then micritic calcite is free to grow into microspar.

Calcitization and replacement of skeletal grains. Skeletal grains composed of aragonite have mostly been preserved by drusy sparite in ancient limestones, through solution of the aragonite and later precipitation of calcite (Section 4.7.1). In some cases, however, the grains have been replaced by calcite, with no intervening void phase, a process referred to as calcitization. When this has occurred, features to note are: (i) relics of the internal structure of the shell, preserved through inclusions of organic matter, (ii) an irregular mosaic of small and large calcite crystals, with wavy, curved or straight intercrystalline boundaries, and (iii) a brownish colour to the neomorphic spar, due to residual organic matter, which imparts a pseudopleochroism to the crystals (Fig. 4.48, Hudson, 1962; Bathurst, 1964; 1975). Skeletal grains originally composed of high Mg calcite may be preferentially replaced by ferroan calcite during later diagenesis (Richter & Füchtbauer, 1978).

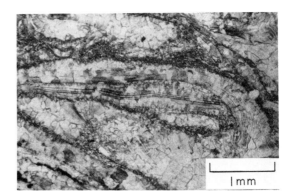

Fig. 4.48 Bivalve shells, originally composed of aragonite, which have been calcitized. The internal structure of the shells is preserved; the calcite crystals are pseudopleochroic and lack a drusy fabric. Plane polarized light. Purbeck limestone, Jurassic; Dorset, England.

1 mm

138

Neomorphic fibrous calcite. Some fibrous calcites, such as those with radiaxial fabrics (Section 4.72. Fig. 4.45) possess evidence of having replaced an earlier cement. This is suggested by features such as undulose extinction, inclusion patterns unrelated to the fibrous calcite crystal fabrics, non-planar intercrystalline boundaries and non-competitive growth fabrics. These fabrics must be interpreted with care though, since similar features occur in speleothems (stalactites) and other vadose carbonates (Kendall & Broughton, 1978).

Neomorphic replacement of cements and skeletal material is thought to proceed by a thin solution film mechanism, whereby a fluid film less than a few microns thick moves through the grain or cement, dissolving the carbonate (usually aragonite) in front and precipitating calcite behind. Some features of the original carbonate are retained within the replacement calcite, but are unrelated to the fabrics of that calcite; inclusion-defined growth lines and micron-sized relics of the original carbonate are examples. In addition, some fabrics of the replacement calcite are controlled by the original carbonate; undulose extinction in a fibrous calcite for example may reflect the systematic variation in orientation of the acicular crystals that it has replaced. The replacement of Pleistocene corals through thin solution films has been described by James (1974) and Pingitore (1976), and the replacement of acicular cement by fibrous calcite, also involving this process, by Kendall & Tucker (1973).

Degrading neomorphism

The opposite of aggrading neomorphism is degrading neomorphism where large crystals of $CaCO_3$ are replaced by smaller calcite crystals. This process of 'crystal diminution' is very rare in limestones and where documented (e.g. Wardlaw, 1962; Tucker & Kendall, 1973; Bathurst, 1975) it is in limestones which have been subjected to tectonic stress or very low grade metamorphism. Micritization of skeletal grains by endolithic algae is not a neomorphic process of course, but does result in a fine-grained mosaic.

4.7.4 STYLOLITES

Stylolites are thin seams of clay and insoluble material which mostly run parallel to bedding in a limestone (or sandstone). (They are present in Fig. 9.8, Chapter 9.) They are frequently sutured, others are less regular. Stylolites result from pressure solution, dissolution of the limestone along planes as a result of overburden or tectonic pressure (Wanless, 1979). Fossils or other grains can often be seen partly missing, adjacent to stylolites. The clay which fills the stylolite seam is the insoluble residue derived from limestone solution. Stylolites, or less well-defined solution planes sometimes referred to as flasers, (Fig. 4.60) commonly develop preferentially in clay-rich horizons. Solution is also common at limestone-shale junctions (as in nodular limestones) and along bedding planes. Calculations show that the amount of limestone dissolved through pressure solution is considerable; sequences can be reduced by up to 40%. In view of this, pressure solution is often cited as one of the main sources of $CaCO_3$ for limestone cementation, specifically as a source of the late diagenetic, often ferroan, sparitic calcite cement (Oldershaw & Scoffin, 1967; Hudson, 1975). Early stylolite formation may take place before precipitation of any cement, while a later phase may cut all diagenetic fabrics. In fact, an early diagenetic cement tends to inhibit subsequent pressure solution (Purser, 1978). Tectonic stylolites, which are developed normal to stress directions, cut and displace late diagenetic solution planes.

4.8 Dolomitization, dedolomitization and silicification

It is not uncommon to find that ancient limestones have been partially or even completely dolomitized. The dolomite so-formed can be replaced by calcite in the process of dedolomitization (Section 4.8.5). Limestones can also be silicified to various degrees (Section 4.8.6). These allochemical diagenetic processes frequently result in an obliteration of sedimentary and petrographic details.

4.8.1 DOLOMITES

The conversion of $CaCO_3$ minerals into dolomite, $CaMg(CO_3)_2$, may take place soon after the sediments have been deposited, i.e. penecontemporaneously and during early diagenesis (formerly referred to as syngenetic dolomitization), or a long time after deposition, usually after cementation, during the later stages of diagenesis (epigenetic dolomitization). The term primary has often been applied to dolomite but this word should only refer to a direct precipitate from water; in fact the majority of dolomites have formed by replacement of pre-existing carbonate minerals. The word dolomite is used for both the mineral and rock-type; for the latter the term dolostone is also used. Carbonate rocks are divided on the basis of dolomite content into:

limestone: 0 to 10% dolomite
dolomitic limestone: 10 to 50% dolomite
calcitic dolomite: 50 to 90% dolomite
dolomite: 90 to 100% dolomite

To convey an indication of grain or crystal size in a dolomite, the terms dolorudite, dolarenite, dolosparite and dolomicrite can be used. In many cases, if the original structure has not been completely destroyed the dolomites can be described in terms of Folk's or Dunham's classification, preceded by the word dolomitic, or prefixed by dolo-, as in dolobiosparite. For the detailed description of textures and fabrics the scheme of Table 4.2 should be used.

The distribution of dolomites in the stratigraphic record is not even, but increases back through time. The limestone to dolomite ratio in the Mesozoic is around 10:1, in Palaeozoic 3:1 and in the Precambrian 1:3. The preferential occurrence of dolomites in the Precambrian has led to the suggestion that seawater had a different composition so that dolomites could be precipitated directly. Alternative views are that dolomite-forming environments were more prevalent through palaeogeographic and palaeoclimatic considerations, or that the older limestones have had more time in which to come into contact with solutions capable of causing dolomitization.

140

Dolomites at the present time are forming in four main locations: the Bahamas and Florida (Shinn *et al.*, 1965; Atwood & Bubb, 1970; Gebelein *et al.*, 1979), the Arabian Gulf (Illing *et al.*, 1965; Hsü & Siegenthaler, 1969), Bonaire Island and San Andres Island in the Caribbean (Deffeyes *et al.*, 1965; Kocurko, 1979) and the Coorong, South Australia (Von der Borch, 1976). In the Bahamas, Florida, Bonaire and Arabian Gulf, dolomite is forming hard surface crusts in the supratidal zone and occurring within carbonate sediments to a depth of around 1 metre. The dolomite consists of 1 to 5 micron-sized rhombs. There is much evidence, in the form of dolomitized gastropods and pellets, for a replacement, early diagenetic origin. In the case of Bonaire and the Arabian Gulf, evaporites, chiefly gypsum, are associated with the dolomitic sediments. In all these cases Mg/Ca ratios of porewaters where dolomites are being precipitated are 3–4 times that of seawater. In the Bahamas, dolomites have recently been discovered in the sediments of depressions between beach ridges on Andros Island (Gebelein *et al.*, 1979). The dolomite here is being precipitated where meteoric water, located in a freshwater lens beneath the ridges, mixes with marine-derived groundwater beneath the depressions (Fig. 4.49). This occurrence could be significant geologically, as will emerge later. On

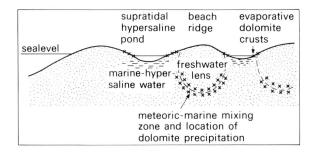

Fig. 4.49 Precipitation of dolomite in the Bahamas. Dolomite crusts (evaporative dolomite) are formed around hypersaline ponds and dolomite is precipitated in the subsurface where freshwater beneath beach ridges mixes with marine-hypersaline water.

San Andres Island, Colombia, cliff-forming Pleistocene limestones are being dolomitized by downward percolating brines with a high Mg/Ca ratio, which are formed from the evaporation of seawater in the supratidal spray zone. The Coorong in Australia is a series of coastal lagoons and lakes where dolomite is being precipitated, but not in association with evaporites. The dolomite is apparently forming where continental groundwater, modified by evaporation, discharges into shallow lakes. Dolomite may precipitate directly from lake waters where the chemical conditions are appropriate (Müller *et al.*, 1972). Only one record of marine subtidal dolomite is known (Behrens & Land, 1972), from Baffin Bay, a shallow often hypersaline lagoon, with restricted connection to the Gulf of Mexico.

Dolomite is a highly-ordered mineral consisting of calcium and magnesium ions in separate layers, alternating with carbonate ions. Because of its precise structure, a number of characteristic superstructure or ordering reflections are obtained with X-ray diffraction. Dolomite precipitates with great difficulty and in the laboratory its synthesis has often required non-geological conditions (Graf & Goldsmith, 1956; Lippmann, 1973; Chilingarian *et al.*, 1979). A synthetic

CaMg carbonate approximating to dolomite but lacking the ordering reflections on X-ray traces has been referred to as protodolomite (Graf & Goldsmith, 1956; but see Gaines, 1977; 1978). Dolomite forming at the present day frequently has poorly-developed ordering reflections. It often has an excess of calcium over magnesium (causing displacement of the principal dolomite peak) so that the composition is in the range $Ca_{52-55}Mg_{48-45}$; these can be called calcian dolomites.

Supratidal and hypersaline dolomite is produced through evaporation and precipitates at elevated Mg/Ca ratios. The raising of this ratio is brought about by the precipitation of aragonite and/or gypsum, which removes Ca from the porewater. In the geological record there are many documented occurrences of penecontemporaneous or early diagenetic supratidal dolomite (e.g. papers in Ginsburg, 1975). Examples include the Ordovician of Maryland (Matter, 1967), the Mississippian of eastern Canada (Schenk, 1969) and the Hauptdolomit (Upper Triassic) of the Alps (Müller-Jungblath, 1968). The penecontemporaneous dolomite is best recognized by its close association with intertidal-supratidal features, such as fenestrae (birdseyes), stromatolites, desiccation cracks and flakes, evaporites, and perhaps a restricted range of fossils indicating hypersalinities. Petrographic features are a fine grain size and some preservation of fossils and structures. Such dolomite occurs in Fig. 4.33.

4.8.3 LATER DIAGENETIC DOLOMITISATION

This type of dolomitization tends not to be related to a particular facies in a limestone or to particular beds, but will often cut across depositional units and structures. The dolomitization may be connected with unconformities, occuring in rocks immediately below, or in some cases it is related to tectonic structures, such as joints, faults or fold axes. Later dolomitization in most cases destroys the textural details of the original limestone. The dolomitization may be pervasive, where all the limestone is converted to dolomite and there are only relics of the original components (Fig. 4.50). Alternatively it may be selective, where, depending on factors such as porewater and sediment chemistry and crystal size, only the matrix or grains of a particular mineralogy are replaced.

Dolomite also occurs as randomly distributed rhombs in limestones and it may occur as a cement in cavities (Fig. 4.51). Many late stage dolomite crystals,

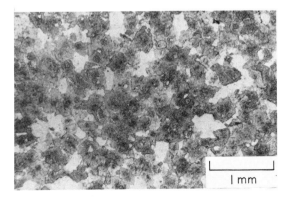

Fig. 4.50 Dolomitized limestone with relics of ooids (spherical structures with quartz nuclei). Some dolomite crystals have a rhombic shape, and the rock as a whole has a high porosity. Plane polarized light. Upper Permian, Nottinghamshire, England.

1 mm

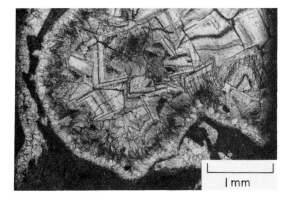

Fig. 4.51 Large, zoned ferroan dolomite crystals in a gastropod chamber. Plane polarized light. Torquay limestone, Middle Devonian; Devon, England. Courtesy of A. Buglass.

both replacements and cements, show a chemical or inclusion-defined zonation. Staining reveals that many of these later dolomites are iron-rich (ferroan), so that weathering of these rocks typically results in a light buff to dark brown colour, due to oxidation of the iron.

4.8.4   THE ORIGIN OF DOLOMITES

The origin of dolomite rocks is still a major problem in carbonate sedimentology (see review of Chilingarian *et al.*, 1979). One mechanism proposed for large scale dolomitization is that of seepage-reflux (Adams & Rhodes, 1960), thought to have been taking place on Bonaire Island by Deffeyes *et al.* (1965). In this mechanism, seawater drawn into supratidal areas by capillary action and flooding is concentrated by evaporation to produce porewaters with a high Mg/Ca ratio. These then descend or reflux seawards through the subsurface sediments, causing their dolomitization (Fig. 4.52). Unfortunately, it is now known that seepage-reflux is not operating at Bonaire to the extent envisaged and, apart from the San Andres case (Section 4.8.2; Kocurko, 1979) it has not been demonstrated on a geological scale elsewhere. For dolomitization in supratidal areas such as the sabkhas of the Arabian Gulf, Hsü & Siegenthaler (1969) proposed the mechanism of 'evaporative pumping', whereby seawater is drawn laterally from the lagoon into the supratidal sediments by the intense

Fig. 4.52 Seepage-reflux mechanism for dolomitization. Seawater seeps through sediments in intertidal-supratidal zone and into supratidal lakes if present. As a result of intense evaporation and precipitation of gypsum (which raises the Mg/Ca ratio), dense porewaters are produced which sink downwards and flow seawards, supposedly dolomitizing the calcareous sediments through which they pass.

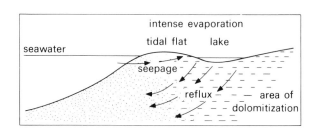

143

evaporation over the sabkhas (it is not a capillary process, sediment pores are water-filled). The intense evaporation also raises the Mg/Ca ratio of the porewaters (through the precipitation of aragonite and gypsum) and so causes dolomitization. Evaporative-supratidal dolomitization operating during a major regressive phase could produce a widespread dolomite horizon.

An alternative process is thus required for the many ancient dolomites, often reefs or subtidal shelf carbonates, which contain no sign of there having been intertidal or supratidal conditions during sedimentation. The only real alternative mechanism is one involving groundwater, in particular the mixing of freshwater and seawater, as considered by Hanshaw *et al.* (1971), Badiozamani (1973) and Folk & Land (1975). Seawater is supersaturated with respect to dolomite but because of the precise Ca-Mg ordering required, and the relative ease of aragonite and high Mg calcite precipitation, dolomite does not precipitate unless the Mg/Ca ratio is raised (as in the hypersaline solutions of supratidal flats) and then the dolomite is forced out of the solution. Dolomite precipitation is therefore more likely to take place from diluted solutions (when there would be fewer interfering ions present) and at slow crystallization rates (Fig. 4.53). Calculations have suggested that the mixing of meteoric

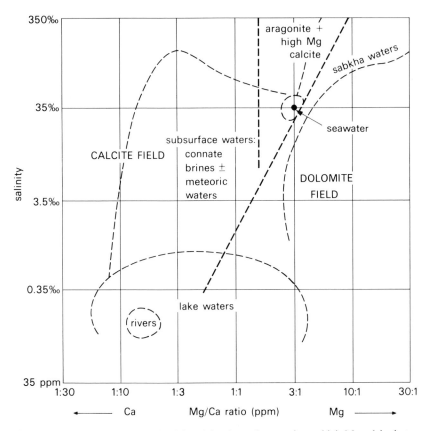

Fig. 4.53 The precipitational fields of calcite, dolomite and aragonite + high Mg calcite in terms of salinity and Mg/Ca ratio as envisaged by Folk & Land (1975). The fields of naturally-occurring waters are also shown.

groundwaters with up to 30% seawater would cause undersaturation with respect to calcite but an increasing saturation for dolomite (Badiozamani, 1973). Dolomitization has thus been predicted for seawater-meteoric water mixing zones, where salinities are decreased but Mg/Ca ratios maintained (Fig. 4.53). Situations where this could occur are in the near-surface and shallow sub-surface of evaporitic environments where hypersaline brines come into contact with fresh continental waters, and the shallow to deep burial phreatic environment where freshwater mixes with seawater buried with the sediments. The occurrence of mixing-zone dolomite in the subsurface sediments on Andros Island (Fig. 4.49) may thus be the modern analogue of many ancient dolomites.

From Fig. 4.56 it is seen that dolomite could also be precipitated through an increase in the Mg/Ca ratio of the porewater, with no salinity increase. Two possible ways in which this can take place are (i) through leaching of Mg from high Mg calcite grains in the limestone, and (ii) through leaching of Mg adsorbed on to clays in mudrock horizons (Kahle, 1965). Scattered dolomite rhombs in a limestone could reflect a local source of Mg from high Mg calcite and the local dolomitization of some thin limestones in thick mudrock sequences could reflect Mg from the adjacent clays.

An important consequence of dolomitization, particularly of the later diagenetic type, is that porosity is increased. Dolomite has a more compact crystal structure than calcite so that theoretically the complete dolomitization of a limestone results in a porosity increase of 13%, as long as there is no subsequent compaction or cementation. This feature of dolomitization is important for hydrocarbon reservoir potential. Many oil fields of western Canada for example are in dolomitized Devonian reef limestones.

4.8.5 DEDOLOMITIZATION

Dolomite may be replaced by calcite to produce a limestone again. This calcitization process is referred to as dedolomitization and predominantly takes place through contact with meteoric waters. Calcite replacement of dolomite can also occur in association with the solution of gypsum-anhydrite, a near-surface phenomenon as well (Sect. 5.5). Recognition of 'dedolomites' is similar to that of replaced evaporites, a question of noting dolomite crystal shapes, usually rhombohedra, occupied by calcite (pseudomorphs), or calcite crystals with replacement fabrics (see neomorphic spar, Sect. 4.7.3) containing small relic inclusions of dolomite. In some cases the original limestone texture is partially regenerated on dedolomitization, in other instances layers and concretions of fibrous calcite randomly and completely replace the dolomite. Dedolomitization has been described and discussed by Shearman *et al.* (1961), Evamy (1967) and Chilingarian *et al.* (1979).

4.8.6 SILICIFICATION

Silicification, like dolomitization, can take place during early or late diagenesis. It takes the form of selective replacement of fossils or the development of chert nodules and layers (Section 9.4). Silica also occurs as a cement in some limestones. The main types of diagenetic silica in limestones are: (i) euhedral

145

quartz crystals, (ii) microquartz, (iii) megaquartz, and (iv) chalcedonic quartz. They are described in Section 9.2 and shown in Figs. 9.1, 9.2 and 9.9. Both length fast and length slow chalcedonic quartz occur and the latter may indicate the former presence of evaporites (Section 5.5). Sponge spicules are the main source of silica, together with diatoms and radiolarians (Section 9.4). Descriptions of silicification in limestones include Banks (1970), Orme (1974), Meyers (1977) and Robertson (1977).

4.9   Porosity in carbonate sediments

The porosity of carbonate sediments shortly after deposition is very high: sand-sized sediments around 50%, lime mud around 80%. Porosity is lost or reduced through cementation, compaction and pressure solution, and gained through solution, dolomitization and tectonic fracturing (Murray, 1960; Choquete & Pray, 1970). Porosity in limestones can be divided into two main types: primary (depositional) and secondary (diagenetic-tectonic). Three common types of primary porosity are (i) framework porosity, formed by rigid carbonate skeletons such as corals, stromatoporoids and algae, especially in reef environments, (ii) interparticle porosity in carbonate sands, dependent on grain size distribution and shape, and (iii) porosity in carbonate muds provided by fenestrae (birdseyes) and Stromatactis (Section 4.6.3). Secondary porosity includes: (i) moulds, vugs and caverns formed by solution of grains and rock, often through leaching by meteoric groundwaters, (ii) intercrystalline porosity produced through dolomitization (Section 4.8.4), and (iii) fracture porosity, formed through tectonic movements and pressures, and through collapse and brecciation of limestone as a result of solution (such as of interbedded evaporites).

Primary porosity, and often the secondary too, is commonly facies controlled. Certain facies, such as reefs and fore-reefs, have high primary porosities, while others have low porosities, lagoonal micrites and outer shelf carbonates for example, unless affected by the diagenetic-tectonic processes leading to porosity development. Studies of carbonate facies distributions, cementation patterns and diagenesis, in particular dolomitization, coupled with porosity-permeability measurements are thus all required to detect any reservoir potential. Examples of carbonate hydrocarbon reservoirs are: the Upper Jurassic Arab-D Formation of Saudi Arabia with a primary intergranular porosity; Middle and Upper Devonian reef and fore-reef limestones of western Canada and the Ordovician Trenton Limestone of north-eastern U.S.A., both with a porosity at least in part a result of dolomitization; and the Upper Cretaceous Chalk of the North Sea and the Tertiary Asmari Limestone of Iran, both with fracture porosity.

4.10   Carbonate depositional environments and facies

4.10.1   NON-MARINE CARBONATE SEDIMENTS

Lacustrine limestones: Lacustrine carbonates are on three principal types: (i) inorganic precipitates, (ii) algal sediments and (iii) skeletal sands. Inorganic

146

precipitation, producing lime muds, mostly takes place through evaporation, but CO_2 loss, as a result of plant photosynthesis or pressure-temperature changes, and mixing of fresh stream water with saline lake water, also cause carbonate precipitation. The mineralogy of the carbonate precipitated depends largely on the Mg/Ca ratio of the water. Aragonite, calcite (high and low Mg) and dolomite may all be precipitated (Müller et al., 1972). Precipitation in shallow agitated zones may produce ooids, as in Great Salt Lake (Sandberg, 1975; Halley, 1977).

Lime muds may also be produced through the activities of algae, and from phytoplankton blooms. The main role of algae, however, is in the formation of stromatolites, common in modern lakes (e.g. Great Salt Lake; Green Lake New York, Eggleston & Dean, 1976) and in many ancient lake sequences (e.g. the Green River Formation, Wyoming and Utah, Bradley, 1929; Surdam & Wray, 1976; the Pliocene Ridge Basin of California, Link & Osborne, 1978). Oncolites also occur, those from Lake Constance being especially well-documented (Schäfer & Stapf, 1978). Skeletal sands contain fragments of calcareous algae, such as *Chara,* bivalves and gastropods.

Lacustrine carbonates are arranged in a similar facies pattern to their marine counterparts. Stromatolite 'reefs' and ooid shoals occur in more agitated, shallow waters, with lime muds occurring shoreward on littoral flats and in protected bays, and in the central deeper parts of lakes. One characteristic feature of lake basin deposits is a rhythmic lamination, consisting of carbonate-organic matter couplets, often interpreted as seasonal in origin.

Apart from the lacustrine limestone sequences noted above, others occur in the Triassic of South Wales (Tucker, 1978) and Greenland (Clemmensen, 1978) and in the Devonian Orcadian Basin of N.E. Scotland (Donovan, 1975).

Calcrete or caliche: In many parts of the world where rainfall is between 200 and 600 mm per year, and evaporation exceeds this precipitation, calcareous soils are formed. They are typically seen in river floodplain sediments but they also develop in other continental sediments (aeolian, lacustrine and colluvial deposits), and in marine carbonate sediments, should they become subaerially exposed. Many terms are applied to these pedogenic carbonates but calcrete and caliche, the latter chiefly in the U.S.A., are widely used. Calcrete occurs in several forms, from nodules to continuous layers, with massive, laminated and pisolitic textures. It is generally considered that calcrete forms through a *per descensum* process, of solution of carbonate particles in the upper, A-horizon of the soil profile and reprecipitation in the lower B-horizon (Goudie, 1973). Calcretes gradually develop in time, initially from scattered to packed nodules (Fig. 4.54), and then to a massive limestone layer. The amount of time involved is several to tens of thousands of years. The characteristic fabric of calcretes is a fine grained equigranular calcite with floating quartz grains through displacive growth (Fig. 4.55). Quartz grains and pebbles may be split or exfoliated in this displacive growth process. Replacement of original particles also takes place. Many calcretes possess spar-filled tubules, formerly occupied by rootlets. Calcrete pisolites can easily be confused with algal oncolites. Unless they have been reworked, calcrete pisolites can be distinguished by evidence for in-situ growth: dominant downward-directed laminae, reverse-grading and a fitted, often polygonal arrangement of pisolites. Reviews on calcretes have been

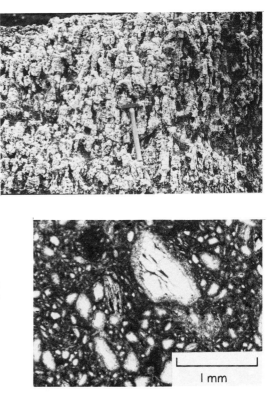

Fig. 4.54 Calcrete, consisting of closely packed elongate nodules which have grown in a river floodplain sediment. Old Red Sandstone, Devonian; Gloucestershire, England.

Fig. 4.55 Photomicrograph of modern calcrete showing dense micritic calcite which has displaced the quartz grains so that the latter are now not in contact. There is also an envelope of coarser calcite crystals around each quartz grain, this is a typical feature of calcretes. Plane polarized light. Recent calcrete, Almeria, Spain.

presented by Goudie (1973) and Reeves (1976); calcretes in continental settings have been described by Reeves (1970), Steel (1974), Allen (1974) and many others; calcretes in marine-coastal settings have been reported by Multer & Hoffmeister (1968), James (1972), Scholle & Kinsman (1974) and Read (1974).

4.10.2 MARINE CARBONATES

Many ancient limestone sequences were deposited on extensive shelves and platforms, in many cases bordering deeper basins. Carbonate facies are typically arranged in a broad pattern from the shore-zone out to the shelf-break, and into the basins too, although in any one sequence or period of carbonate sedimentation, not all facies may be developed. A model of the typical and complete carbonate facies pattern is shown in Fig. 4.56. Briefly, in the nearshore intertidal-supratidal area, carbonate mud flats dominate, passing landwards to evaporite deposits of sabkhas and salinas (Section 5.6) if in an arid climatic belt. In the shallow to deep shelf area, carbonate skeletal sands and muds accumulate. Shallow areas of higher energy on the shelf or along the shelf margin are the sites of ooid formation, and oolites together with skeletal sands may form barriers, beaches and shoals. Carbonate tidal deltas, also sites of ooid formation, may develop along barrier coastlines at the mouths of major tidal channels which connect lagoons and the open shelf. Along the shelf margin, reefs and other carbonate buildups commonly develop. These may also form barriers leading to the formation of quiet-water lagoons, perhaps of restricted

148

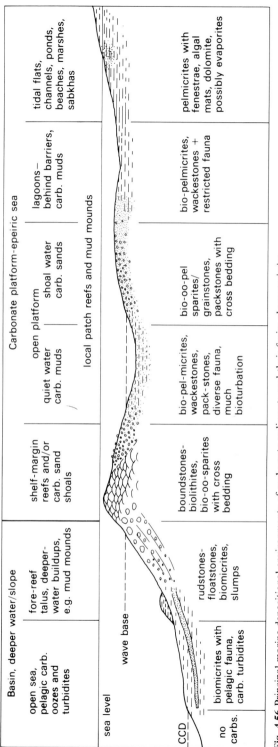

Fig. 4.56 Principal marine depositional environments of carbonate sediments and their facies characteristics.

circulation, on the shelf behind the reefs. Small patch reefs often form on the shelf and within open lagoons. Carbonate detritus, derived from the reefs and shoals along the shelf margin, may be transported into adjacent basins through debris flows and turbidity currents. Periods of low terrigenous input into the basins may permit sedimentation of pelagic carbonates, particularly on submarine rises there. From this model of carbonate sedimentation the main carbonate facies can be related to seven major depositional sites : (i) intertidal-supratidal flats, (ii) lagoons and restricted bays, (iii) intertidal-subtidal shoal areas, (iv) open shelves and platforms, shallow to deep, (v) reefs and carbonate buildups, (vi) starved basins and other sites of pelagic carbonate sedimentation, and (vii) carbonate turbidite basins. The features of these depositional sites and their sediments are summarized below, with pertinent references.

Intertidal-supratidal flats are vast areas, regularly to rarely covered by water, dominated by weak currents and wave action. Many tidal flats occur shoreward of lagoons behind barriers. Tidal flat carbonates are dominantly micrites, often pelleted, although local lenses of coarser sediment (grainstone) may represent tidal channel fills. Fenestrae (Section 4.6.3) are the characteristic structure, giving rise to the distinctive birdseye limestone (Fig. 4.32). The fauna may be restricted in diversity; gastropods in particular may abound, together with ostracods, Foraminifera and bivalves. Thin coarse layers of subtidal skeletal grains may occur, transported on to the tidal flat by storms. Algal mats and stromatolites (Section 4.3.3) are typical of tidal flat deposits. Many are simple planar varieties, showing desiccation cracks and laminoid fenestrae. Small domes may develop and in higher energy areas, columnar stromatolites. Bioturbation is common, and rootlets may occur. Synsedimentary cementation of tidal flat sediments can produce surface crusts, which may break up to form flakes, and tepee structures (Section 4.7.1). Penecontemporaneous dolomitization may take place, giving fine-grained dolomite mosaics (Section 4.8.2). In arid climatic areas, the evaporite minerals gypsum-anhydrite, and possibly halite, will develop in the sediment. They may be preserved as pseudomorphs (Section 5.5). Slight uplift and contact with meteoric waters may result in palaeokarstic surfaces, laminated crusts and calcretes. Modern carbonate tidal flats have been described by Shinn *et al.,* (1969), Davies (1970a), in Ginsburg (1975) and Hardie (1977); tidal flat limestones have been described by Fischer (1964), Matter (1967), Schenk (1969), Laporte (1967) and in Ginsburg (1975).

Lagoons and restricted bays are subtidal areas located behind barriers, which may be reefs or carbonate sand shoals. Organisms living in lagoons, and therefore the sediments accumulating in these predominantly quiet water areas, depend largely on the degree of restriction and permanence of the barrier. Lagoons may be normal in terms of salinity, as in lagoons of atolls; brackish where there is much freshwater run-off, as in the inner part of Florida Bay for example; or hypersaline to a greater or lesser extent, such as the inner part of the Great Bahama Bank, Shark Bay (W. Australia), and the lagoons of the Trucial Coast (Arabian Gulf). The sediments are variable in grain size, although many are carbonate muds, rich in peloids. Aggregates are common. Towards the barrier the muds pass into coarser sediments, and coarse skeletal debris may be derived from small coral patch reefs which often grow in normal lagoons. The lagoon floor is dominated by molluscs, green algae and

150

Foraminifera. The green algae in particular are a major sediment contributor (Section 4.3.3), and endolithic algae play a significant role in skeletal breakdown and the production of micritized grains. Algal mats may occur, as in the Bahamas and Shark Bay (Section 4.3.3), and sea-grasses too. Bioturbation is often intensive, largely through the activities of crustaceans. Sedimentary structures may be poorly developed, although thin, vaguely-graded beds of coarser grains may be formed through periodic storm reworking. For descriptions of modern lagoonal carbonates see Milliman (1974), Logan *et al.* (1974) and Bathurst (1975).

Lagoonal limestones are common in the geological record, particularly in back-reef and back-barrier situations, passing shorewards into tidal flat facies. Many are wackestones, biomicrites and pelmicrites, in some cases with organisms preserved in growth position. Examples are the calcisphere-rich 'porcelanites' of the Carboniferous, the *Amphipora* wackestones of Devonian back-reefs in western Canada, Europe and Australia, and the thick-shelled magalodont bivalve facies of Triassic back-reefs (Wilson, 1975).

Intertidal-subtidal shoal areas occur in belts of high tidal current and wave activity and include barriers, beaches, shoals and tidal deltas. Depths of deposition are less than 5–10 m. In many cases these shoal areas are located along the seaward margin of carbonate platforms. The sediments are carbonate sands, composed of ooids and rounded and sorted skeletal grains. The latter are fragments of normal marine organisms: corals, bivalves, algae, echinoderms and brachiopods (in the Mesozoic-Palaeozoic). Sedimentary structures are ubiquitous, chiefly cross bedding on all scales, perhaps with herring-bone cross bedding through tidal current reversals (Fig. 2.24), also planar bedding (in truncated sets if a beach), scours and channels. The sand bodies may be linear, localized patches, or blankets if associated with a sealevel rise.

Descriptions of modern carbonate sand bodies include Ball (1967), Loreau & Purser (1973), Bathurst (1975) and Hine (1977). Ancient equivalents are biosparites, oosparites and grainstones, occurring in most thick carbonate sequences.

Open shelves and platforms with carbonate sediments are poorly developed at the present time; examples include the Yucatan Shelf, Mexico (Logan *et al.*, 1969) and the eastern Gulf of Mexico. In the geological record extensive epeiric seas covered shelves and intracratonic platforms thousands of square kilometres in area. Carbonate deposition takes place in seawater of normal salinity and oxygenation. The sediments are grainstones in nearshore and shoalwater areas, passing offshore into wackestones and biomicrites where deposition is largely below wave-base. Periodic storms may effect these sediments to produce shell-lags and storm layers, the latter consisting of graded units a few centimetres thick. The sediment is typically bioturbated. Common skeletal components are molluscs, chiefly bivalves, Foraminifera and coralline algae, and brachiopods and echinoderm debris in the Mesozoic and Palaeozoic. Mud mounds and patch reefs may occur within these open shelf areas, particularly towards the outer shelf margins. Terrigenous silt and clay are often an important constituent of shelf limestones, frequently as thin intercalations.

Epeiric-sea limestones are often part of long-term transgressive or regressive phases of sedimentation, or they are part of shorter-term, oft-repeated

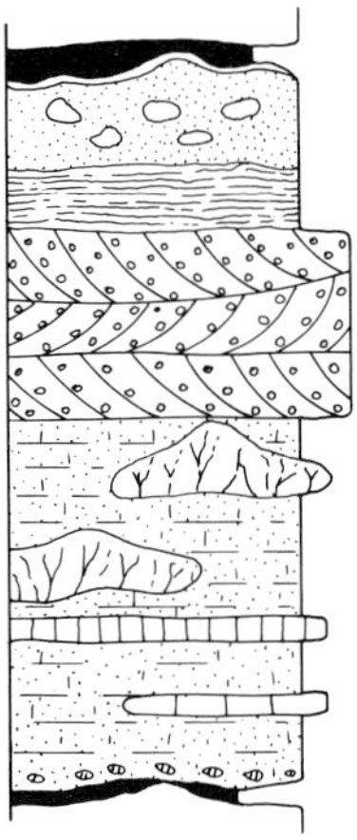

Sediments	Interpretation
palaeosoil ± palaeokarstic surface ± calcrete	supratidal/emergence
fenestral biopelmicrites wackestones + stromatolites	tidal flat
biopel- & oo-sparites/grain-stones, + cross bedding (etc.)	low intertidal-shallow, agitated subtidal
local bioherms + biostromes	
biopelmicrites/wackestones ± terrigenous clay, much bioturbation, storm beds	deeper-water subtidal
basal intraformational conglom.	reworking during transgression

Fig. 4.57 Shoaling-up limestone sequence; there are many variations on this general theme, depending largely on the energy level of the shoreline and on the climate. Sequences are typically several to many metres thick.

depositional cycles. Two common types of the latter are (i) wholly carbonate cycles, consisting of a shoaling-up sequence capped by a palaeokarstic surface (Fig. 4.57) and (ii) carbonate-siliciclastic cycles, where the siliciclastic sediments result from periodic deltaic advances across a carbonate shelf (Fig. 2.56). Epeiric-sea carbonates have developed over vast areas of stable cratons, such as the North American and Russian platforms, at various times during the Phanerozoic. They are particularly well-developed in the Carboniferous. Examples are the limestones of the Yoredale cycles in Britain and N.W. Europe, and the Pennsylvanian and Wolfcampian shelf sequences of southwestern U.S.A., both interbedded with deltaic sediments (see review of Wilson, 1975).

Reefs and carbonate buildups. Although coral reefs are one of the most familiar and most studied of modern carbonate environments, there are many other types of reef developing at the present time and preserved in the geological record. Such *carbonate buildups,* a term now widely used for locally-formed limestone bodies which had original topographic relief, are a common feature of many carbonate sequences going back to the Precambrian. Many buildups are important reservoir rocks for oil and gas. The literature on modern and ancient reefs is vast; reviews and compilations include Laporte (1974), Heckel (1974), Bathurst (1975), Wilson (1975), Frost *et al.* (1977) and Sellwood (1978).

The term reef itself is best restricted to a carbonate buildup which possessed a wave-resistant framework constructed by organisms, but to be clear the term ecologic reef or organic framework reef should be used for this. Specific types of such ecological reef are *patch reef,* small and circular in shape; *pinnacle reef,* conical; *barrier reef,* separated from coast by a lagoon; *fringing reef,* attached to coast; and *atoll,* enclosing a lagoon. Other terms frequently used are *bioherm* for local in-situ organic growth with or without framework; *biostrome* for

152

laterally extensive in-situ growth with or without framework; *organic bank* or *loose skeletal buildup* for an accumulation of mostly skeletal sediment, chiefly through trapping or baffling; *mud mound, mud bank* or *lime mud buildup* (formerly reef knoll) for an accumulation of mostly lime mud, micrite, probably by trapping and baffling.

Many different organisms can be and have been involved in the construction of reefs. At the present time, the main reef builders are corals and coralline algae; others of limited importance are sponges, serpulids, oysters and vermetid gastropods. In the past, practically all invertebrate groups have at one time or another contributed to reef growth. Special mention can be made of: blue-green algae (stromatolites) in the Precambrian and Cambrian, stromatoporoids in the Ordovician to Devonian, rugose corals in the Silurian to Carboniferous, scleractinian corals from the Triassic, phylloid algae in the Carboniferous-Permian, sponges in the Triassic-Jurassic, and rudistid bivalves in the Cretaceous (see Heckel, 1974 for a review). Organisms in reefs take three roles: the frame-builders, those providing a skeletal framework (corals at the present time); the frame-binders and encrusters, organisms which consolidate the framework, such as calcareous algae and bryozoans; and the reef-users, such as boring bivalves and algae, predatory fish and echinoderms. With many ancient reefs it is clear that there was no true solid framework, but much in-situ organic growth. This last feature gives rise to the two typical features of reef limestones, a massive appearance with no stratification (Figs. 4.58, 4.59) and the prevalence of organisms in growth position. Many reef limestones would be classified as biolithites, framestones and bindstones. Primary cavities are a feature of reefs, although they are often infilled with skeletal debris and cement. Early cementation is a feature of many modern and ancient reefs (Section 4.7.1).

There are many factors controlling the growth of modern coral reefs (Stoddart, 1969) and it is likely that these same factors exerted an influence on coral and other reefs in the past. For coral reef growth, these factors are: (i) water temperature, optimum growth is around 25°C, (ii) water depth, most growth takes place within 10 m of the surface, (iii) salinity, corals cannot tolerate great fluctuations and (iv) turbidity and wave action, coral growth is favoured by intense wave action and an absence of terrigenous silt and clay. The majority of reefs occur along shelf margins, an agitated zone where waves and currents of the open sea first impinge on the seafloor. Smaller patch reefs (Fig. 4.59)

Fig. 4.58 Devonian reef limestones in the Canning Basin, western Australia. The reef consists of massive unbedded limestone (central part of cliff); in front of the reef (to the left) occurs a talus slope of reef debris, these fore-reef limestones have an original dip; behind the reef (to the right) are flat-bedded limestones of the back-reef lagoon. Courtesy of C.T. Scrutton.

Fig. 4.59 Small patch reef consisting largely of rugose and tabulate corals, surrounded by thin-bedded shelf limestones. Wenlock Limestone, Silurian; Shropshire, England. Courtesy of C.T. Scrutton.

develop on the shelves, often in the open lagoons behind linear shelf margin reefs. Reefs, often atolls, are also developed on submerged volcanic islands within the ocean basins. The configuration and morphology of some present-day reefs is a reflection of karstic solution of earlier reefs during glacial low stands of sealevel (Purdy, 1974).

Many modern reefs, but particularly those in shelf-margin situations, show a characteristic three-fold division into fore-reef or reef-front, reef-flat, and back-reef. The reef-front is a steep slope, vertical in places, with organisms constructing reef in the upper part, passing down to a talus slope of coarse reef debris. Reef-derived carbonate turbidites may be present in an adjoining basin. A prominent system of surge channels gives a spur and groove morphology along the reef-front, extending onto the reef-flat. On the reef-flat, covered by no more than 1–2 m of water, there is prolific organic growth, of corals and algae on modern reefs. The back-reef area consists of reef debris adjacent to the reef-flat, passing shoreward to a quiet water lagoon. This same facies pattern is seen in many ancient buildups (Fig. 4.58). Classic examples include the Permian Capitan reef in Texas (Newell *et al.*, 1953), Devonian reefs of western Canada, Europe and Australia (Playford & Lowry, 1966; Klovan, 1974; Krebs, 1974), and the Triassic reefs of the Northern Calcareous Alps (Zankl, 1969).

Mud mounds are massive accumulations of micrite, tens to hundreds of metres across which pass laterally into well-bedded limestones. The mud mounds only contain scattered fossils, those of some possible significance are crinoids, bryozoans and algae. Stromatactis cavities (Section 4.6.3) are common in many mud mounds. Mud mounds occur on shelves and in shelf-margins and a deeper water setting is often indicated. The best developed mud mounds occur in the Palaeozoic, in the Carboniferous of northwest Europe where they are referred to as Waulsortian 'reefs' (Bathurst, 1959; Lees, 1964), the Carboniferous of New Mexico and Montana (Pray, 1958; Cotter, 1965) and in the Devonian of New York (Heckel, 1972a) and Europe.

The origin of mud mounds is still a problem. Processes which could be involved are (i) the entrapment and/or precipitation of micrite by algae, (ii) entrapment of mud through the baffling action of bryozoans and crinoids and

154

(iii) concentration of mud into mounds by currents. Comparisons have been made with the mud banks of Florida (Turmel & Swanson, 1976) and Shark Bay (Davies, 1970b), where accumulations of micrite are forming through the baffling and sediment-trapping action of sea-grasses and algae. The sediment itself is derived from green algal disintegration (Section 4.3.4) and the breakdown of larger skeletal grains.

Starved basins, shelves and submarine rises are the sites of pelagic carbonate sedimentation. Where water depth is too great for benthic organisms to flourish, in excess of some 50–100 m, then carbonate sediments composed of pelagic organisms will accumulate in the absence of clay. The maximum depth of accumulation is controlled by the rate of carbonate solution. The ocean is saturated with respect to $CaCO_3$ in the upper 100–200 m and below this undersaturated. $CaCO_3$ thus dissolves below 200 m but it is not until greater depths that the rate of $CaCO_3$ solution increases substantially. The depth at which the rate of solution is balanced by the rate of supply is known as the carbonate compensation depth (CCD). This depth varies in the oceans, its position being controlled by calcareous plankton productivity which itself depends largely on nutrient supply and water temperature. In the tropical regions of the world's oceans, the CCD for calcite is between 4500 and 5000 m; the CCD for aragonite is about 1000 m less. Calcareous oozes can accumulate on seafloor which is shallower than the CCD; siliceous oozes and red clays are present below this depth. Fluctuations in the CCD back into the Cainozoic and Mesozoic are now well documented (Ramsay, 1974).

Modern pelagic carbonates are composed of pteropods (aragonitic), coccoliths (Fig. 4.18) and Foraminifera (calcitic), and are found on outer continental shelves, continental slopes and ocean floors starved of terrigenous clay, and on submarine rises, drowned reefs and volcanoes (seamounts and guyots) rising up from the ocean floor. Ancient pelagic limestones occur in the Mesozoic of the Alpine region, the Ammonitico Rosso for example (Garrison & Fischer, 1969; Bernoulli & Jenkyns, 1974) and in the Devonian and Carboniferous of Hercynian Europe, the Cephalopodenkalk and Griotte (Tucker, 1974). They were deposited on submerged reefs and carbonate platforms, and on submarine rises within the basinal areas. Characteristic features of pelagic limestones, apart from a dominantly pelagic fauna, are their condensed nature and evidence for synsedimentary cementation in the form of hardgrounds, lithoclasts, sheet cracks and neptunian dykes. Many pelagic limestones are nodular (Fig. 4.60 and Jenkyns, 1974) and some contain ferromanganese nodules and crusts (Section 6.7).

The Cretaceous chalks of N.W. Europe and S.E. U.S.A. are composed largely of coccoliths and are thus pelagic limestones. Deposition took place at depths of around 50–150 m, so that there is a significant benthic macrofauna of echinoids, bivalves and brachiopods. Hardgrounds are common within the chalks and these are frequently mineralized with phosphate and glauconite (Kennedy & Garrison, 1975a).

Papers on pelagic limestones are contained in Hsü & Jenkyns (1974).

Turbidite basins filled with carbonates are not as common as those with siliciclastic turbidites (Section 2.10.7). These basins are adjacent to carbonate platforms which supply the sediment. In the slope area between shelf and basin,

Fig. 4.60 Pelagic limestone: a micritic, nodular limestone which contains a pelagic fauna. Pressure solution effects are common and result in the clay seams (flasers) between nodules. Griotte, Upper Devonian; Montagne Noire, France.

5 cm

slump-folded limestones and debris-flow breccias are common. In the basins themselves, limestone turbidites are usually interbedded with hemipelagic dark shales (Fig. 4.61). Sole structures, graded and planar bedding, and cross lamination are all developed in limestone turbidites, just as in siliciclastic examples.

Limestone breccias and turbidites from slopes and basins have been described from the Lower Palaeozoic of the Appalachians, Rocky Mountains and Newfoundland (papers in Cook & Enos, 1977), the Devonian of western Europe (Meischner, 1964; Tucker, 1969) and the Cretaceous of the Alps (Scholle, 1971).

Fig. 4.61 Limestone turbidites composed of crinoidal and other skeletal grains interbedded with hemipelagic mudrocks. Sequence is inverted. This is the typical appearance of turbidites (both carbonate and siliciclastic) in laterally continuous beds with interbedded deepwater mudrocks. Devonian; Cornwall, England.

4.10.3 CARBONATE FACIES SEQUENCES

Carbonate facies sequences are generated by depositional processes of progradation and buildup, and external processes of sealevel change and tectonic movements. With constant sealevel in a period of tectonic stability, the two main depositional processes of progradation are the building out of intertidal-supratidal flats over adjacent subtidal sediments, producing a shoaling-upward cycle (Fig. 4.57) which may be capped by evaporites (Section 5.6), and the building out of shelf-margin reefs over the fore-reef talus. Other processes are the migration of carbonate sand shoals and barriers over lagoonal or quietwater shelf sediments. A eustatic fall in sealevel, if sufficient to terminate marine sedimentation on the shelf, will result in meteoric processes affecting the sediments, in particular the formation of palaeokarstic surfaces and calcrete. Solution porosity may develop in sediments at this time. A less drastic sealevel fall will produce more widespread shallow water and tidal flat facies on the shelf. Sealevel rises will produce landward migration of facies belts, the development of deeper water facies over shallow water ones, and a termination of reef growth if the sealevel rise is too great and too quick.

Simple tectonic processes of vertical uplift or subsidence will change water depths on shelves and have the same effects as eustatic sealevel changes. Other contemporaneous tectonic movements are chiefly along faults and these acting during sedimentation may give rise to rapid facies changes across the fault lines, the localized development of reefs, and the formation of narrow to broad fault-bounded basins within intracratonic areas. Facies patterns, sequences and cycles for carbonates are discussed by Irwin (1965), Duff *et al.* (1967), Heckel (1972b), Wilson (1975) and James in Walker (1979).

Further reading

Textbooks giving in-depth reviews of carbonate sediments and limestones are:

Bathurst, R.G.C. (1975) *Carbonate Sediments and Their Diagenesis*, pp. 658. Elsevier, Amsterdam.
Wilson, J.L. (1975) *Carbonate Facies in Geologic History*, pp. 471. Springer-Verlag, Berlin.
Milliman, J.D. (1974) *Marine Carbonates*, pp. 357. Springer-Verlag, Berlin.
Flügel, E. (1978) *Mikrofazielle Untersuchungs-methoden von Kalken*, pp. 454. Springer-Verlag, Berlin.

For a review of carbonate diagenesis, including dolomitization see:

Larsen, G. & Chilingarian, G.V. (1979) *Diagenesis in Sediments and Sedimentary Rocks*, pp. 579. Elsevier, Amsterdam.

For the identification of fossils in thin section see:

Horowitz, A.S. & Potter, P.E. (1971) *Introductory Petrography of Fossils*, pp. 302. Springer-Verlag, Berlin.
Scholle, P.A. (1978) A color illustrated guide to Carbonate Rock Constituents, Textures, Cements, and Porosities. *Mem. Am. Ass. Petrol. Geol.* **27**, pp. 241.

Collections of relevant papers include:

Bricker, O.P. (Ed.) (1971) *Carbonate Cements*, pp. 376. Johns Hopkins Press, Baltimore.
Friedman, G.M. (Ed.) (1969) *Depositional Environments in Carbonate Rocks*, pp. 209. *Spec. Publ. Soc. econ. Paleont. Miner.*, **14**, Tulsa.
Laporte, L.F. (Ed.) (1974) *Reefs in time and space*, pp. 256. *Spec. Publ. Soc. econ. Paleont. Miner.*, **18**, Tulsa.
Pray, L.C. & Murray, R.C. (Eds.) (1965) *Dolomitization and limestone diagenesis*, pp. 180. *Spec. Publ. Soc. econ. Paleont. Miner.*, **13**, Tulsa.

5

Evaporites

5.1   Introduction

Evaporites are mainly chemical sediments which have been precipitated from water following the evaporative concentration of dissolved salts. The principal evaporite minerals are gypsum ($CaSO_4.2H_2O$) and anhydrite ($CaSO_4$) and halite ($NaCl$). There are many other naturally-occurring evaporite minerals and of these, the potassium and magnesium salts sylvite, carnallite, polyhalite, kainite and kieserite are important constituents of some marine salt deposits (formulae given in Table 5.1).

Table 5.1 The common marine and non-marine evaporite minerals.

Common marine evaporite minerals		Non-marine evaporite minerals	
halite	$NaCl$	halite, gypsum, anhydrite	
sylvite	KCl	epsomite	$MgSO_4.7H_2O$
carnallite	$KMgCl_3.6H_2O$	trona	$Na_2CO_3.NaHCO_3.2H_2O$
kainite	$KMgClSO_4.3H_2O$	mirabilite	$Na_2SO_4.10H_2O$
anhydrite	$CaSO_4$	thenardite	$NaSO_4$
gypsum	$CaSO_4.2H_2O$	bloedite	$Na_2SO_4.MgSO_4.4H_2O$
polyhalite	$K_2MgCa_2(SO_4)_4.2H_2O$	gaylussite	$Na_2CO_3.CaCO_3.5H_2$
kieserite	$MgSO_4.H_2O$	glauberite	$CaSO_4.Na_2SO_4$

Evaporites are of great economic importance and have a wide range of uses and applications. Evaporite beds are an essential component of many oilfields of the world, commonly being the cap rocks to carbonate reservoir rocks (as in the Middle East and western Canada) or effecting structural traps through salt diapirism (see papers in Buzzalini *et al.*, 1969). Geologically, evaporites are useful in the studies of palaeoclimatology since they are generally restricted to arid areas of low latitude, where temperatures are very high, relative humidity is low and evaporation far exceeds any rainfall. The location of the world's major evaporite deposits are shown in Fig. 5.1.

5.1.1   EARLY GEOCHEMICAL WORK

Studies of evaporites go back to the last century when chemists such as Usiglio and Van't Hoff undertook experiments to evaporate seawater and to synthesize evaporite minerals with a view to determining stability fields and precipitational controls. Starting with seawater, composition given in Table 5.2, Usiglio established that gypsum was precipitated when seawater had been evaporated to about 19% of its original volume, and that halite appeared when the volume

158

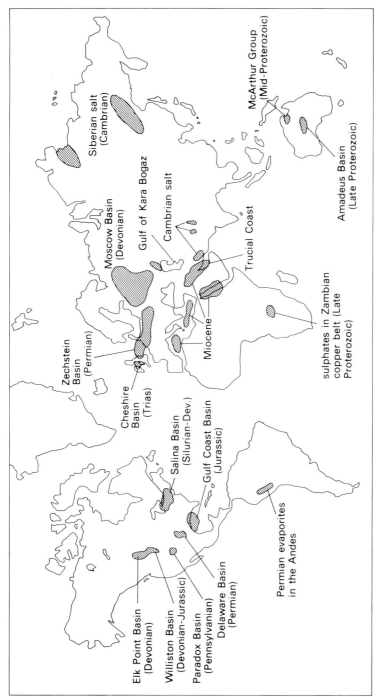

Fig. 5.1 Map showing location (and age) of the major evaporite deposits of the world. Also shown is the location of the Trucial Coast, Arabian Gulf, where sabkha sulphates are forming today, and of the Gulf of Kara Bogaz, on the eastern side of the Caspian Sea, the nearest modern analogue to a barred basin.

Map labels:

Siberian salt (Cambrian)
Moscow Basin (Devonian)
Gulf of Kara Bogaz
Cambrian salt
Zechstein Basin (Permian)
Cheshire Basin (Trias)
Salina Basin (Silurian–Dev.)
Gulf Coast Basin (Jurassic)
Elk Point Basin (Devonian)
Williston Basin (Devonian–Jurassic)
Paradox Basin (Pennsylvanian)
Delaware Basin (Permian)
Permian evaporites in the Andes
Miocene
Trucial Coast
sulphates in Zambian copper belt (Late Proterozoic)
Amadeus Basin (Late Proterozoic)
McArthur Group (Mid-Proterozoic)

Table 5.2 The composition of seawater expressed in parts per million and percentage of total dissolved species. For comparison, the composition of world average river water, with a salinity of around 120 p.p.m. is also given. Data from Krauskopf (1979).

Dissolved species	Seawater		River water
	p.p.m.	% of total	
Cl^-	18,000	55.05	7.8
Na^+	10,770	30.61	6.3
SO_4^{2-}	2,715	7.68	11.2
Mg^{2+}	1,290	3.69	4.1
Ca^{2+}	412	1.16	15.0
K^+	380	1.10	2.3
HCO_3^-	140	0.41	58.4
Br^-	67	0.19	0.02
H_3BO_3	26	0.07	0.1
Sr^{2+}	8	0.03	0.09
F^-	1.3	0.005	0.09
H_4SiO_4	1	0.004	13.1

was reduced to 9.5%. Continued evaporation produced magnesium and potassium minerals, but some naturally-occurring salts, such as kieserite and polyhalite were not obtained. Van't Hoff and co-workers showed that the minerals precipitated depended on the rate of evaporation and on the composition of the water, in particular whether equilibrium between metastable salts and solution was maintained. Following such experimental studies, it was soon noted that there were discrepancies between volumes of the various evaporite minerals found in marine salt deposits, such as the Permian Zechstein of N.W. Europe, and those predicted theoretically from the evaporation of seawater (Table 5.3). The discrepancies were explained by early replenishment of the barred-basin or lagoon with normal seawater and incomplete evaporation. Details of the physical chemistry of salt precipitation and reviews of earlier work can be found in Stewart (1963), Borchert & Muir (1964), Braitsch (1971) and Krauskopf (1979).

Table 5.3 The theoretical thickness of salts precipitated from seawater compared with the thickness of these salts in the Permian Zechstein of Germany, expressed as 100 metres of precipitated evaporite. Note that there is much more $CaSO_4$ and much less Mg and K salts in the Zechstein deposits compared with the theoretical (after Borchert & Muir, 1964). Also shown is the approximate thickness of the various salts produced by the evaporation of a column of seawater 1000 m high.

Component	Mineral	Thickness in 100 m of evaporite		Salt thickness from 1000 m of seawater
		from seawater	Permian Zechstein	
$MgCl_2$	in bischofite and carnallite	9.4	0.5	1.5
KCl	sylvite and in carnallite	2.6	1.5	0.4
$MgSO_4$	in kieserite	5.7	1.0	1.0
NaCl	halite	78	78	12.9
$CaSO_4$	anhydrite	3.6	16	0.6
$CaCO_3$	calcite	0.4	3	0.1
$CaMg(CO_3)_2$	dolomite			

160

Thick evaporite sequences, in some cases reaching a 1000 m or more, commonly fill large intracratonic sedimentary basins. Examples include the Permian Zechstein Basin of N.W. Europe, and the evaporite basins of North America (Fig. 5.1). Other evaporite sequences interdigitate with non-evaporitic sediments, limestones and marls especially, and occur on stable platforms and shelves and in subsiding basins. Evaporites deposited in lakes or marine embayments may be located in intracratonic rifts (fault-bounded basins), which in a few cases then developed into intercratonic rifts (continental margins) through seafloor spreading. The Dead Sea, a site of modern salt precipitation, the Triassic salt basins of Britain and Tertiary salt deposits of France and Germany are examples of evaporites formed in intracratonic rifts. The Tertiary evaporites on either side of the Red Sea and the Cretaceous evaporites known to exist along the eastern and western continental margins of the Atlantic were deposited in rifts which subsequently became the sites of ocean crust formation (Kinsman, 1975).

Salt deposits are commonly cyclic. Some consist of many thin evaporite beds, a few to tens of metres thick, typically of gypsum-anhydrite with little or no halite, alternating with limestone and marls. The very thick salt deposits of intracratonic basins often consist broadly of gypsum-anhydrite passing up into the more soluble halite, with thin beds of the highly soluble bittern salts (potassium and magnesium chlorides and sulphates) at the top. This sequence or cycle may be repeated several times, as in the Permian Zechstein of N.W. Europe.

There has been much discussion on the depositional environments of salt deposits, partly because there are so few modern sites of evaporite precipitation and none on the scale of those which clearly existed in the past. Evaporite depositional environments are shown diagrammatically in Fig. 5.2, and are discussed in Dean & Schreiber (1978) and Kendall (1979).

Two principal modes of evaporite deposition are recognized: (i) subaqueous precipitation from a shallow to deep standing body of water on to the floor of a lake, arm of the sea or marine barred-basin and (ii) subaerial precipitation within sediment or in very shallow brine pools and salinas. The subaqueous precipitation of evaporites is basically a simple 'evaporating dish' process, with occasional replenishment by fresh or marine water. At the present time, evaporites are being precipitated directly on to the floor of salt lakes but there are no truly marine barred-basins where evaporites are accumulating. An oft-quoted case is the Gulf of Kara Bogaz on the eastern side of the Caspian Sea, where halite, epsomite $(MgSO_4.7H_2O)$ and astrakhanite $(Na_2Mg(SO_4)_2.2H_2O)$ are being precipitated on the bay floor (papers in Kirkland & Evans, 1973).

For many years, evaporation of standing bodies of seawater in relatively deep barred-basins was considered the only viable mechanism for the formation of the world's major evaporite bodies (Hsü, 1972). However, an important alternative mode of formation was provided by the discovery in the early 1960's of gypsum-anhydrite forming within sediments of high intertidal–supratidal flats, called *sabkhas*, along the Trucial Coast of the Arabian Gulf (Shearman,

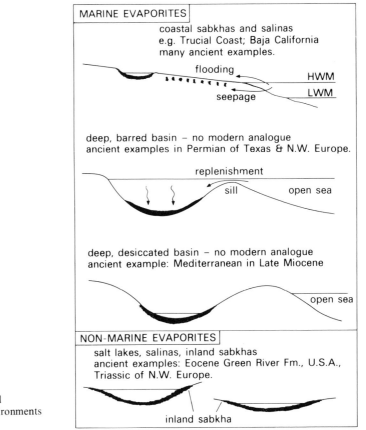

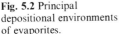

Fig. 5.2 Principal depositional environments of evaporites.

Contents of the figure:

MARINE EVAPORITES

coastal sabkhas and salinas
e.g. Trucial Coast; Baja California
many ancient examples.

flooding
HWM
seepage
LWM

deep, barred basin – no modern analogue
ancient examples in Permian of Texas & N.W. Europe.

replenishment
sill
open sea

deep, desiccated basin – no modern analogue
ancient example: Mediterranean in Late Miocene

open sea

NON-MARINE EVAPORITES

salt lakes, salinas, inland sabkhas
ancient examples: Eocene Green River Fm., U.S.A.,
Triassic of N.W. Europe.

inland sabkha

1966). The evaporites are precipitated from sediment porewaters in the vadose and upper phreatic zones. Evaporites are also forming within the exposed sediment around salt lakes and playas (inland or continental sabkhas). The subaerial precipitation of evaporites has taken place on the floor of deep desiccated marine basins; this occurred during the late Miocene when the Mediterranean dried up after being cut off from the Atlantic (Section 5.6).

Evaporites may develop in soils of desert areas as crusts and indurated horizons. Gypcrete or gypsite is not uncommon in North Africa, the Middle East and India.

5.2 Gypsum and anhydrite

Rocks of gypsum-anhydrite possess distinctive structures and textures and are susceptible to replacement, recrystallization and solution. Geological evidence and present day occurrences show that both gypsum and anhydrite may be precipitated at the Earth's surface, subaqueously in shallow and deep water and subaerially in coastal and inland sabkhas. However, on burial to depths greater then several hundred metres all $CaSO_4$ is present as anhydrite and on uplift, anhydrite is invariably converted to gypsum (secondary gypsum). The many studies of gypsum-anhydrite have shown that the stable phase is determined by the activity of water (related to salinity) and temperature (Hardie, 1967).

Petrographically, gypsum and anhydrite are easily distinguished on their optical properties. Gypsum has low relief and weak birefringence and belongs to the monoclinic crystal system; anhydrite has moderate birefringence, higher relief and is orthorhombic. Both may show a prominent cleavage.

5.2.1 SABKHA SULPHATE AND NODULAR ANHYDRITE

The main site of marine sulphate precipitation today, where the early part of the gypsum-anhydrite 'cycle' can be observed, is in the high intertidal and supratidal zones of such areas as the Texas and Trucial coasts. Gypsum is being precipitated displacively within the sediments as discoidal, rosette, selenite and twinned crystals from less than 1 mm to more than 25 cm in size (Masson, 1955; Kerr & Thomson, 1963; Shearman, 1966; Kinsman, 1969; Butler, 1970). Sediment pore waters are largely derived from surface flooding of seawater, a process referred to as flood recharge, and evaporative pumping (Section 4.8.4; Hsü & Siegenthaler, 1969). Dolomitization of carbonate particles is commonly associated with gypsum precipitation, as a result of the high Mg/Ca ratio (Section 4.8), and this releases calcium ions for further gypsum precipitation.

If the evaporation is sufficiently intense, as along the Trucial Coast, then with increasing concentration of pore fluids across the sabkhas, the gypsum crystals are replaced by a fine mush of equant and lath-shaped anhydrite crystals. This takes place when chlorinities are in excess of 145‰, i.e. a concentration of seawater by a factor of 7.5 to a salinity of around 260‰. The shape of the gypsum crystals may be retained or pseudomorphed by the anhydrite if the host sediment is cohesive (Fig. 5.3). Continued precipitation of anhydrite results in closely-packed nodules with host sediment restricted to thin stringers. The nodular texture produced is referred to as chicken-wire anhydrite and this is the typical texture of many ancient sulphate deposits (Fig. 5.4). Anhydrite is also precipitated as thin beds or layers of coalesced nodules in the more landward parts of the sabkhas. These beds are commonly irregularly contorted and buckled, forming the so-called enterolithic texture, also common in ancient sulphate sequences. In the most landward part of the sabkha, some re-hydration of anhydrite to gypsum may occur from contact with fresh

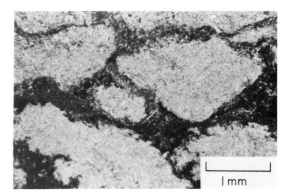

Fig. 5.3 Nodules of aphanitic anhydrite, partly retaining shape of original gypsum crystals. Dolomitic host sediment between nodules. Crossed polars. Triassic, N.W. England.

1 mm

163

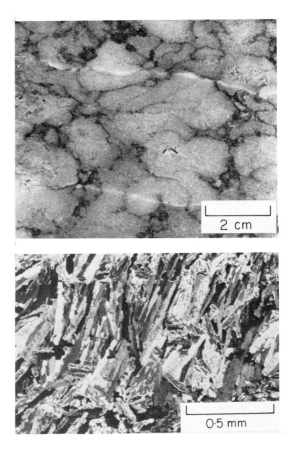

Fig. 5.4 Chicken-wire anhydrite. Closely-packed nodules of anhydrite with thin stringers of sediment between. Upper Permian, N.E. England.

2 cm

Fig. 5.5 Laths of anhydrite (the crystals have moderate birefringence, second order colours). Crossed polars. Triassic, N.W. England.

0·5 mm

continental groundwaters.

The formation of anhydrite requires an arid climate with high mean annual temperatures (above 22°C) and with seasonal temperatures in excess of 35° (Kinsman, 1969). Where the climate is less arid, then primary nodules of gypsum crystals may develop within the sediment. This is happening in sabkhas along the Mediterranean coast of Egypt (West *et al.*, 1979).

Primary textures exhibited by anhydrite in both modern and ancient nodular deposits include fine equant mosaics (aphanitic) and felted and parallel-subparallel arrangements of laths (Fig. 5.5). Recrystallization of equant and lath anhydrite may take place to produce coarse granular mosaics ('pile of bricks' texture), large fibrous crystals and fibro-radiating aggregates (Holliday, 1973).

5.2.2 LAMINATED SULPHATE

Laminated anhydrite (or gypsum) consists of thin sulphate laminations alternating with laminae of a different composition, often calcite, organic-rich calcite or organic matter (Fig. 5.6). Thin gypsum-anhydrite laminae in halite is one type of layered halite (Section 5.3). Anhydrite-calcite and anhydrite-organic matter couplets are typically less than several millimetres in thickness, but they may comprise sequences hundreds of metres thick. These laminated

164

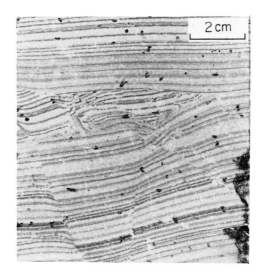

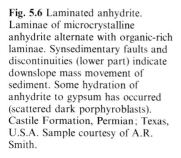

Fig. 5.6 Laminated anhydrite. Laminae of microcrystalline anhydrite alternate with organic-rich laminae. Synsedimentary faults and discontinuities (lower part) indicate downslope mass movement of sediment. Some hydration of anhydrite to gypsum has occurred (scattered dark porphyroblasts). Castile Formation, Permian; Texas, U.S.A. Sample courtesy of A.R. Smith.

anhydrites occur in the lower sections of thick basinal evaporite sequences, as in the Middle Devonian of the Elk Point Basin (Davies & Ludlam, 1973) or they constitute most of the basin fill, as in the Permian Castile Formation of the Delaware Basin (Anderson et al., 1972). Although the anhydrite laminae are chiefly planar (Fig. 5.6), they may be contorted and buckled. The latter in fact superficially resemble sabkha-type anhydrite and they have been interpreted as such in the Elk Point Basin case (Shearman & Fuller, 1969). However, one very important feature of laminated anhydrite is the lateral persistence of the laminae; frequently individual laminae can be correlated over vast distances, of tens to hundreds of kilometres (Anderson & Kirkland, 1966; Davies & Ludlam, 1973). This feature, indicating uniform conditions over a wide area, necessitates direct precipitation of the sulphate from water in a relatively deep basin, at least below wave base. Seasonal (annual) changes in water chemistry and temperature have been invoked to account for the laminations and the term varved has been applied to the couplets. The organic matter is probably sapropelic in origin, derived from seasonal phytoplankton blooms within the photic zone.

5.2.3 RESEDIMENTED SULPHATE

Once gypsum-anhydrite has been precipitated, either subaqueously or subaerially, it may be transported and resedimented, or reworked by wind, wave and current processes, to produce clastic deposits. Interbedded with laminated gypsum-anhydrite for example, there occasionally occur thicker graded units of clastic sulphate with sole structures, which are interpreted as turbidites. Horizons of small- and large-scale folded, contorted and brecciated gypsum-anhydrite also occur (Fig. 5.6) and these are regarded as slumps, slides and debris flows. Examples of this resedimented sulphate occur in the Elk Point Basin (Davies & Ludlam, 1973), the Permian Zechstein (Schlager & Bolz, 1977) and the Messinian (Upper Miocene) of the Northern Apennines and Sicily (Schreiber et al., 1976). Downslope movement of evaporites into a basin is clearly indicated and this further demonstrates the subaqueous, relatively deep water origin of the interbedded anhydrite laminites.

Reworking of intertidal-supratidal gypsum produces cross-laminated sulphates, intraformational conglomerates of gypsum crystal fragments and gypsum stromatolites where detrital grains are trapped by algal mats. Such shallow water-tidal flat clastic gypsum beds are well documented from the Messinian of Sicily (Hardie & Eugster, 1971; Schreiber *et al.,* 1976).

5.2.4 SECONDARY AND FIBROUS GYPSUM

Uplift of anhydrite sequences, perhaps a long time after their formation and burial, results in the generation of secondary gypsum, as the anhydrite comes into contact with fresh near-surface ground water. Secondary gypsum consists of two varieties, porphyroblastic and alabastrine gypsum. Gypsum porphyroblasts are large crystals, typically several millimetres across or larger, which occur scattered through the anhydrite (Fig. 5.7). Alabastrine gypsum consists of small to large, often poorly-defined interlocking crystals, many with irregular extinction (Fig. 5.8; Holliday, 1970). In spite of gypsification, the original nodular or laminated texture of anhydrite is invariably preserved.

Veins of fibrous gypsum (satin spar) are commonly parallel or subparallel to the bedding, having a displacive (intrusive) relationship. Usually they are a few mm or cm in thickness and consist of vertically-arranged fibres (Fig. 5.9). It is

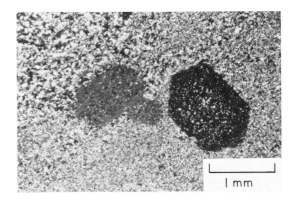

Fig. 5.7 Gypsum porphyroblasts replacing aphanitic anhydrite. Crossed polars. Permian, N.E. England.

Fig. 5.8 Alabastrine gypsum, formed by replacement of anhydrite. Crossed polars. Gachsaran Formation, Miocene, Iraq. (Courtesy of M.G. Shawkat.)

166

Fig. 5.9 Veins of fibrous gypsum (satin spar) in gypsiferous mudrock. Crossed polars. Permian, N.W. England.

thought that the fibrous gypsum grew under pressure in water-filled veins induced by hydraulic fracture (Shearman *et al.*, 1972). However, the source of the gypsum is uncertain. One view is that it derives from the increased volume of sulphate resulting from anhydrite hydration while an alternative is that it comes from residual sulphate-rich pore fluids.

5.3   Halite

Halite commonly infills large sedimentary basins and makes up the greater part of evaporite cycles. Halite is the main evaporite mineral of many saline lakes.

In halite deposits, the rock salt may be massive, bedded, layered or intimately intermixed with siliciclastic sediments. Rarely, the halite is reworked into ripples or shows cross bedding. In bedded halite, the bedding is on a scale of several centimetres and it is usually produced by colour changes due to variations in the amount of clay in the salt. Some bedded halite deposits, such as the Triassic Cheshire halite of Britain, show spectacular large-scale contraction polygons and cracks. *Layered halite* commonly consists of alternations of halite with anhydrite. Thick sequences of this halite type occur in the Permian Zechstein of N.W. Europe, the Devonian Prairie Formation of western Canada and the Permian Salado Formation of the Delaware Basin. Halite laminae are typically several centimetres in thickness and the anhydrite laminae are several millimetres. Layering in halite may also be defined by thin silty laminae or planar dissolution surfaces which truncate the underlying crystal fabrics. Where intermixed with siliciclastic sediment, halite consists of randomly-oriented euhedral crystals, scattered or densely-packed.

167

In thin section, halite is isotropic (since cubic); it often shows strong cleavage and fracture planes and possesses fluid inclusions (Fig. 5.10). Recrystallization of halite can take place relatively easily during diagenesis and through deformation. Layered halite consists chiefly of an opaque variety with abundant liquid inclusions (termed chevron halite from the arrangement of the inclusions, Fig. 5.11, and a clear transparent variety (Dellwig, 1955; Wardlaw & Schwerdtner, 1966). The opaque, chevron halite is regarded as the original salt precipitate while the clear halite is a later precipitate, having formed by replacement of the opaque variety and infilling of dissolution hollows and cavities.

Halite is being precipitated at the present time in coastal and inland salinas and evidence from ancient halite sequences suggests that it can also be precipitated in shallow to deep water lakes and marine basins. Bedded halite with polygonal crack patterns clearly formed in salinas which periodically suffered prolonged subaerial exposure. Dissolution planes and thin silt laminae within such halite have been interpreted as the result of wind deflation. Where halite is intimately intermixed with clastic sediment, it is probable that the halite has grown displacively within the silt in an analogous fashion to the growth of gypsum crystals and anhydrite nodules (Smith, 1971).

The origin of layered halite is unresolved. Halite-anhydrite alternations have been related to seasonal changes in the temperature and/or composition of the basin waters in a similar way to laminated anhydrite. The term Jahresringe,

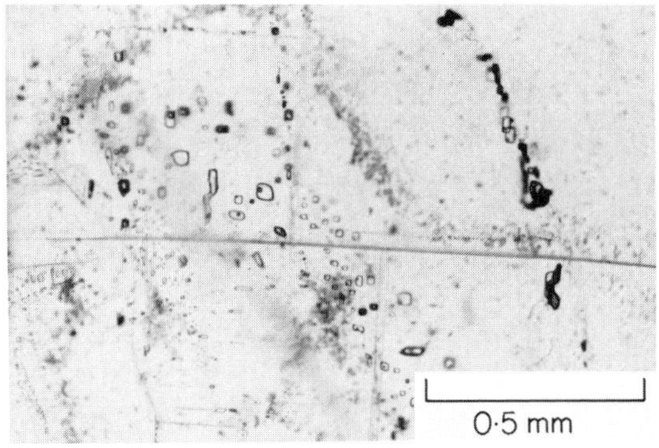

Fig. 5.10 Halite with fluid-filled inclusions. Plane polarized light. Cheshire Halite, Triassic, N.W. England. Courtesy of R.M. Tucker.

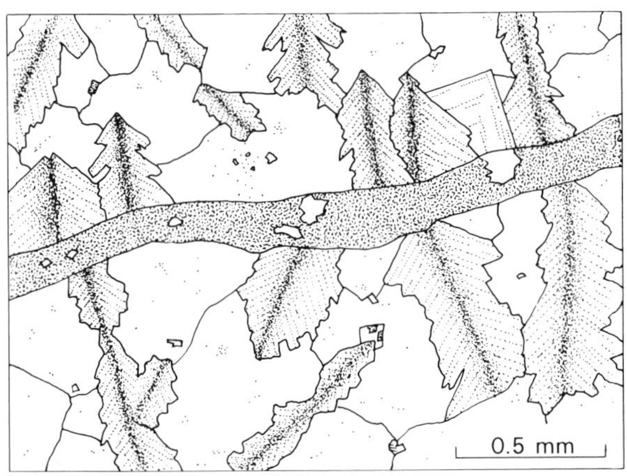

0·5 mm

Fig. 5.11 Sketch of chevron and clear halite with an anhydrite-clay-carbonate layer. Plane polarized light. Prairie Formation, Devonian; Alberta, Canada. After Wardlaw & Schwerdtner (1966) and Schmalz (1969).

0.5 mm

168

implying an annual rhythmicity has been applied to halite-anhydrite couplets in the Permian Zechstein (Richter-Bernberg, 1950). An annual interpretation of the layering requires high rates of salt precipitation, the order of 500 m in 8000 years. It also requires deposition in deep basins, deeper than the thickness of the salt, since the subsidence rates for a shallow water origin are improbably great. With a deep basin, periodic influxes of brine or seawater into the basin are necessary to give the great thickness of salt (the evaporation of 1000 m of seawater only gives 12.9 m of halite, Table 5.3). Variations in bromine content across couplets in the Devonian Prairie Evaporite Formation, are thought to indicate such seasonal influxes, although seasonal is not equated with annual (Wardlaw & Schwerdtner, 1966). A type of layered halite has been discovered forming in shallow brine-filled depressions on supratidal flats in Baja California (Shearman, 1970). In this modern example halite alternates with gypsum and although only a thin deposit (25 cm) has developed so far, there is clearly the potential for thick deposits of layered halite to form in supratidal salinas, if there is a background of continuous subsidence.

The NaCl for many halite deposits is derived from the evaporation of seawater, which of course contains a vast reserve of NaCl. NaCl can also be concentrated from fresh continental waters, as in many salt lakes, or derived from the solution and recycling of older evaporites. Halite of marine origin can be distinguished by its high bromine values (more than 50 ppm) compared with non-marine and recycled halite which has much lower Br concentrations (Holser, 1966a).

5.4   Other evaporite minerals and their occurrence

5.4.1   POTASSIUM AND MAGNESIUM SALTS

Potassium and magnesium salts which occur in some marine evaporite deposits include the sulphates, kieserite and kainite, and the chlorides sylvite and carnallite. These highly soluble salts are the last to form in the evaporation of seawater, hence they tend to occur in the uppermost parts of evaporite cycles and rarely attain a great thickness. In view of their highly soluble nature, diagenetic mineral changes through contact with residual brines and fresh groundwater are inevitable. It is likely that the mineral assemblages of these final precipitates are not original and some minerals may be entirely secondary in origin. Sylvite for example has probably formed through incongruent dissolution of carnallite and some polyhalite has probably formed through alteration of kainite by solutions rich in calcium ions. Polyhalite has been reported from supratidal flats in Baja California formed by reaction of gypsum with bitterns, solutions enriched in K^+ and Mg^{2+} (Holser, 1966b). This modern occurrence and the association of ancient potassium salts with intertidal and shallow-water sediments (as in the Permian Zechstein of Yorkshire, Smith, 1973) suggest that most of these deposits formed in salinas and sabkhas.

5.4.2   LACUSTRINE EVAPORITES

Apart from halite and gypsum-anhydrite, there are a number of evaporite minerals which in large concentration are only found in saline lakes (Table 5.4).

Table 5.4 The water chemistry of five salt lakes (in p.p.m.). The main point to note is the great variability in the relative concentrations of the constituents between lakes.

	Dead Sea	Great Salt Lake, Utah	Mono Lake California	Borax Lake California	Gulf of Kara Bogaz, U.S.S.R.
Cl^-	208,020	112,900	15,100	5,945	142,500
SO_4^{2-}	540	13,590	7,530	22	46,900
HCO^-_3	240	180	26,430	6,668	—
Na^+	34,940	67,500	21,400	6,199	81,200
K^+	7,560	3,380	1,120	322	
Ca^{2+}	15,800	330	11	nil	4,900
Mg^{2+}	41,960	5,620	32	30.7	19,900
total salinity	315,040	203,490	71,900	> 19,400	> 293,000

The evaporite mineralogy of salt lakes can vary considerably from one region to the next, since the geochemistry of the lake waters, which determines the evaporites precipitated, is itself very variable (Table 5.4), being largely dependent on the local geology. There is commonly a zonation of evaporite minerals within a salt lake, of the least soluble occurring around the edge and the most soluble being precipitated in the lake centre (Hardie *et al.*, 1978). Salt lakes often possess hard surface crusts with polygonal cracks and efflorescences. Salt lakes such as the Dead Sea and Great Salt Lake (Utah), with dominant halite, are distinguished from bitter lakes such as Mono Lake (California) and Carson Lake (Nevada), where sodium carbonate and sulphate minerals dominate. Ancient lacustrine evaporites are well-developed in the Eocene Green River Formation of Wyoming and Utah.

5.5 Evaporite solution and replacement

Since evaporites are composed of relatively to highly soluble minerals, it is not uncommon to find that at outcrop and in the shallow subsurface evaporite beds have been dissolved away by other minerals. Evaporite solution leads to the collapse and brecciation of overlying strata (Blount & Moore, 1969). Accompanying sulphate solution, associated dolomites are commonly replaced by calcite (dedolomitization: Section 4.8.4).

The recognition of replaced evaporites is based on several lines of evidence. The replacement crystals may possess relics of the original evaporite mineral. This is particularly the case where silica has replaced gypsum-anhydrite. The crystal shape of the original evaporite mineral or the nodular shape in the case of anhydrite, is usually retained on replacement to produce a pseudomorph. Pseudomorphs of halite are easily recognized by their cubic, often hopper shapes and pseudomorphs of gypsum have a characteristic lozenge (Fig. 5.12) and swallow-tail shape. The fabrics of the replacement mineral may also be a guide. Two forms of silica, length slow chalcedonic quartz and lutecite are commonly associated with former evaporite beds (Folk & Pittman, 1971). Replaced evaporites are common in many intertidal-supratidal carbonate sequences and have been described by West (1964), Chowns & Elkins (1974), Tucker (1976), Milliken (1979) and others.

170

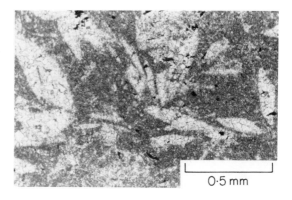

Fig.5.12 Calcite pseudomorphs after gypsum in a micritic limestone; typical of high-intertidal facies. Plane polarized light. Lower Carboniferous, eastern Scotland.

0·5 mm

5.6 Evaporite sequences and discussion

Following the recognition of sabkhas and supratidal flats as important environments of evaporite, particularly sulphate, precipitation, many ancient examples have been described. The typical facies are nodular (chicken-wire) and enterolithic anhydrite, although as noted in Section 5.2.2, these textures can develop in deep water anhydrites (Dean *et al.*, 1975). The key features for the identification of sabkha evaporites are the shallow water and intertidal sedimentary structures contained within associated carbonates (Section 4.10.2). As a result of net deposition upon the sabkha surface, the sabkha gradually progrades seawards over the intertidal sediments. A sabkha cycle of supratidal evaporites overlying intertidal and subtidal carbonates is produced (Fig. 5.13), which may be repeated many times in an evaporite sequence. Ancient sabkha sequences have been described by Arthurton & Hemingway (1972), Bosellini & Hardie (1973), Shearman (1966) and Wood & Wolfe (1968).

For the precipitation of evaporites within deep, water-filled basins, the most important factor, apart from an extremely arid climate and periodic

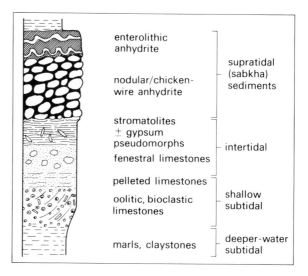

Fig. 5.13 A sabkha cycle. Such cycles typically range from several to several tens of metres in thickness.

171

replenishment of seawater, is the barrier which gives near-complete isolation from the main mass of seawater. The barrier may be structural, such as a fault-bounded ridge, or sedimentary, such as a carbonate reef or sand bar. In many cases, carbonate sediments and reefs deposited before evaporite precipitation contributed towards an isolation of the basins, and defined sub-basins. To account for the deficiencies of halite and/or potassium salts in some thick evaporite sequences, reflux of dense, bottom-flowing brines out of the basin has been postulated.

Although many evaporite sequences and textures are being re-examined and re-interpreted in the light of recent studies on sabkhas and salinas, there are several features which can only be explained in terms of the classic relatively deep water, barred-basin model. These are: laminated evaporites with individual laminae which can be correlated over distances of many kilometres; graded beds of anhydrite clasts, interpreted as turbidites, and horizons of contorted and brecciated evaporites, interpreted as slumps. The vertical rock sequence developed through evaporation of a relatively deep, water-filled basin, should begin with sapropelic deposits, representing an initial stagnation of the water as the barrier became effective, and be followed by laminated anhydrite as evaporation proceeded (Fig. 5.14; Schmalz, 1969). When evaporation is well-advanced, halite would be deposited, perhaps with anhydrite laminae, until the basin was nearly filled. Final evaporation of brines and bitterns in salinas and sabkhas might then produce K^+ and Mg^{2+} salts.

The two main sites of evaporite precipitation, sabkhas-salinas and deep water-filled basins, are not mutually exclusive. One may pass laterally into the other or develop into the other with time. Sabkhas and salinas are quite likely to be developed on the marginal, shallow water-subaerial regions around a deep basin in which evaporites are also being precipitated. Infilling of a basin with salt will lead to salinas and sabkhas over the whole area of the basin. With very wide sabkhas, differential subsidence could lead to the development of brine-filled depressions, in which layered halite and anhydrite or laminated anhydrite could be precipitated subaqueously.

The interpretation of some evaporite deposits in deep basins has been very controversial. Late Miocene (Messinian) evaporites discovered beneath the

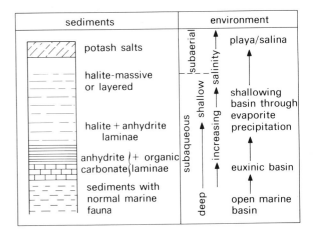

Fig. 5.14 Evaporite sequence formed in an initially deep, marine barred-basin, with periodic replenishment.

172

floor of the Mediterranean have been interpreted as deep water and subaerial. Convincing evidence now shows that the Mediterranean sea dried up, causing salts to be precipitated in sabkhas and salt lakes located on the floor of this deep, desiccated basin (Hsü *et al.*, 1977, 1978). This event is referred to as the Messinian salinity crisis (for information on effects of this event see Cita & Wright, 1979). Another case of controversy centres on some of the Devonian evaporites in the Elk Point Basin of western Canada. Laminated anhydrite, which passes up into thick anhydrite and halite, has been interpreted as shallow water-intertidal (Shearman & Fuller, 1969) and then re-interpreted as the deposits of a deep, stratified water body (Davies & Ludlam, 1973).

Further reading

Dean, W.E. & Schreiber, B.C. (Eds.) (1978) Notes for a short course on marine evaporites. *Soc econ. Paleont. Min.* Tulsa, *Short Course* **4.**

Kendall, A.C. (1979) Continental and supratidal (sabkha) evaporites. Subaqueous evaporites. In: *Facies Models* (Ed. by R.G. Walker), pp. 145–174. Geoscience Canada.

Kirkland, D.W. & Evans, R. (Eds.) (1973) *Marine evaporites: origins, diagenesis and geochemistry,* pp. 426. Dowden, Hutchinson & Ross, Stroudsburg.

Braitsch, O. (1971) *Salt deposits, their origins and composition,* pp. 297. Springer, Berlin.

Borchert, H. & Muir, R.O. (1964) *Salt deposits: the origin, metamorphism and deformation of evaporites,* pp. 338. Van Nostrand, London. (Although written before the hey-day of sabkhas).

6

Sedimentary Ironstones and Iron-Formations

6.1 Introduction

Iron is present in practically all sedimentary rocks to the extent of a few percent but less commonly it forms ironstones and iron-formations where the iron content exceeds 15%. Important reserves of iron ore are contained within these sedimentary iron deposits. The element iron occurs in two valence states, a divalent form, ferrous iron (Fe^{2+}) and a trivalent form, ferric (Fe^{3+}). As a result of this, the behaviour of iron and the precipitation of its minerals are strongly controlled by the chemistry of the surface or diagenetic environment. The common iron minerals in sedimentary rocks are given in Table 6.1.

Table 6.1 The iron minerals of sedimentary rocks.

Oxides	hematite α-Fe_2O_3
	magnetite Fe_3O_4
	goethite α-FeO.OH
	limonite $FeO.OH.nH_2O$
Carbonate	siderite $FeCO_3$
Silicates	chamosite $Fe_3Al_2Si_2O_{10}.3H_2O$
	greenalite $FeSiO_3.nH_2O$
	glauconite $KMg(FeAl)(SiO_3)_6.3H_2O$
Sulphides	pyrite FeS_2
	marcasite FeS_2

The majority of sedimentary iron deposits were formed under marine conditions and many of the Phanerozoic contain normal marine fossils. There are important differences between those which formed in the early-middle Precambrian and those of the Phanerozoic. The former, referred to as *iron-formations* or *banded iron formations,* are typically thick sequences of various iron minerals interbedded with chert, deposited in large intracratonic basins (Section 6.5.1). The *Phanerozoic ironstones* are usually thin sequences, often oolitic in character, which were deposited in localized areas (Section 6.5.2). One complicating factor for the interpretation of many iron-rich sedimentary rocks is that there are no modern analogues for comparison.

The only iron deposits forming to any extent at the present time are the bog-iron ores of mid to high latitude lakes and swamps (Section 6.6). In addition, ferromanganese nodules and crusts, and metalliferous sediments are forming on the seafloor (Section 6.7). Neither of these modern developments are very

174

significant geologically, and indeed in many ways they are unrelated to the ironstones and iron-formations of the geological record.

6.2 Source and transportation of iron

Traditionally, two sources of iron have been considered: continental weathering and contemporaneous volcanicity (see papers and discussion in Lepp, 1975). Contemporaneous volcanicity as a source has few adherents today, although there are some deposits, notably the Archaean iron-formations, which are closely associated with volcanics. In these cases iron could be supplied by hydrothermal exhalations and submarine weathering of extrusives. Continental weathering is now generally accepted as the source of iron for most sedimentary iron deposits. Intense weathering under a humid tropical climate releases the iron from mafic and heavy minerals in igneous and other rocks and produces iron-rich lateritic soils. Through erosion and transportation, the iron is carried to the sea by rivers but the manner in which the iron travels has not been fully resolved.

Iron in true solution in river water and ground water is in very low concentrations (less than 1 ppm) and in seawater the concentration is around 0.003 ppm. The low values arise from the fact that in the pH and Eh ranges of most natural surface waters (Fig. 6.1), iron is present as the highly insoluble

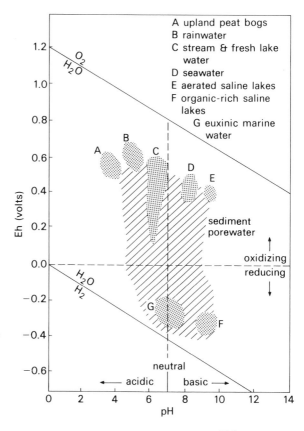

Fig. 6.1 Eh-pH diagram showing the fields of some naturally occurring waters. Redox potentials of natural solutions are limited by reactions involving water, which are dependent on pH. The upper limit of Eh is determined by the oxidation of water to oxygen (the upper diagonal line) and the lower limit of Eh is the reduction of water to hydrogen (lower diagonal). After Baas-Becking *et al.*, 1960.

175

ferric hydroxide (also the main constituent of laterite). Three mechanisms of iron transportation have been suggested. Ferric hydroxide readily forms a colloidal suspension, which is stabilized in the presence of organic matter. Iron could be transported by rivers in this form and then precipitated in the sea as the colloidal suspensions flocculated. Iron can be transported by adsorption and chelation on to organic matter. And in a similar manner, iron can be carried by clay minerals, either as part of the clay structure or of greater importance, as oxide films on the surface of clays (Carroll, 1958). During sedimentation and early diagenesis the iron is released to the porewaters and reprecipitated as iron minerals. A more radical hypothesis asserts that iron formations are diagenetic in origin, having been formed by the replacement of calcareous sediments (Kimberley, 1979, 1980; Dimroth, 1979). The source of the iron is thought to be interbedded organic-rich muds or pyroclastic material.

Ironstone formation is favoured where there are low rates of sedimentation, both of siliciclastic material and carbonates. Continental areas supplying the iron through deep tropical weathering are thus assumed to be low-lying with negligible relief.

In spite of the foregoing, it is considered by many people that the quantity of iron in Proterozoic and Archaean iron-formations is still too high to be explained by continental weathering as occurs today. An atmosphere with less oxygen and more carbon dioxide has thus been proposed (Garrels et al., 1973); this would enable iron to be leached and transported more efficiently, particularly as Fe^{2+} in solution (Section 6.5.1).

6.3   The formation of the principal iron minerals

The change from one oxidation state of iron to the other is dependent on changes in the Eh and pH of the environment. Eh is a measure of the oxidizing or reducing nature of the solution, basically whether an element like iron will gain or lose electrons; pH is a measure of the acidity or alkalinity, that is the hydrogen ion concentration. Fe^{3+} is stable under more oxidizing and more alkaline conditions whereas Fe^{2+} is stable under more reducing and more acidic conditions. In fact, in the pH-Eh range of natural environments, Fe^{3+} is present as the highly insoluble $Fe(OH)_3$, whereas Fe^{2+} is present in solution. Apart from Eh and pH, two other factors controlling the precipitation of iron minerals are (i) the activity (i.e. effective concentration) of carbonate ions, which can be measured by the partial pressure of carbon dioxide, Pco_2 and (ii) the activity of sulphur, frequently represented by $p S^{2-}$, the negative logarithm of the activity of the sulphide ion (Berner, 1971).

One of the main factors affecting the Eh of natural aqueous environments is the amount of organic matter present, since its decomposition, mainly brought about by bacteria, consumes oxygen and creates reducing conditions. Normal seawater and other surface waters have a positive Eh, as do the porewaters in surficial sediments on the seafloor (Fig. 6.1). However, organic matter deposited in the sediments soon decomposes so that a reducing environment is formed some tens of centimetres below the sediment-water interface. If there is a high rate of organic sedimentation and/or restricted circulation, then reducing conditions may develop in the bottom waters and create a euxinic, anoxic environment (Fig. 3.7).

The stability fields of the common iron minerals are plotted on Eh-pH, Eh-pS^{2-} and Eh-Pco_2 diagrams in Figures 6.2 and 6.3. These diagrams are constructed from thermodynamic data and so have their limitations when applied to natural systems. From Figures 6.2 and 6.3, it can be seen that hematite is the stable mineral under moderately to strongly oxidizing conditions. It is generally believed that hematite forms diagenetically from a hydrated ferric oxide precursor approximating to goethite, by an ageing process involving dehydration (Berner, 1969). The formation and preservation of hematite in a sediment require a low original organic content.

For the ferrous minerals, pyrite, siderite and magnetite, stable under conditions of negative Eh, the stability fields are strongly dependent on the Pco_2 and pS^{2-} of the solution (Fig. 6.3). Pyrite forms where the sulphide activity is high (i.e. low pS^{2-}), whereas siderite forms where the sulphide activity is low and the carbonate activity high. Since most surface environments are oxidizing, these minerals are usually precipitated within sediments, during early diagenesis, where reducing conditions have developed through bacterial decomposition of organic matter. The iron for these minerals is Fe^{2+} in the porewaters, mostly liberated from clays and ferric oxides in the sediment by the negative Eh.

The sulphide for the precipitation of pyrite comes mainly from the bacterial reduction of dissolved sulphate in porewaters; this produces H_2S which reacts

Fig. 6.2 Eh-pH diagram showing the stability fields of ferrous and ferric iron, hematite, siderite, pyrite and magnetite. The diagram shows that hematite is the stable mineral in all environments which are moderately to strongly oxidizing. For the minerals pyrite, siderite and magnetite, stable in reducing environments, the mineral stability fields are strongly dependent on the concentrations of carbonate and sulphide in the solution (see Figs. 6.3a and 6.3b), as well as pH. Fig. 6.2 shows the stability fields for the condition of high carbonate and low sulphide. If sulphide is greatly in excess of carbonate, then the pyrite field expands to occupy nearly all the lower half of the diagram. When both sulphide and carbonate are in low concentrations, then the magnetite field expands into near-neutral environments. After Krauskopf, 1979.

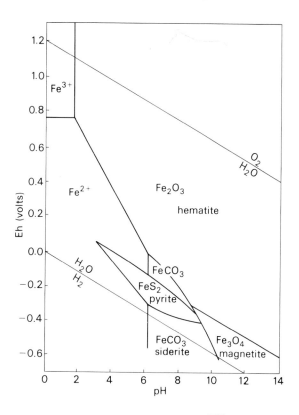

177

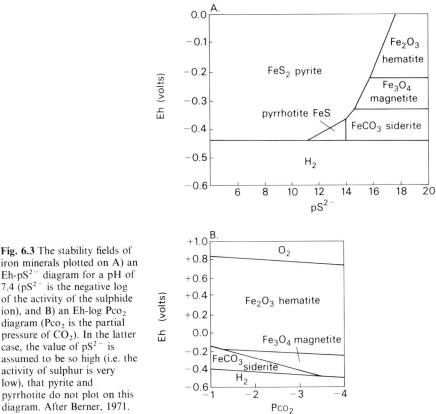

Fig. 6.3 The stability fields of iron minerals plotted on A) an Eh-pS^{2-} diagram for a pH of 7.4 (pS^{2-} is the negative log of the activity of the sulphide ion), and B) an Eh-log Pco$_2$ diagram (Pco$_2$ is the partial pressure of CO$_2$). In the latter case, the value of pS^{2-} is assumed to be so high (i.e. the activity of sulphur is very low), that pyrite and pyrrhotite do not plot on this diagram. After Berner, 1971.

with the Fe^{2+} in solution (Berner, 1970). The sulphate-reducing bacteria require organic matter and they are instrumental in the latter's decomposition. Seawater contains much dissolved sulphate and so pyrite is the typical authigenic mineral of organic-rich marine muds. Dissolved sulphate is generally present in low concentrations in freshwater so that pyrite is less common in non-marine sediments. Pyrite can also form in association with the bacterial reduction of sulphate in gypsum. In all these cases the initial precipitates are black, finely-crystalline metastable iron sulphides: machinawite and greigite. In a relatively short time these are transformed to pyrite. If there is excessive bacterial production of H$_2$S, through high rates of organic matter sedimentation and minimal circulation, H$_2$S will occur in waters overlying the bottom sediments. The metastable iron sulphides can then be precipitated directly.

For the formation of siderite, the requirement of low sulphide activity is rarely attained in marine sediment porewaters because of the abundant dissolved sulphate. Siderite is thus more common in non-marine sediments. If there is insufficient Fe^{2+} relative to Ca^{2+} in the porewaters, then calcite forms in preference to siderite. The formation of magnetite is favoured by low activities of both sulphide and carbonate, together with negative Eh and neutral pH (Figs. 6.3). Conditions such as these are rare in nature and so magnetite formation is not common.

The formation of the iron silicate minerals chamosite, greenalite and glauconite is poorly understood. Being ferrous minerals, it is suspected that conditions of low Eh are necessary for greenalite and chamosite to form but since they are absent and poorly represented, respectively, in modern sediments, their interpretation to some extent relies on geological data (Section 6.4.4).

The formation of iron minerals and the construction and use of Eh-pH and other such diagrams are considered at length in Garrels & Christ (1965), Curtis & Spears (1968). Berner (1971) and Krauskopf (1979).

6.4   Occurrence and petrography of the iron minerals

6.4.1   IRON OXIDES

Hematite is present in both Precambrian iron-formations and Phanerozoic ironstones, and of the latter it is more common in Palaeozoic developments. In the Precambrian cases the hematite is chiefly present as thin beds and laminae, alternating with chert, but it also occurs as massive, pelletal and oolitic forms. In the Phanerozoic ironstones the hematite is mainly present as ooids and impregnations and replacements of fossils (Figs. 6.4, 6.5). Later diagenetic migrations and replacements of calcareous host sediments and calcite cements by hematite are not uncommon. Although the hematite itself may be a primary mineral, probably precipitated via an amorphous hydrated ferric oxide, there is often petrological evidence that the oxide has formed by replacement of

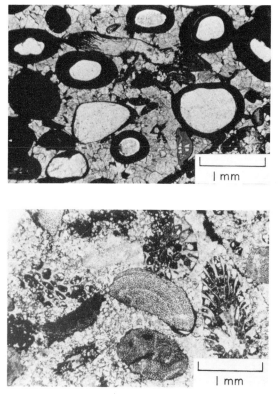

Fig. 6.4 Hematite oolite. Ooids have quartz nuclei and a cortex of hematite; fossils are also impregnated by hematite (a brachiopod in upper part). The matrix is a cement of ferroan calcite. Plane polarized light. Clinton Oolite, Silurian; Appalachians, U.S.A.

Fig. 6.5 Hematite-impregnated crinoid ossicles and bryozoan fragments in a ferroan calcite cement. Plane polarized light. Rhiwbina Iron Ore, Lower Carboniferous; Glamorgan, Wales. Courtesy of T. P. Burchette.

179

chamosite. In some cases this is a syn-sedimentary replacement, arising from the reworking of chamosite grains into a more oxidizing environment where hematite or its precursor is stable.

Hematite in thin section is opaque and typically cryptocrystalline. It can be recognized by its red colour in reflected light (quickly checked by raising the microscope lamp so that light shines down on to the rock slice).

Goethite is absent from Precambrian iron deposits, but it is a major constituent of Phanerozoic ones, in particular those of the Mesozoic. In many cases goethite appears to be a recent weathering product, having formed by oxidation and hydration of other iron minerals. However there are ironstones where the goethite appears to be primary, or at least synsedimentary. The goethite may form ooids (Fig. 6.6; James & Van Houten, 1979) and

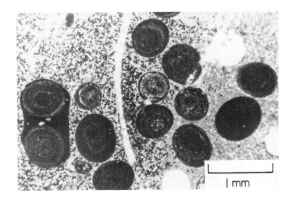

Fig. 6.6 Goethite oolite, with a matrix of goethite and calcite. A compound ooid is present and a bivalve shell fragment. Plane polarized light. Claxby Ironstone, Lower Cretaceous; Lincolnshire, England.

occasionally these may consist of alternations of goethite and chamosite. The goethite here could have formed through seafloor oxidation of the chamosite. Ferriferous spherules largely consisting of goethite are forming at the present time in Lake Chad, W. Africa (Lemoalle & Dupont, 1973).

Goethite in section is a yellow to brown colour and generally appears isotropic.

Limonite is a poorly-defined hydrated form of iron oxide, containing goethite, other materials such as clay and adsorbed water. The term is best restricted to the yellow-brown amorphous product of subaerial weathering of iron oxides and other minerals.

Magnetite is abundant in the Precambrian iron-formations where it is interlaminated with chert. It is a minor component of Phanerozoic ironstones but in some instances it is important. It generally occurs as small replacement crystals or granules within oolitic ironstones. It is distinguished from hematite by its steel-grey colour in reflected light. (It is also magnetic of course!)

6.4.2 IRON CARBONATES

Siderite is a major constituent of both Precambrian and Phanerozoic iron-rich sediments. It is the groundmass to many Phanerozoic chamosite oolites and it can replace ooids and skeletal grains (Fig. 6.7). Siderite is common in non-

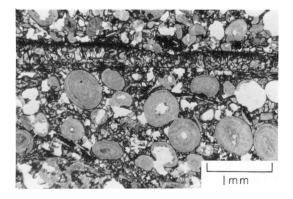

Fig. 6.7 Sideritic chamosite oolite. Ooids are composed of chamosite, matrix of siderite. A bivalve shell fragment has been replaced by siderite. Plane polarized light. Dogger Ironstone, Middle Jurassic; Yorkshire, England.

marine, organic-rich mudrocks, either as small disseminated crystals or as nodules and rounded masses. At the present time, siderite is forming in muds of deltaic and lacustrine environments and siderite nodules are forming in sediments of the Atchafalaya River Basin, U.S.A. (Ho & Coleman, 1969).

Siderite crystals as seen in thin section are of three types: coarse crystals up to several millimetres across, similar to other carbonates such as calcite in terms of high birefringence and rhombohedral cleavages; a very fine grained variety of equant-rhombic crystals a few microns in diameter, and a fibrous variety which forms spherulites. The first type occurs predominantly in the matrix of oolitic and bioclastic ironstones (Fig. 6.7) and can usually be recognized from a yellowish-brown oxidation zone ('limonite') along crystal boundaries and twin and cleavage planes. The fine grained variety occurs interbedded with cherts in the Precambrian iron-formations and constitutes most siderite nodules. Aggregates of fibrous spherulites form the rock known as sphaerosiderite (Fig. 6.8). Siderite can vary considerably in chemical composition, with Ca^{2+}, Mg^{2+} and Mn^{2+} substituting for Fe^{2+} by as much as 10%.

Fig. 6.8 Sphaerosiderite: spherulites of fibrous siderite. 6.8a: plane polarized light. 6.8b, crossed polars. Coal Measures, Upper Carboniferous; Yorkshire, England.

(a)   (b)

6.4.3   IRON SULPHIDES

The iron sulphides, in particular pyrite, are constituents of many iron-rich (and other) sediments, but they rarely form the major part. Pyrite and its metastable precursors are forming within estuarine and tidal flat sediments, and they are being precipitated on the floor of the Black Sea.

Pyrite is distinguished from other opaque iron minerals by its yellowish colour in reflected light. Pyrite is present as disseminated grains and crystals (cubic); it is rarely a replacement of skeletal fragments. Aggregates of spherical micro-concretions of pyrite are known as framboids. *Marcasite* is a dimorph of pyrite which is rarely found in ironstones but forms nodules in chalks and coal measure sediments.

6.4.4 IRON SILICATES

The three most important iron silicates are chamosite, greenalite and glauconite.

Chamosite typically occurs as ooids in Phanerozoic ironstones, within a matrix of siderite or calcite (Fig. 6.7). It also forms flakes and is finely disseminated in chamositic mudrocks. In many cases there are no nuclei to the ooids or they may have grown around chamosite flakes or fragments of broken ooid. One of the features of chamosite ooids which distinguishes them from the more familiar aragonite ooids is the evidence that they were originally soft. The ooids are commonly flattened and distorted and in some instances flattened ooids have formed the nuclei to later ooids, showing that the ooid deformation took place on the seafloor (Fig. 6.9). The term spastolith has been applied to these squashed ooids.

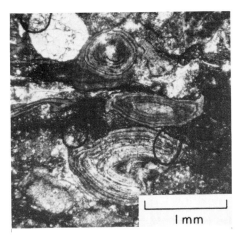

Fig. 6.9 Flattened and distorted chamosite ooids (spastoliths). Plane polarized light. Raasay Ironstone, Lower Jurassic, western Scotland.

1 mm

Chamosite is green in colour and has a low birefringence. It has a variable composition and structure but it is related to chlorite and kaolinite and possesses a 7 Å basal reflection.

Chamosite is a ferrous silicate and theoretical considerations of its stability field indicate that it should form under reducing conditions (Curtis & Spears, 1968). However, chamosite ironstones often contain a rich marine benthic fauna which required normal oxidizing conditions on the seafloor. In view of this it is generally accepted that chamosite is precipitated as a mixed gel of $Fe(OH)_3$, $Al(OH)_3$ and $SiO_2.nH_2O$, which is stable at positive Eh. Conversion of this gel to chamosite would take place after burial within the reducing environment beneath the sediment-water interface. The chamosite or its gel

182

precursor could be oxidized to hematite (or its precursor) through current reworking of the sediment on the seafloor. An alternative origin of the chamosite ooids is through diagenetic replacement of calcareous ooids (Kimberley, 1979, 1980) with the iron derived from interbedded organic-rich muds. However, the evidence that many chamosite ooids were soft is consistent with precipitation as a gel; calcareous ooids are never soft (Section 4.3.1). One point in favour of a diagenetic origin is the absence of chamosite ooids in modern seas. In the few known marine occurrences, off the Orinoco and Niger deltas, and on the Sarawak shelf, the chamosite is a poorly-ordered ferrous-iron clay forming diagenetically within faecal pellets (Porrenga, 1967). Interestingly, some chamosite pellets are being oxidized to goethite around the edges. Chamosite also occurs in a restricted marine Scottish loch (Rohrlich *et al.*, 1969).

The hydrated ferrous silicate *greenalite* is interbedded with chert and constitutes beds and lenses in Precambrian sedimentary iron deposits. Greenalite occurs as rounded to subangular pellets, with little internal structure. It is a green isotropic mineral related to chamosite and chlorite. Greenalite is considered a primary mineral in iron-formations but it is not certain if it is an original precipitate. It ·may well have formed by diagenetic replacement of detrital iron-rich particles or via an iron-silicate gel, as with chamosite.

Glauconite is a potassium iron aluminosilicate with a high Fe^{3+}/Fe^{2+} ratio. From X-ray diffraction studies several types of glauconite can be recognized depending on the structure and degree of ordering (Burst, 1958). It typically occurs as light to dark green pellets and aggregates up to 1 mm in diameter (Fig. 6.10). It is present in many sandstones and may be the major constituent, forming greensands, well-known in the Cretaceous of Britain and eastern U.S.A. Glauconite is being formed on many modern continental shelves at depths from a few tens to hundreds of metres, but it is invariably a poorly-ordered form (Bell & Goodell, 1967; McRae, 1972). It tends to occur in areas with low sedimentation rates. The conditions of formation of glauconite are not fully understood; it is often associated with organic matter, which would create local reducing conditions, but the overall environment is aerobic. Much modern glauconite forms within tests of Foraminifera and in faecal pellets.

Fig. 6.10 Glauconite pellets. Some quartz and phosphate grains are also present. 6.10a: plane polarized light. 6.10b: crossed polars. Upper Greensand, Cretaceous; Sussex, England.

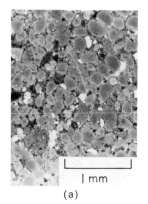

(a)

(b)

1 mm

1 mm

In section, glauconite is a light-green colour, often pleochroic. Pellets are microcrystalline and show an aggregate polarization pattern (Fig. 6.10).

6.5   Precambrian iron-formations and Phanerozoic ironstones

The separation of iron-rich sedimentary rocks into the two groups, Precambrian iron-formations and Phanerozoic ironstones is based on major differences in mineralogy, geochemistry, sedimentology and stratigraphy.

6.5.1   PRECAMBRIAN IRON-FORMATIONS

In spite of the tremendous economic importance of these rocks and the vast amount of data available they still arouse much discussion and controversy (see papers in *Economic Geology*, **68** (7) for 1973 and Lepp, 1975). Iron-formations are present within the cratonic shields of most continents (Fig. 6.11).

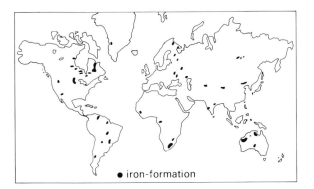

Fig. 6.11 Distribution of Precambrian iron-formations.

● iron-formation

From studies of iron-formations in Canada two groups are recognized: Algoma-type, lenticular deposits which are closely associated with volcanic rocks and greywackes in greenstone belts, and Superior-type where the formations are more extensive units deposited on stable shelves and in subsiding basins. With a few problematic exceptions Superior-types are restricted to the Early-Middle proterozoic (1900–2500 Ma), whereas Algoma-types are mostly Archaean in age (2500–3000 Ma), with some younger examples.

On the basis of the dominant early diagenetic iron mineral present, four sedimentary iron facies are recognized within iron-formations: oxide, silicate, carbonate and sulphide facies (James, 1954). Chemical factors control iron mineral precipitation (Section 6.3) and these largely relate to the sediment's carbon content, determined by organic productivity and degree of agitation. In Archaean iron-formations of the Canadian Shield, oxide-silicate facies pass into carbonate and then sulphide facies with increasing depth and distance from shore (Goodwin, 1973).

Facies in iron-formations can also be defined on sedimentary characteristics and the best known is the laminated facies (Fig. 6.12), consisting of the fine interbedding of chert with hematite or magnetite, less commonly with siderite

184

Fig. 6.12 Banded cherty iron-formation consisting of hematite interbedded with chert (sometimes referred to as jaspilite). Some disruption of the bedding by slumping has occurred in the central part. Early Proterozoic, Michigan, U.S.A.

5 cm

or greenalite. Laminations on the meso-scale (10–50mm) and micro-scale (0.2 –2mm) can be distinguished and in the Brockman Iron-Formation of the Hamersley Basin, Australia, individual mesobands can be traced over 30,000 sq km (Trendall, 1973). Laminated iron facies were deposited in deep-water basins, on shelves below wave base, and in lagoons, the lamination probably reflecting seasonal changes in the environment. Sediments of shallow, agitated waters consist of ooids, pisolites, pellets, granules and intraclasts, giving rise to iron facies exactly comparable to those of limestones (Section 4.10). Such non-laminated facies are well-developed in the Biwabik and Gunflint Iron-Formations of the Lake Superior region and in the Sokoman Iron-Formation of the Labrador Trough (Dimroth, 1979). Ooids generally consist of hematite and chert and granules are of greenalite. Sedimentary structures present include cross bedding channels, ripples and desiccation cracks, and stromatolites (Dimroth, 1968).

Points of discussion with the Precambrian iron-formations centre on the source of the iron, the origin of the chert and the depositional environment. A volcanic source for the iron has been suggested, particularly for the Archaean Algoma-types which are clearly associated with contemporaneous volcanic rocks. For many of the Proterozoic iron-formations, where there was no contemporaneous volcanicity deep weathering of continental rocks is assumed to have provided the iron. To facilitate leaching of iron and its transportation it has often been suggested that the Earth's atmosphere had a higher carbon dioxide content at that time, prior to 1900 Ma (Garrels *et al.*, 1973; Cloud, 1973; Drever, 1974). A greater partial pressure of CO_2 would have the effect of lowering the pH of surface waters, leading to a greater efficiency in iron leaching and transportation. Seawater with a lower pH could itself have been a major reservoir of Fe^{2+} in solution. A biochemical mechanism of iron deposition suggested by Cloud (1973) involves precipitation of the Fe^{2+} as ferric oxides and hydroxides through reaction with oxygen derived from primitive photosynthesizing organisms. Direct chemical precipitation of silica and seasonal blooms of silica-secreting organisms have been suggested for the chert layers. The occurrence of microfossils in the cherts (La Berge, 1973) indicates at least some biological activity.

There is growing evidence to suggest that the early Precambrian atmosphere did contain some oxygen and as a result of this a completely different

185

hypothesis of diagenetic eluviation-replacement has been put forward by Kimberley (1979, 1980) and Dimroth (1979). It is postulated that iron-formations are replaced limestones. The source of the iron and silica is unknown, but is thought to be leaching of interbedded terrigenous clay and/or pyroclastic material.

For the environment of deposition, most workers have assumed or proposed that the environments were marine although locally with lagoons and restricted basins. As an alternative, there is a suggestion that cherty iron formations were deposited in lakes and playa-lakes (Eugster & Chou, 1973).

6.5.2 PHANEROZOIC IRONSTONES

Phanerozoic ironstones vary considerably in grade, lithology and iron minerals present. The important types are hematite ironstones, chamosite oolites, sideritic mudrocks and nodular siderites, and sulphide ironstones. With oolitic ironstones a common Phanerozoic type, the mineralogy of ooids and matrix is often different and so in naming the rock it is customary to refer to these components separately; thus a sideritic chamosite oolite consists of chamosite ooids in a matrix of siderite.

Hematite Ironstones A well-documented example is the Clinton ironstone, which developed along the eastern shelf of the Central Appalachian Basin in the Middle Silurian (Hunter, 1970). The ironstone consists mainly of hematite oolite (Fig. 6.4), but chamosite oolite also occurs. Skeletal grains in the sediment are frequently impregnated and replaced by hematite. Facies analysis has shown that hematite oolites were deposited in shallower water and closer to the shoreline than the chamosite facies. A glauconite facies is recognized farther offshore. These ironstones developed during periods with low rates of siliciclastic influx and they occur at horizons of major lithologic and faunal breaks in the sequence, interpreted as representing periods of shallowing.

The Lower Ordovician Wabana ironstone of Newfoundland is similar to the Clinton in being dominantly oolitic and of shallow water aspect. Ooids consist of alternations of hematite and chamosite, with the former apparently replacing the latter. Hematitic ironstones occur within limestone sequences in the Devonian and Carboniferous of Britain, continental Europe and the U.S.A. They mostly take the form of hematite-impregnated and replaced skeletal debris in a carbonate cement (Fig. 6.5). In some cases, such as the Rhiwbina iron ore of South Wales, the ironstones are concentrates produced by the current reworking of grains impregnated with chamosite, which was then converted to hematite.

Chamosite oolites These form thin sequences, the order of 1–10 m, which are condensed in thickness relative to other contemporaneous sediments. The ironstones frequently only occur in localised areas, passing laterally into typical marine sediments. Oolitic ironstones may show cross bedding and other structures indicating shallow agitated waters, and they are commonly interbedded with chamositic and sideritic mudrocks. Fossils found within chamosite oolites are often diverse, indicating normal marine conditions. Examples are the Northampton (Taylor, 1949) and Frodingham Ironstones

186

(Hallam, 1966) of the British Jurassic. In these cases the ironstones probably formed within local embayments away from the diluting effects of terrigenous mud and sand. Shoals within a pro-delta environment have been suggested as a likely location for chamosite ooid growth, with iron-bearing waters being carried beyond the main area of terrigenous sedimentation (a 'clastic trap' mechanism). Other oolitic ironstones contain a restricted fauna. The Westbury Ironstone (Jurassic, Southern England) for example contains an abundance of oysters (Talbot, 1974). In these cases it is likely that the chamosite facies formed in coastal lagoons, sometimes hyposaline, separated from normal marine areas by sand bars and barriers. In the Miocene of Columbia, thin beds of chamosite and goethite ooids accumulated in interdistributary bays of a prograding delta at times of low sediment input (James & Van Houten, 1979).

Sideritic mudrocks and nodular siderites Mudrocks with finely disseminated siderite have commonly been deposited in deltaic, lagoonal, estuarine and other environments where iron-bearing muds were deposited and the necessary chemical conditions existed in the sediment. The sediments are usually grey or bluish-grey but they take on a brown colour with weathering through oxidation of the siderite. Sideritic mudrocks are common in Jurassic ironstone sequences, associated with chamosite oolites; examples include the Northampton and Westbury ironstones of Britain cited above.

Mudrocks containing flattened nodules or more continuous beds of siderite constitute the clay ironstones, black band ironstones and sphaerosiderites. Nodular siderites are particularly common in coal measure sequences, such as the Carboniferous of Britain, the Appalachians and the U.S.S.R. The clay ironstones and black band ironstones mainly formed within fresh and brackish water swamp and lagoon sediments, as they are forming at the present time (Ho & Coleman, 1969). Black band ironstones have a higher organic content than clay ironstones. Sphaerosiderites, with their radiating-spherulitic crystal fabric (Fig. 6.8) chiefly formed in the lower horizons of fossil soils (seatearths and underclays), with the iron derived from leaching in the higher part of the soil profile.

Sulphide ironstones Pyrite rarely makes up a large part of Phanerozoic sedimentary ironstones but it is disseminated within many black carbonaceous shales and some limestones. In some ironstones pyrite occurs as aggregates and replaces other iron minerals and skeletal fragments to form thin beds of high sulphide content. Such horizons occur within Ordovician chamosite and hematite oolites of Wabana, Newfoundland and in the Sulphur Bed of the Cleveland Ironstone (Jurassic), England.

6.6   Bog Iron Ores

The only modern environment where significant iron ores are at present forming is in swamps and lakes of mid to high latitudes such as northern America, Europe and Asia. The ores range from hard oolitic, pisolitic and concretionary forms to earthy and soft types. Iron minerals present are frequently difficult to identify because of their amorphous or poorly-crystalline form. However, goethite is most common, siderite less so, and vivianite ($Fe_3P_2O_8.H_2O$) an accessory. Manganese oxide content may reach 40% (in most ironstones Mn is less than 1%. The formation of bog ores occurs when acidic groundwater seeps into oxygenated marshes and swamps. The rise in Eh and pH causes ferrous iron in solution to be precipitated, mainly as ferric hydroxides. This seepage of groundwater has been suggested as a mechanism for some ancient ironstones.

6.7   Ferromanganese nodules and crusts, and metalliferous sediments

Ferromanganese nodules and crusts and metalliferous sediments of the seafloor have received much attention in recent years, chiefly in view of their possible economic significance.

Nodules and encrustations of iron and manganese oxides are widely distributed on the seafloor, but particularly in parts of the Atlantic, Pacific and Indian oceans. They are especially well-developed in areas of negligible sedimentation, where there are strong bottom currents, and often at depths of several thousand metres. In oceanic settings, they commonly form on the flanks of active mid-ocean and aseismic ridges, on seamounts and on abyssal plains. They also form in sediment-starved continental-margin areas, such as on the Blake Plateau, off the east coast of the U.S.A. Somewhat similar manganese nodules are found in some temperate lakes and shallow marine areas.

The marine manganese nodules and crusts vary considerably in chemistry and mineralogy from one part of the seafloor to another. Both Mn-rich/Fe-poor, and Mn-poor/Fe-rich varieties occur. One of the main points of interest is in the relatively high concentrations of metals including Co, Ni, Cu, Cr and V (Table 6.2), such that the submarine 'mining' of these deposits is now considered a viability. Although the minerals constituting the nodules are frequently X-ray amorphous, todorokite $(Mn^{2+}R^{+}R^{2+})$ $(Mn^{4+}Mn^{2+})$ $O_6.3H_2O$ (Where R represents other metals) is one manganese oxide which commonly occurs, together with the hydrated iron oxide goethite. Ferromanganese nodules and pavements, being hard substrates, are often encrusted with organisms such as Foraminifera, bryozoans, serpulids and ahermatypic corals. Phosphate nodules and phosphatization of sediment may occur in association with marine manganese deposits.

In the geological record, ferromanganese nodules are not common, but they do occur in Jurassic and Devonian condensed pelagic limestones of the Alpine

Table 6.2 Average concentration of Fe, Mn, Cu, Co and Ni (in percent) of shallow and deepwater sediments and ferromanganese nodules from three seafloor settings. From data in Glasby, 1977.

	Near-shore sediments	Deep-sea sediments, Atlantic	Nodules from seamounts	Abyssal nodules	Active ridge nodules
Fe	4.83	5.74	15.81	17.27	19.15
Mn	0.0850	0.3980	14.62	16.78	15.51
Cu	0.0048	0.0115	0.058	0.37	0.08
Co	0.0013	0.0039	1.15	0.256	0.40
Ni	0.0055	0.0079	0.351	0.54	0.31
Depth			1900m	4500m	2900m

and Hercynian fold-belts in Europe, and also in Cretaceous red clays of Timor. The origin of these ferromanganese deposits has given rise to much debate. A hydrothermal-volcanic origin has been favoured by some workers and this could well be the case where there is a close association with volcanism, as in active mid-ocean ridge occurrences. However, in many instances ferromanganese nodules are in areas remote from any sort of volcanic activity, so that some form of direct or indirect precipitation of the metal oxides from seawater is required. Although the rate of accretion of the nodules varies considerably, and in some cases tens of years is sufficient, the majority appear to form at incredibly slow rates of around $1mm/10^6yrs$.

Apart from nodules and crusts, metalliferous sediments also occur in oceanic settings, especially in the vicinity of active spreading ridges where they develop on top of ocean-floor basalts. These basal sediments are typically enriched in Fe, Mn, Cu, Pb, Zn, Ni, Co, Cr and V. Metal-rich muds have been recovered from the mid-Atlantic ridge and East Pacific Rise during deep-sea drilling and from the spreading ridge of the Red Sea, where Zn-Cu-Pb enriched sediments are associated with hot, metal-rich brines. The fluids effecting these metal enrichments are thought to derive either from mantle magmatic sources or from the interaction of basalt with seawater.

Sediments enriched in metals are found in association with many ancient pillow lavas and ophiolite suites. The Cretaceous umbers (Fe and Mn-rich) and ochres (Fe-rich/Mn-poor) of the Troodos Massif, Cyprus can be compared with the basal sediments of present-day active mid-ocean ridges.

Ferromanganese deposits are considered in detail in Glasby (1977) and in the context of modern and ancient pelagic sedimentation by Jenkyns (1978).

Further Reading

For a useful collection of papers on iron deposits, including some referred to in this chapter, see:
Lepp, H. (Ed.) (1975) Geochemistry of Iron. pp. 464. Benchmark Papers in Geology; Dowden, Hutchinson & Ross, Stroudsburg.

Papers concerned with Precambrian Iron-Formation are contained in:
Economic Geology volume 68, number 7 (1973).
Ferromanganese deposits are discussed in Glasby, G.P. (Ed.) (1977) *Marine Manganese Deposits,* pp. 523. Elsevier, Amsterdam.

7

Sedimentary Phosphate Deposits

7.1   Introduction

Sedimentary phosphate deposits or phosphorites are important natural resources. Phosphates are one of the chief constituents of fertilizer and they are used widely in the chemical industry. In addition, phosphorites commonly contain relatively high concentrations of useful elements such as uranium, fluorine and vanadium.

Phosphorus is one of the essential elements for life and is present in all living matter. Although only forming a minor part of plants and animal soft parts, phosphate is the major constituent of all vertebrate skeletons (bones) and some invertebrate hard parts. In the marine environment phosphate is a primary nutrient so that it is a control of organic productivity. In seawater, it is mainly present as dissolved 'orthophosphate' and particulate phosphate, with the phosphate in the latter case chiefly contained within or adsorbed on to organic detritus. In the oceans, most organic productivity utilizes dissolved 'orthophosphate' and takes place in the upper levels through phytoplanktton growth. The concentration of phosphate in coastal waters, such as within estuaries, is often higher than surface waters offshore. High phosphate concentrations also occur in the waters of anoxic basins. Phosphorus in the marine environment is discussed in Riley & Skirrow (1975).

Many sedimentary rocks contain a few per cent calcium phosphate in the form of grains of apatite (a heavy mineral: Section 2.5.5), bone fragments or coprolites. Rocks largely composed of phosphate on the other hand are relatively rare. Sedimentary phosphate deposits are here discussed in three categories: nodular and bedded phosphorites, where upwelling and organic productivity have played a major role in their formation (Section 7.3); bioclastic and pebble bed phosphorites where sediment reworking has been of paramount importance (Section 7.4) and guano and its derivatives (Section 7.5).

7.2   Mineralogy

The most common sedimentary phosphate minerals are varieties of apatite. The apatite of igneous rocks is chiefly fluorapatite, $Ca_5(PO_4)_3F$, but in sedimentary apatite replacement of the phosphate by carbonate may reach several per cent and fluorine may be replaced by hydroxyl or chlorine ions. In addition, the calcium ions may be substituted by sodium, strontium, uranium and rare earths. Most sedimentary phosphates are carbonate hydroxyl fluorapatites which can be represented by the formula $Ca_{10}(PO_4,CO_3)_6F_{2-3}$. Two specific sedimentary apatite minerals, best identified by X-ray diffraction and chemical

190

analysis, are *francolite* with more than 1% fluorine and appreciable carbonate, and *dahllite,* a carbonate hydroxyapatite with less than 1% F. These two minerals are anisotropic. The term *collophane* is often loosely applied to sedimentary apatite of cryptocrystalline form for which the precise composition has not been established. It is isotropic.

7.3   Nodular and bedded phosphorites

7.3.1   RECENT–SUBRECENT OCCURRENCES

Marine phosphate deposits have been known from the ocean floors since the 'Challenger' oceanographic expedition of the 1870s dredged up phosphate nodules and slabs from the continental shelf and slope off southern Africa. Since then seafloor phosphorites have been recorded off many other continents (Fig. 7.1), in particular the west coast of North and South America (those off California are especially well-documented, Dietz *et al.,* 1942), the east coast of the U.S.A., and the shelves off northwest Africa and Japan (Tooms *et al.,*

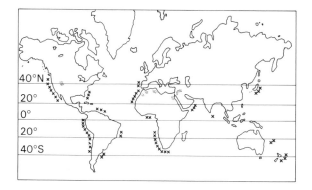

Fig. 7.1 Distribution of marine phosphorites, although in many cases the deposits are not actually forming at the present time. Also shown are the locations of the Permian Phosphoria Formation of northwestern U.S.A. and the Cretaceous–Tertiary phosphorites of the Middle East–North Africa.

1969). The marine phosphorite occurs in areas of negligible sedimentation, on outer continental shelves and slopes, particularly on the tops and sides of local ridges and banks, on fault scarps and the flanks of submarine canyons. Phosphate nodules and crusts generally occur at depths from 60–300 m.

The nodules are usually several centimetres in diameter but may reach a metre or more across. They range in shape from flat slabs to spherical nodules and irregular masses. The internal structure of the nodules varies from homogeneous to concentrically-laminated and conglomeratic and many contain pellets and 'ooids'. Vertebrate skeletal debris, especially of fish, and coprolites are often associated. In some situations pebbles derived from local seafloor exposures of Tertiary limestone are phosphatized, as well as the in-situ bedrock surface itself (Burnett & Gomberg, 1977). Submerged reefs forming seamounts in the oceans frequently have crusts of phosphorite upon them and skeletal grains have been replaced by carbonate-apatite (Marlowe, 1971). The principal mineral comprising the phosphorites is collophane but francolite and dahllite are also present. Glauconite (Section 6.4.4) and ferromanganese oxides (Section 6.7) are frequently associated with phosphate nodules.

It has recently been established from uranium isotope data (Kolodny & Kaplan, 1970; Burnett & Gomberg, 1977) that many marine phosphorites such as those off California and Florida were formed in pre-Holocene time, in some cases as far back as the Miocene. Two areas where Recent accumulation of phosphorites has been demonstrated are the Walvis Bay region of South West Africa and along the continental margin of Peru and Chile. On the South West African shelf, the phosphorite occurs within diatom oozes as dispersed phosphate, biogenic fragments, coprolites, pellets and phosphatic nodules (Baturin, 1970). The nodules vary from soft and gelatinous to friable or to hard and massive forms. The phosphate content increases with increasing induration. Diatoms, originally composed of opaline silica, have been replaced by cryptocrystalline phosphate within nodules. Off Peru and Chile the phosphorite is developing on the upper continental slope as replacements of benthic Foraminifera (Manheim *et al.*, 1975). Obliteration of the foraminiferal structure by the replacement leads to the production of phosphorite pellets.

7.3.2 ORIGIN OF MARINE PHOSPHORITES

Two factors control the formation of marine phosphorites: upwelling and low sedimentation rates (Fig. 7.2). The importance of upwelling was first appreciated by Kazakov in 1937 (Gulbrandsen, 1969), who noted that many 'modern' phosphorites are located in areas where cold, nutrient-rich waters rise from the depths towards the surface. Upwelling currents lead to high organic productivities and phytoplankton growth in surface waters which in turn results in organic-rich (and so phosphate-enriched) sediments and oxygen-deficient waters overlying the seafloor. Mass mortalities of fish occasionally take place in areas of upwelling, particularly as a result of poisoning by phytoplankton blooms. More organic matter, with its combined phosphorus, and skeletal phosphate (bones) are thus contributed to the seafloor during these events.

Although it was once thought that phosphorite was precipitated directly from seawater, perhaps as some type of colloid, data from sites of active phosphorite formation indicate that it is being formed within the surficial sediments largely by replacement (Baturin, 1970; Manheim *et al.*, 1975; Birch, 1979). The breakdown of organic matter in the sediment liberates phosphate which is precipitated in pellets and coprolites, and replaces siliceous and

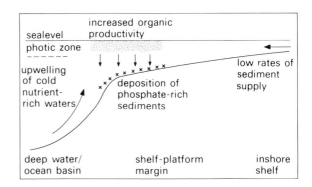

Fig. 7.2. Model for formation of marine phosphorites.

192

calcareous skeletons and lime mud, eventually giving rise to nodular masses of phosphorite. The role of the phytoplankton is crucial in transporting the phosphate from the upwelling currents to the seafloor.

A further stage in the formation of extensive marine phosphorites is reworking. Ocean currents and severe storms remove much of the fine, unphosphatized sediment from the seafloor, leaving a concentrate of nodules, pellets and coprolites in various stages of induration and phosphate impregnation, which are then further phosphatized. Even greater reworking of bottom sediment is achieved during sealevel changes; transgressive–regressive events across the shelf off southern Africa have been important in the formation of the phosphorites there (Birch, 1979).

There is a latitudinal control on the formation of phosphorites: they occur in tropical or subtropical arid zones within 50° of the equator (Cook & McElhinny, 1979). The paucity of phosphorites actually forming at the present time, and their greater development in the Tertiary, has been related to slightly higher seawater temperatures at that time (Tooms *et al.*, 1969).

7.3.3   ANCIENT PHOSPHORITE SEQUENCES

Phosphate deposits which in many cases can be related to upwelling currents and high organic productivities are known from the Precambrian onwards. One of the most famous is the Phosphoria Formation of Permian age occurring in Idaho, Wyoming and Montana, U.S.A. (McKelvey *et al.*, 1959). Phosphorite occurs in two members (Meade Peak and Retort) separated by chert (Rex Chert Member). The phosphate members consist of interbedded carbonaceous mudrock, phosphatic mudrock, pelletal, oolitic, pisolitic, nodular and bioclastic phosphorite and phosphatic dolomite and limestone.

The phosphorites developed in an outer shelf-upper slope setting and pass laterally eastwards into nearshore carbonates, evaporites and red beds, and westwards into deeper water sediments. The situation is thought to have been directly analogous to modern phosphorite accumulations, with phosphate derived initially from upwelling currents and contributed to the sediments via phytoplankton growth in surface waters (Fig. 7.2). Much of the phosphorite consists of pellets up to several millimetres in diameter composed of collophane, in a matrix of collophane with smaller pellets (Fig. 7.3). The pellets could well

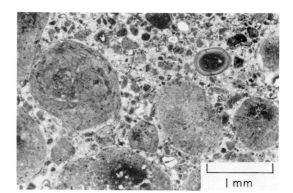

Fig. 7.3. Phosphorite pellets in matrix of smaller pellets and cryptocrystalline collophane. An ooid of collophane is also present. Plane polarized light. Phosphoria Formation, Permian; Idaho, U.S.A.

193

Fig. 7.4 Bioclastic phosphorite consisting largely of fish scales cemented by darker collophane. Plane polarized light. Phosphoria Formation, Permian; Idaho, U.S.A.

be of faecal origin. Ooids of phosphorite (Fig. 7.3) are present in some beds but their origin is less clear. Skeletal phosphate occurs in the form of fish scales and bones and inarticulate brachiopolds (Fig. 7.4).

Other Palaeozoic phosphorites in North America are found in the Mississippian and Ordovician of Utah, Idaho and adjacent states. Along the western margin of North America, bedded phosphorites occur in the Miocene Monterey Formation of California and Jurassic La Caja Formation of Mexico. Extensive and valuable phosphate deposits occur in the Upper Cretaceous to Lower Tertiary of North Africa and the Middle East, from Morocco and the Spanish Sahara to Iraq and Turkey (Sheldon, 1964). These bioclastic and pelleted phosphorites, also related to upwelling, accumulated on the continental margin around the southern shore of Tethys.

Some ancient bedded phosphorites have been interpreted as formed within confined basins, lagoons or estuaries, where high phosphate values, promoting high organic productivities, do occur. Examples are Tertiary phosphorites from the Atlantic coastal plain of the U.S.A. (Pevear, 1966; Rooney & Kerr, 1967) and some probable Tertiary phosphorites from Agulhas Bank, S. Africa (Parker, 1975). Phosphate precipitation and replacement associated with non-sequences and hardground surfaces occur in ancient pelagic sediments, such as the Cretaceous Chalks of Britain (Kennedy & Garrison, 1975a).

7.4 Bioclastic and pebble bed phosphorites

Vertebrate skeletal fragments are locally concentrated to form bone beds, often with fish scales. The depositional processes instrumental in their concentration are mainly current and wave reworking of sediments and winnowing of finer material, such that the phosphatic grains are left as lag deposits. Environments where these processes may take place include transgressive and regressive shelf and shore zones and fluviatile and intertidal channels. Bioclastic phosphorites can form in association with phosphate deposits related to upwelling (Section 7.3), the bone beds off Florida for example (Burnett & Gomberg, 1977). In the geological record, classic bioclastic phosphorites are the Rhaetic Bone Beds, Upper Triassic of southwest Britain and the Ludlow Bone Bed (Upper Silurian) of the Welsh Borderland.

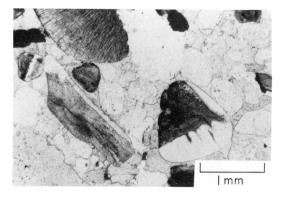

Fig. 7.5 Photomicrograph of bone fragments showing internal structure, in calcite-cemented sandstone. Plane polarized light. Rhaetic Bone Bed, Upper Triassic; Gloucestershire, England.

1 mm

In thin section, skeletal phosphate is distinguished by its light yellow to brown colour and presence of a microstructure of regularly-arranged canals (canaliculi) and growth lines (Fig. 7.5). Bone phosphate is often isotropic, or anisotropic with weak, irregular, patchy or undulose extinction.

Coprolites are often present in bioclastic phosphorites. They are generally spherical to elongate faecal pellets, up to 20 mm in diameter, composed of collophane. They may be homogeneous or show a vague concentric lamination (Amstutz, 1958).

During diagenesis, bone fragments in bioclastic phosphorites are further enriched in phosphate and their initially low fluorine content increases with age. The bone fragments may be cemented by collophane and phosphate nodules may nucleate around grains. Diagenetic phosphate can be precipitated within limestones, mudrocks and sandstones in the form of nodules, cements or replacements of calcareous skeletal grains. In most cases this phosphate would be derived from the decay of disseminated organic matter. A neutral or slightly acid pH facilitates the replacement of $CaCO_3$. The chemical and physical factors controlling phosphate precipitation and replacement have been discussed by Ames (1959) and Gulbrandsen (1969).

Once formed, phosphate nodules and phosphatized limestone fragments, as well as phosphatic fossils, are very resistant to weathering. They are easily reworked into succeeding beds and may form concentrates, again of economic value. Examples are the phosphate pebble beds in the Pliocene Bone Valley Formation in Florida (Riggs, 1979), the Palaeocene Brightseat Formation of Maryland (Adams *et al.,* 1961), the Cambridge Greensand and Glauconitic Marl (Cretaceous) of England (Kennedy & Garrison, 1975b) and the Middle Devonian of New York (Baird, 1978). Phosphorites on the North African continental shelf are largely reworked from Tertiary phosphate deposits.

7.5   Guano

The excrement of birds, and to a lesser extent bats, may in certain circumstances form thick phosphate deposits of economic significance. Leaching of the fresh guano leaves an insoluble residue composed mainly of calcium phosphate. Thick accumulations of bird guano are found on some small oceanic islands in

195

the eastern Pacific such as Ocean Island and the Banabans, along the Pacific coast of South America, and in the West Indies. Many of these deposits are comparatively old and are not being formed at the present day.

Geologically, guano itself is not significant. However, downward percolation of solutions derived from guano may cause phosphatization of underlying carbonate sediments and rocks (Braithwaite, 1968).

Further Reading

There is no comprehensive review of phosphatic sediments but for further information on marine phosphorites see initially the following papers:

Birch, G.F. (1979) Phosphatic rocks on the western margin of South Africa. *J. sedim. Petrol.* **49**, 93–110.
Manheim, F., Rowe, G.T. & Jipa, D. (1975) Marine phosphorite formation off Peru. *J. sedim. Petrol.* **45**, 243–251.
Tooms, J.S., Summerhayes, C.P. & Cronan, D.S. (1969) Geochemistry of phosphate and manganese deposits. *Oceanogr. Mar. Biol. Ann. Rev.* **7**, 49–100.

For a description of the classic Phosphoria Formation see:

McKelvey, V.E., Williams, J.S., Sheldon, R.P., Cressman, E.R., Cheney, T.M. & Swanson, R.S. The Phosphoria, Park City and Shedborne formations in the western phosphate field. *Prof. Pap. U.S. geol. Surv.* **313-A**, 1–47.

8
Coal, Oil Shale and Petroleum

8.1   Introduction

Organic matter in Recent and ancient sediments is a reflection of the original organic input and subsequent processes of diagenesis and metamorphism. However, much of the organic matter which comprises living organisms decays and breaks down in the presence of oxygen, eventually into carbon dioxide and water. This organic degradation takes place in most subaerial environments and many subaqueous environments. Where there is a deficiency of oxygen, the organic decomposition is incomplete and arrested, and quite stable organic compounds can develop and be preserved in the rock record. Organic matter is best preserved in anoxic environments, such as in stagnant lakes and marine basins, swamps and bogs, and in situations when anaerobic conditions set in early during diagenesis.

The fundamental process in the production of both plant and animal organic matter, and in fact the very basis of all life, is *photosynthesis*. Through this plants manufacture carbohydrates from carbon dioxide and water, using sunlight for energy and chlorophyll as a catalyst. Although it is a complicated process it can be represented by:

$$6 \, CO_2 + 6 \, H_2O \rightarrow C_6H_{12}O_6 + 6 \, O_2$$

Decomposition of organic matter in the presence of oxygen, aerobic decay, is basically the reverse of photosynthesis with various intermediate steps. When there is little or no oxygen present anaerobic decay takes place to produce hydrocarbons and other more complex organic compounds.

Since the majority of natural environments are oxidizing, most sediments contain only small quantities of organic matter. On average, sandstones contain 0.05% organic matter, limestones around 0.3% and mudrocks 2%. The principal organic deposits are hard coals, brown coals (lignite), their modern analogue peat, and oil shales. These deposits and the oil and gas which is derived from organic matter in sediments, comprise the fossil fuels, of immense importance to man. The now highly analytical and specialized disciplines of organic geochemistry and organic petrology are doing much to improve our understanding and exploitation of fossil fuels, all the more so in view of the expected depletion of petroleum reserves in the 21st century.

8.2   Modern organic deposits

The three main types of organic deposit accumulating at the present time are humus, peat and sapropel. *Humus* is fresh, decaying and decayed organic matter occurring mainly in the upper part of soil profiles. Decay products are mainly humic acids which are capable of leaching rock fragments and clays. In time, most humus is completely oxidized and so does not form organic deposits. It may, however, be important in the downward transportation of ions in the soil profile (eluviation) and in the formation of iron pans and other soil precipitates.

Peat forms as a result of two main factors: growth of plants which are composed of organic compounds resistant to decay, and the development of anaerobic conditions which prevent the oxidation and bacterial decomposition of the organic matter. Also peats are developed chiefly under colder climates where the rate of decay is much slower. The principal plant groups involved in peat formation are those tolerant of acid conditions: the mosses, especially *Sphagnum* and *Hypnum,* heathers, rushes, sedges and horsetails, and some trees such as pine, birch and willow. Peat is largely developed in freshwater swamps and bogs and on moorland areas in temperate and polar regions. Vast areas of Canada, northern Europe and the U.S.S.R. are covered by Pleistocene to Recent peat deposits. Peat is also accumulating in some mangrove swamps of the humid tropics.

Sapropel refers to organic material which accumulates subaqueously in shallow to deep marine basins, lagoons and lakes. The organic matter is largely derived from phytoplankton which live in the upper water levels, but fragments of higher plants can be an important constituent. Planktonic algae, including *Botryococcus* which contains globules of oil, proliferate in the photic zone of some lakes and give rise to hydrocarbon-rich sediments. Organic matter derived from such algae can accumulate along shorelines to form significant deposits. This occurs in the Coorong Lagoon, South Australia, and in Lake Balkash, U.S.S.R. Fine-grained terrigenous clastic sediment can be deposited along with the sapropelic organic matter so that all transitions occur from pure sapropel through organic-rich muds (Section 3.7.2) to organic-free muds.

8.3   Ancient organic deposits

Organic deposits can be divided into two broad groups: those formed through in-situ organic growth, as peat and humus, and those consisting of organic matter which has been transported or deposited from suspension, as with sapropels. Many brown and hard coals belong to the first, humic group while oil shales, and some coals, are examples of the second, sapropelic group. Lignites and coals have less than 33% inorganic material; (usually less than 10%); this mainly consists of clay and quartz silt and sand (Section 8.5.2). Oil shales contain more than 33% inorganic material; the organic content is kerogen (see below) and hydrocarbons are given off on distillation (Section 8.8). Organic-rich mudrocks with up to 10% organic matter have been considered in Section 3.7.2. Tar sands contain solid and semi-solid hydrocarbons which have invariably migrated into the deposit. Organic matter in sediments may simply be the in-situ alteration products of original material, the residue left after petroleum generation, or material which has itself migrated from a nearby or distant source rock. The term *bitumen* is loosely used for any organic material in sediments but more specifically it is used for liquid or solid hydrocarbons which are soluble in organic solvents. Asphalt is a solid or semi-solid bitumen, strictly one derived from an oil rich in cycloparaffin hydrocarbons. *Kerogen* refers to organic matter which is largely insoluble in organic solvents; it is a geopolymer consisting of long-chain hydrocarbons of high molecular weight. *Petroleum* consists of crude oils, chiefly short and long-chain hydrocarbons, and gas, mainly methane, which in most cases has migrated into porous reservoir rocks from source rocks.

8.4 Coals and the coal series

Most coals are *humic coals,* formed from the in-situ accumulation of woody plant material; *sapropelic coals* are those formed from algae, spores and comminuted plant debris. Humic coals form a natural series from peat through brown coals and bituminous coals to anthracite. Coalification (also carbonification) refers to all the processes which take place when peat is converted into coal. Since most of the changes are controlled by temperature of burial, the term organic metamorphism is also used. Various microbiological, physical and chemical processes operate during coalification, all contributing towards the *rank* of the coal. The rank is basically a measure of the degree of coalification or level of organic metamorphism. With increasing rank, the carbon content increases, the volatile content decreases and other properties are affected (Table 8.1). Humic coals are divided on their rank into a number of categories (Table 8.1): these are (i) peat, (ii) soft brown coal (iii) hard brown coal (iv) bituminous hard coal and (v) anthracite. Apart from the level of organic metamorphism, coals are also divided into various *types*, determined by their organic constituents (Section 8.5.1).

Table 8.1 The clasification of coal on the basis of rank, with the approximate values of the various parameters used to estimate rank.

Rank stages		Carbon content % (dry ash free)	Volatile content % (approx.)	Calorific value Kcal/kg	Vitrinite reflectance
peat					
		60			
soft brown coal					
		53	4000		0.3
hard brown coal	lignite				
		71	49	5500	
	sub-bitum. coal				
		77	42	7000	0.5
bitum-inous hard coal	low rank ⎫ high rank ⎬ rank				
		87	29	8650	1.1
		91	8	8650	2.5
anthracite					
		100	0		

The initial stages of coalification take place during peat formation. The processes are mainly biological and bacterial, with little alteration of the original plant material. In *soft brown coals* many plant fragments are readily discernible, together with their original cell structure. A process of 'gelification' during the formation of *hard brown coal* causes homogenization and compaction of plant cell walls and leads to the formation of vitrinite, one of the constituents of bituminous coals (Section 8.5.1). The brown coals are relatively young, mostly occurring in the Tertiary and Mesozoic. Important reserves of lignite and sub-bituminous coals occur in the Tertiary and Cretaceous of Germany, North America and the Far East.

Bituminous hard coals contain less volatiles and moisture and consist of some 78 to 90% carbon. They are black, hard and bright in comparison with the soft, earthy and dull brown coals. They typically break into rectangular lumps, defined by prominent joint surfaces known as cleats. A distinctive bedding-parallel banding is usually developed, consisting of bright and dull layers of differing petrological composition. *Anthracite* contains more than 91% carbon although its calorific value is similar to high rank bituminous coals; it burns with a smokeless flame. Anthracite is a bright shiny rock with conchoidal fracture. Bituminous coals and anthracites are generally older coals, widely developed in the Carboniferous of western Europe, North America and the U.S.S.R.

8.4.2 SAPROPELIC COALS

Cannel coal and boghead coal are the two common sapropelic coals. Cannel is a massive unlaminated coal with an even and fine-grained texture, and conchoidal fracture. It consists of altered plant debris, spores and some algae. Boghead coal is similar to cannel coal but it is largely derived from algae, particularly the oil-bearing type such as *Botryococcus*. Sapropelic coals formed in ponds and shallow lakes within the areas of swamp where the vegetation grew for the formation of the humic coals.

8.5   Coal Petrology

8.5.1 ORGANIC CONSTITUENTS

Most studies of coal now use polished surfaces of the coal and reflected light microscopes, with oil-immersion objectives for increased contrast. Marie Stopes in 1919 recognized four visible (macroscopic) ingredients of bituminous coals. These *lithotypes* are vitrain, fusain, clarain and durain and comprise the millimetre-thick bands of humic coals. Stopes (1935) later introduced *maceral* terms, analogous to the minerals of igneous rocks, for the constituents of coal which can be recognized with the microscope. These terms form the basis of the 'Stopes-Heerleen System of Classification', which is now used by all coal petrologists. Microscopic associations of macerals, forming bands thicker than 50 μm are referred to as microlithotypes. The various macerals and microlithotypes used in coal petrology, together with the original lithotypes, are shown in Tables 8.2, 8.3. The macerals are divided into three groups: vitrinite, inertinite and liptinite (formerly exinite), and their appearance in reflected light is shown in Fig. 8.1.

Table 8.2. The principal macerals of coals.

maceral group	macerals	origin
vitrinite	collinite telinite	plant debris, esp. wood and bark
inertinite	fusinite semifusinite sclerotinite micrinite	woody tissues fungi fine organic debris
liptinite	sporinite cutinite resinite alginite	spores cuticle resin algae

Table 8.3. The lithotypes and microlithotypes of coal, together with the principal macerals comprising the microlithotypes.

lithotypes (hand specimen scale)	microlithotypes (microscopic scale)	principal macerals in microlithotype
vitrain	vitrite	vitrinites
fusain	fusite	inertinites, esp. fusinite
durain	durite	liptinites + inertinites
clarain	clarite	vitrinites + liptinites

The vitrinite macerals are collinite and telinite. Telinite is derived from plant cell-wall material of wood and bark and so has a cellular structure. Collinite is structureless, often being the organic infill of cell cavities. Vitrinite is derived from plant fragments which accumulated in stagnant water and were soon buried.

The inertinite macerals fusinite and semi-fusinite are derived from wood tissues so that cell structures are frequently preserved (Fig. 8.1c). Sclerotinite forms from fungal remains and micrinite from fine-grained organic detritus and from the thermal decomposition of resin. It is generally held that the inertinites have formed through some degree of oxidation of the woody tissue before or during the peatification stage such as some subaerial exposure of the plant debris before burial. With fusinite, however, which possesses a charcoal-like character, formation through forest fires has been suggested and much contested.

The liptinite group consists of macerals derived from megaspores, microspores, cuticles, resins and algae, as the various names suggest. These macerals can be recognized by their shape and structure, although in many

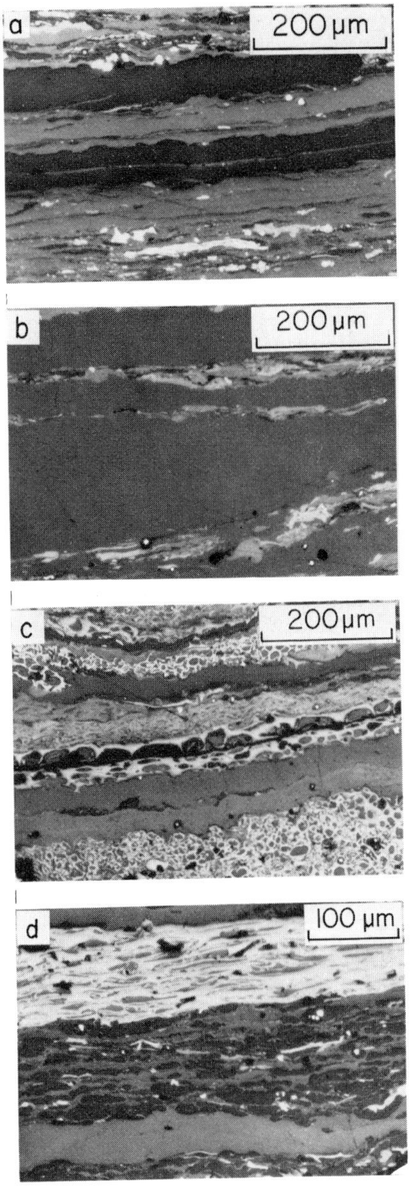

Fig. 8.1 Photomicrographs of bituminous coal showing the various macerals. Liptinite has a low reflectance and so forms the darkest areas; vitrinite has medium reflectance and so appears medium grey, and inertinite has a high reflectance and so has a very bright, light grey appearance. Fig. 8.1a Liptinite (the variety sporinite), vitrinite and inertinite. Liptinite comprises the two elongate dark areas which are compressed megaspores and the small dark elongate streaks, which are compressed microspores. Fig. 8.1b Vitrinite and thin streaks of inertinite (lighter grey colour). Fig. 8.1c Inertinite (semifusinite) and vitrinite. Fig. 8.1d Liptinite consisting of compressed microspores, vitrinite and inertinite (semifusinite, the very light band). Courtesy of D.G. Murchison. All reflected light, oil immersion; Upper Carboniferous, Northumberland, England.

cases the original components were compacted and squashed in the peat stage. Alginite is the main constituent of boghead coals.

With increasing rank, the whole coal tends to become homogeneous and the macerals lose their identities.

The four common microlithotypes forming the microscopic bands and layers in coal and composed of the various macerals described above are vitrite, fusite, clarite and durite (Table 8.3). The lithotypes of Stopes, which can be recognized in hand specimen, are composed of these microlithotypes. *Vitrain* layers are bright and glassy-looking. They are brittle, have a conchoidal fracture and are largely composed of the microlithotype vitrite consisting of vitrinite macerals. *Fusain* is a soft charcoal-like material which gives coal its 'dirty' character. It is chiefly composed of fusite, consisting of fusinite. *Durain* is a dull variety of coal with no lustre and an irregular fracture. It is composed of durite, a mixture of liptinite and inertinite macerals. *Clarain* is a laminated coal with a silky lustre and smooth fracture. It consists mainly of clarite, formed of a fine alternation of vitrinite and liptinite macerals.

Coal petrology is treated at length in Stach *et al.* (1975).

8.5.2 INORGANIC CONSTITUENTS

A wide variety of inorganic minerals may be present in coal, some being sedimentary, while others are early or late diagenetic. Quartz grains, clay minerals and some heavy minerals are the principal sedimentary inorganic constituents of coal. Early diagenetic minerals and concretions of siderite, ankerite, dolomite, calcite (including coal balls, see below) and pyrite. Pyrite is widespread in coals and is largely derived from the activities of sulphate-reducing bacteria (Section 6.3). Later diagenetic minerals typically form in veins parallel to the cleat, and in cavities. These minerals include ankerite, quartz and sulphides of Fe, Pb, Zn and Cu.

8.5.3 COAL BALLS

Of particular interest to the origin of coal is the occurrence of spherical concretions known as coal balls. These are mainly composed of calcite or dolomite but their significance lies in the well-preserved plant material which they contain (Fig. 8.2). The coal balls are early diagenetic nodules which formed

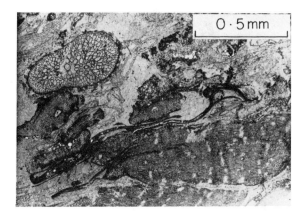

Fig. 8.2 Photomicrograph of a coal ball composed of calcite showing sections through stems and branches of various plants. Plane polarized light. Upper Carboniferous, northern England.

before compaction and alteration of the organic matter which eventually formed the coal. In Carboniferous coal balls fragments of the large trees *Sigillaria* and *Lepidodendron* are abundant, as well as *Calamites*. This indicates that the coal was derived from peat of a forest swamp environment. Modern analogues occur in mangrove swamps of humid tropics such as the Everglades, Florida (Spackman *et al.*, 1969).

8.6 Coal formation and rank

8.6.1 CHEMICAL CONSIDERATIONS

From the sedimentological context of coal seams, coal petrology and coal ball petrology, there is little doubt that humic coals are formed in situ as peat in forest swamps. The formation of brown coals (lignite) from peat requires burial, an absence of air, and time. The conversion of brown coals to bituminous coals in general requires more time, but also increases in temperature and pressure. Changes in the brown coal-bituminous coal range are entirely physical and chemical processes.

Chemically, coals largely consist of humic substances derived from the biochemical and chemical breakdown of the lignin, cellulose and protein of the original plant material. The humic substances have complex structures, high molecular weights and are largely composed of carbon in combination with small amounts of hydrogen, oxygen and nitrogen. In the early stages, lignin is degraded by micro-organisms to humic acids, diphenyl, benzyl and other organic compounds with ring structures and side chains. Then polycyclic aromatic compounds are formed and with increasing rank there is an increase in the aromatic portion, so that anthracite consists principally of condensed benzene rings. The formation of bituminous coals takes place at temperatures between 100 and 200° C; at these temperatures, the processes taking place are decarboxylation, condensation with the elimination of water, and re-arrangement reactions. Free carbon is not thought to be present in coal although it may occur in anthracites. When coal is heated, gases such as methane and hydrogen are given off and coal-tar and oil are formed. Distillation of coal produces coke.

8.6.2 RANK, DEPTH AND TEMPERATURE

The degree of coalification, i.e. the rank of the coal, can be measured by several parameters, including the amount of carbon, hydrogen, oxygen, volatiles and moisture present, the calorific value and the reflectance of the vitrinite. Not all parameters are useful over the whole coal rank spectrum. The reflectance of vitrinite, which increases with increasing rank (Table 8.1), is particularly useful since measurements need not be confined to coal seams, but can be undertaken on other sediments which contain plant fragments (phytoclasts) preserved as vitrinite. Vitrinite reflectance can be used as an indication of burial history (Castano & Sparks, 1974) and to evaluate hydrocarbon potential since it will indicate the burial temperatures reached by the sediments and one can assess whether the temperatures were high enough for petroleum generation (Section 8.9, and Hood et al., 1975; Tissot, 1977).

It has long been known that the rank of coal increases with increasing depth (Fig. 8.3). This has been referred to as 'Hilt's Law' after a 19th century German geologist who first recognized the relationship. The increase in rank is not so much a function of depth as temperature, since it is the latter which effects chemical changes in coal. Rank increasing with depth has been demonstrated for numerous coal basins, mainly from borehole core studies

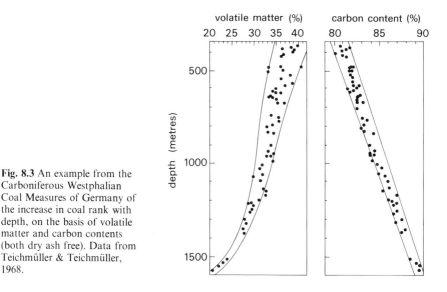

Fig. 8.3 An example from the Carboniferous Westphalian Coal Measures of Germany of the increase in coal rank with depth, on the basis of volatile matter and carbon contents (both dry ash free). Data from Teichmüller & Teichmüller, 1968.

(Teichmüller & Teichmüller, 1968, 1979; Damberger, 1974; Stach *et al.*, 1975). The depth at which coal of a particular rank is converted to one of higher rank does vary between coal basins. If the coals are of the same general age then this is a reflection of variations in the geothermal gradient. Apart from temperature, the length of time that a coal has been subjected to that burial temperature is also important. With coals of different ages buried to the same depths, the older coals have a higher rank. Where coals occur in areas of folding and thrusting then they too are frequently of higher rank. This is the case with anthracite in the South Wales Coalfield and high rank coals associated with thrusts in the Coal Measures of the Ruhr, Germany. Tectonic pressure has been invoked to explain this but a more likely alternative is the frictional heat developed during thrusting. Tectonic pressure is more physical in its effect, producing an anisotropy (preferred orientation) of coal macerals.

8.7   Occurrence of coal

In the geological record, coal-forming conditions were widespread in humid climatic areas from the late Devonian onwards, when land plants evolved and proliferated. The Carboniferous, Jurassic and Lower Tertiary in particular were periods of extensive coal formation. Paralic and limnic coals are distinguished, the paralic ones developed in coastal situations, particularly in deltaic environments, usually as thin (< 3m) but relatively persistent seams. Limnic coals formed in continental basins, associated with lakes, where through rapid subsidence the coal seams may be relatively thick (hundreds of metres even). The majority of the coal seams in the British, German and North American Carboniferous are of the paralic type; limnic coal basins are especially well-developed in the French Carboniferous.

Paralic, humic coals typically occur at the top of coarsening-upward deltaic cycles (Fig. 2.55 and Section 2.10.4). In a typical coal measure cycle, mudrocks with marine fossils (sometimes a marine limestone) give way to non-marine mudrocks then siltstone and sandstone, before the coal seam. Coal seams overlie soil horizons containing roots and rootlets but the nature of the soils varies considerably, depending on the lithology of the sediment in which the plants grew. Where the soils were sandy, they have frequently developed into quartz arenites known as ganisters, through the leaching of less stable minerals by acid porewaters (Section 2.7.1). Other soils are mudrocks of mixed clay mineralogy (Section 3.7.1), often with siderite nodules (Section 6.3), referred to as seatearths or underclays, or if suitable for refractory purposes, then fireclays. Tonstein is a specific term for a kaolinitic palaeosoil, occurring beneath some coal seams but within and above others. Precise chemical conditions are required for its formation if a purely sedimentary origin is accepted (Moore, 1964), but it is frequently held that tonsteins are derived from wind-blown volcanic ash (Spears & Rice, 1973; Section 10.5). Individual coal seams may be very continuous laterally but in most cases they are restricted to a particular coal basin and cannot be correlated from one region to the next. Marine horizons that occur within coal measure sequences, and tonsteins, can frequently be traced over vast areas. The origin of coal measure cycles and the rhythmic sedimentation of which coal seams are part is discussed by Merriam (1964), Duff *et al.* (1967), Westoll (1968) and others.

8.8   Oil Shales

Oil shales are a diverse group of rocks which contain organic material that is mostly insoluble in organic solvents, but which can be extracted by heating (distillation). The organic matter is largely kerogen, but some bitumen may occur. The quantity of oil which can be extracted ranges from about 4% to more than 50% of the weight of the rock, i.e. between 10 and 150 gallons of oil per ton of rock or 50–700 litres per metric ton. Oil shales contain a substantial amount of inorganic material consisting largely of quartz silt and clay minerals. Some oil shales are really organic-rich siltstones and mudrocks (Section 3.7.2), while others are organic-rich limestones.

Much of the organic matter in oil shales is finely disseminated and so altered that the organisms from which it formed cannot be identified. In many oil shales the remains of algae and algal spores are common and so much of the organic matter is assumed to be of algal origin. Fine-grained higher plant debris and megaspores may also be an important constituent. The common sedimentary feature of many oil shales is a distinct lamination, on a millimetre scale, of alternating clastic and organic laminae. Seasonal or annual blooms of planktonic algae are frequently invoked to account for the rhythmic lamination. As with the formation of coal, anaerobic conditions are required to prevent oxidation of the organic matter and to reduce the biological degradation. Many oil shales thus formed in stratified water bodies where oxygenated surface waters permitted plankton growth and anoxic bottom waters allowed the preservation of the organic matter (Fig. 3.7).

The kerogen is derived largely from algal lipid matter (fats and fatty acids) rather than carbohydrates, lignins or waxes. Some kerogen is formed from vascular plant debris, although on distillation this yields little oil.

The formation of kerogen requires the conversion of fatty matter to quasi-hydrocarbon materials. This is brought about by internal biochemical degradation within an increasingly anaerobic environment. Polymerization and condensation are the main reactions taking place, followed by maturation. The formula for kerogen from the classic oil shales of the Eocene Green River Formation of Colorado, Utah and Wyoming, U.S.A., is given as $C_{215}H_{330}O_{12}N_5S$. Kerogen in polished section is typically a yellow to amber, structureless organic material, occurring in bands and stringers, parallel to stratification. Certain metals, such as vanadium, nickel, uranium and molybdenum, are enriched in oil shales. They are complexed with or chelated on to the kerogen.

The majority of oil shales were deposited subaqueously, both in lacustrine and marine situations. Oil shales in the Eocene Green River Formation contain much dolomite and calcite and are rhythmically-laminated or varved. Although formerly regarded as relatively deepwater in origin (Bradley, 1931), deposition is now thought to have taken place in relatively shallow, ephemeral lakes, frequently subjected to desiccation (Surdam & Wolfbauer, 1975). Oil shale formed from a single algal species occurs at several horizons in the Lower Carboniferous of the Midland Valley of Scotland. This formed in freshwater lakes within a deltaic complex where humic coals also developed. Marine oil shales deposited in shallow seas, on continental platforms and basins are known from the Devonian of eastern and central North America, the Jurassic of Europe and the Miocene of California. In some cases they formed in association with upwelling (Section 7.3.2). Since planktonic algae are the principal source of the organic matter and these have a long geological history, Precambrian oil shales do occur. An example is the Nonesuch Shale of Michigan and Wisconsin, around 1100 Ma.

There is considerable interest in oil shales at the present time since they are a source of fossil fuel and they may help to offset the expected exhaustion of petroleum reserves. Extensive deposits of oil shale occur in the U.S.S.R., China and Brazil and low-grade deposits which may become economically viable occur in many other countries of the world (Yen & Chilingarian, 1976). Also of some economic potential are tar sands, such as those of Athabasca, Alberta, Canada (Chilingarian & Yen, 1978). Tar sands are sediments containing much heavy oil (Bitumen and asphalt) which in most cases has migrated from nearby source rocks.

8.9   Petroleum

8.9.1   COMPOSITION AND OCCURRENCE

The generation of petroleum is one of the stages in the alteration of certain types of organic matter buried in sediments. It is formed through increasing burial and temperature and so is part of the general process of organic metamorphism. Petroleum consists of crude oil and gas. Crude oils are largely carbon (around 85%) and hydrogen (12%), with small amounts of oxygen (up to 2%), nitrogen (up to 1.5%) and sulphur (up to 8%). Several hundred organic compounds have been recorded from natural oil. The hydrocarbons are largely n-alkanes (naphthenes) and aromatic hydrocarbons. The oxygen, nitrogen and

207

sulphur occur in such compounds as fatty, napthenic and aromatic acids, pyridines and porphyrins, and thiols and thioalkanes. Low sulphur crude oil ($< 0.5\%$ S) is referred to as sweet, whereas sour oil has a higher sulphur content. Solid and semi-solid organic compounds, (asphalts) are associated with some petroleums and are composed of asphaltenes, resins and other oily constituents.

Gas is invariably associated with crude oil, occurring either in solution, or as a gas cap. Natural gas not associated with oil also occurs. The gas consists predominantly of methane. 'Dry' and 'wet' gases are distinguished on the amount of liquid vapour present. Some higher paraffins such as ethane and propane also occur. CO_2, N_2 and H_2S may form a significant component and helium is also present. Water occurs in most oilfields and is typically a brine, more saline than seawater.

Petroleum can occur in any porous rock but in fact most of the world's oil is located in Tertiary and Mesozoic sedimentary rocks. Petroleum is derived from source rocks (discussed below) and then migrates into reservoir rocks which are typically sands and sandstones and certain types of limestone. The porosity and permeability of the reservoir rocks are obviously very important (Sections 2.9 and 4.9). An impervious seal is required to prevent upward escape of the petroleum from the reservoir and common cap rocks in oilfields are mudrocks and evaporites. To contain the petroleum some form of trap is necessary. Many traps are structural, involving folds (domes and anticlines especially), faults and salt diapirs, while others are depositional, arising from the geometry of the reservoir sandstone body or limestone mass and its overlying cap rock.

At the present time around one third of the world's oil comes from the Middle East: Saudi Arabia, Iran, Iraq, and the Gulf States. There are major oilfields in the U.S.A., Canada, U.S.S.R., Venezuela, Nigeria, Libya, Mexico and Indonesia, while many other countries have smaller oilfields.

8.9.2   FORMATION OF PETROLEUM

Petroleum is largely derived from the maturation of organic matter deposited in fine-grained marine sediments. Many oil source rocks formed at times of high organic productivity of marine plankton, coinciding with transgressive events and high stands of sealevel (Tissot, 1979). The diagenesis of the deposited organic matter produces kerogen and with increasing temperature through depth of burial, this first yields an immature natural gas consisting of methane and some carbon dioxide. In the following catagenesis stage crude oil is generated together with some wet gas (Fig. 8.4). This principal phase of oil generation takes place at temperatures around $70-100°C$; in areas of average geothermal gradient this is at depths of 2–3.5 km. Further increases in temperature through deep or very deep burial (metagenesis stage), produce dry natural gas. The type of kerogen present is also a factor in determining the composition of petroleum generated. Kerogen which is formed from plant cuticle mainly gives rise to gas on organic metamorphism; kerogen from algal organic matter is oil-prone kerogen. Areas of high geothermal gradient tend to favour hydrocarbon generation (Klemme, 1975).

Petroleum can be generated from humic and sapropelic coals if the burial

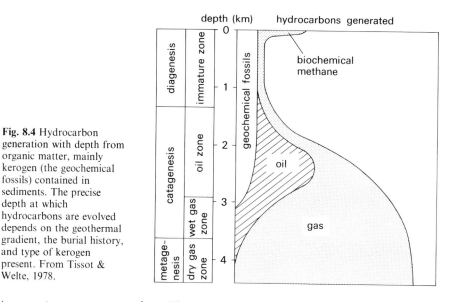

Fig. 8.4 Hydrocarbon generation with depth from organic matter, mainly kerogen (the geochemical fossils) contained in sediments. The precise depth at which hydrocarbons are evolved depends on the geothermal gradient, the burial history, and type of kerogen present. From Tissot & Welte, 1978.

temperatures are appropriate. The crude oils produced are different from those of marine origin, being more waxy.

In the search for petroleum, use can be made of the colour of pollen and spores (palynomorphs) in the source rocks to see if the stage of petroleum generation has been reached. With increasing temperature and level of organic metamorphism, palynomorphs change colour from yellow to brown when crude oil is evolved, and to black when dry gas is generated. An indication of burial temperature can also be obtained from the vitrinite reflectance (Tissot, 1977).

The study of petroleum is a science in itself and for further information reference should be made to standard petroleum geology text-books, such as Levorsen (1967), Hobson & Tiratsoo (1975) or Tissot & Welte (1978).

Further reading

Stach, E., Mackowsky, M.-Th., Teichmüller, M., Taylor, G.H., Chandra, D. & Teichmüller, R. (1975) *Textbook of Coal Petrology,* pp. 428. Gebrüder Borntraeger, Stuttgart.

Murchison, D.G. & Westoll, T.S. (Eds.) (1968) *Coal and Coal-Bearing Strata,* pp. 480. Oliver & Boyd, Edinburgh.

Teichmüller, M. & Teichmüller, R. (1979) Diagenesis of coal (coalification). In: Diagenesis of sediments and sedimentary rocks (Ed. by G. Larsen & G.V. Chilingarian), pp. 207-246. *Dev. in Sed.* **25A.** Elsevier, Amsterdam.

Tissot, B.P. & Welte, D.H. (1978) *Petroleum Formation and Occurrence,* pp. 538. Springer-Verlag, Berlin.

Hobson, G.D. & Tiratsoo, E.N. (1975) *Introduction to Petroleum Geology,* pp. 300. Scientific Press Ltd., Beaconsfield.

9

Cherts and siliceous sediments

9.1   Introduction

Chert is a very general term for fine-grained siliceous sediment, of chemical, biochemical or biogenic origin. It is usually a dense, very hard rock, which splinters with a conchoidal fracture when struck. Most cherts are composed of fine-grained silica, and the majority contain only small quantities of impurities. Certain types of chert have been given specific names. For example, flint is used frequently as a synonym for chert and more specifically for chert nodules occurring in Cretaceous chalks. Jasper refers to a red variety of chert, its colour being due to finely disseminated hematite. Chert of this type is interbedded with iron minerals to form jaspilite in the Precambrian iron-formations (Section 6.5.1). Porcelanite refers to fine-grained siliceous rocks with a texture and fracture similar to unglazed porcelain. The term porcelanite is also used for an opaline-claystone composed largely of opal-CT (Section 9.3.2).

Cherts in the geological record are usually divided into bedded and nodular types. Bedded cherts are frequently associated with volcanic rocks and the 'chert problem' has centered on a volcanic versus biogenic origin of the silica. The modern equivalents of many ancient bedded cherts, the radiolarian and diatom oozes, cover large areas of the deep ocean floors. Nodular cherts are mainly developed in limestones and to a lesser extent in mudrocks and evaporites. For the most part, nodular cherts are diagenetic, having formed by replacement, whereas many bedded cherts are primary accumulations. Siliceous sediments are also being deposited in lakes.

9.2   Chert petrology

Bedded and nodular cherts consist of three main types of silica: microquartz, megaquartz and chalcedonic quartz (Fig. 9.1). Microquartz consists of equant quartz crystals only a few microns across. Megaquartz crystals are larger, reaching 500 μm or more in size; the crystals have unit extinction and often possess good crystal shapes and terminations. Megaquartz is often referred to as drusy quartz as it frequently occurs as a pore-filling cement. Chalcedonic quartz is a fibrous variety with crystals varying from a few tens to hundreds of μm in length. They often occur in a radiating arrangement, forming wedge-shaped, mammillated and spherulitic growth structures (Fig. 9.2). Most chalcedonic quartz is length-fast (chalcedonite) but a length-slow variety, quartzine, also occurs. The latter is rare, but where found it is often in association with replaced evaporites (Section 5.5).

Radiolarians (marine zooplankton with a range of Cambrian to Recent),

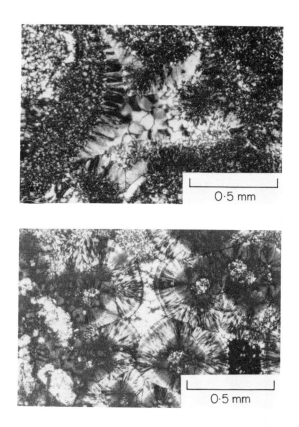

Fig. 9.1 Photomicrograph showing microquartz: finely crystalline mosaic with pin-point extinction, megaquartz: larger quartz crystals in central part, and chalcedonic quartz in a fibrous fringe. In this case, the microquartz is a replacement of carbonate grains and the megaquartz and chalcedonic quartz are pore filling. Crossed polars. Lower Carboniferous; Glamorgan, Wales.

0·5 mm

Fig. 9.2 Photomicrograph of chalcedonic quartz in spherulitic growth structure, also some microquartz and megaquartz. Crossed polars. Chert in limestone, Coral Rag, Jurassic, Yorkshire, England.

0·5 mm

diatoms (marine and non-marine phytoplankton, Triassic to Recent) and siliceous sponges (marine and non-marine, Cambrian to Recent) are composed of opaline silica. This is an isotropic amorphous variety, containing up to 10% water. Opaline silica is metastable so that it decreases in abundance back through time and is absent from Palaeozoic cherts. Radiolarians and diatoms have disc-shaped, elongate and spherical tests with spines and surface ornamentation (Fig. 9.3). They range in size from a few tens to hundreds of μm. Sponge spicules are a similar size, and give circular and elongate sections in rock slices.

9.3   Bedded cherts

9.3.1   SILICEOUS OOZES AND BEDDED CHERTS

Radiolarian and diatom oozes are accumulating on the ocean floors at the present time (Lisitzin, 1972). They occur especially where there is high organic productivity in near surface waters and this is controlled largely by oceanographic factors of upwelling and nutrient supply. Diatoms dominate in siliceous oozes around Antarctica and in the northern Pacific. Radiolarian-rich oozes occur in the equatorial regions of the Pacific and Indian Oceans. It is likely that in pre-late Mesozoic times, radiolarians occupied the niches of diatoms and so were more widely distributed than at present. The siliceous

211

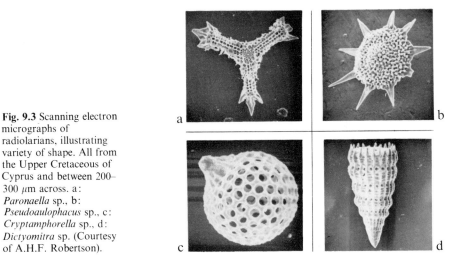

oozes are preferentially deposited in abyssal areas where depths exceed the carbonate compensation depth (CCD, Section 4.10.2), around 4.5 km in the central Pacific. Siliceous oozes form at shallower depths where surface oceanic waters are fertile and there is a paucity of calcareous plankton and terrigenous detrital material. Diatomaceous sediments accumulating at depths of less than 1.5 km in the Gulf of California are of this type (Calvert, 1966). The depth at which silica itself dissolves rapidly, the opal compensation depth, is around 6 km. Because of dissolution during settling, only a small percentage of diatoms and radiolarians actually reach the ocean floor to form sediments. Laminations within siliceous oozes are reflections of seasonal variations in plankton productivity and/or seasonal influxes of detrital clay.

Ancient bedded cherts frequently occur in mountain belts and other zones of deformed rocks. The bedding, which is a characteristic of these cherts, is generally on the scale of several centimetres, with millimetre-thick beds or partings of shale between (Fig. 9.4). Many chert beds are massive with no internal sedimentary structures. Although recrystallization has frequently occurred, structureless beds may well have resulted from the slow and steady deposition of the silica. Some chert beds show graded bedding, parallel and small-scale cross lamination, and basal scour structures (Nisbet & Price, 1974).

Fig. 9.4 Bedded chert, with thin shale partings. Lower Carboniferous, Montagne Noire, S. France.

212

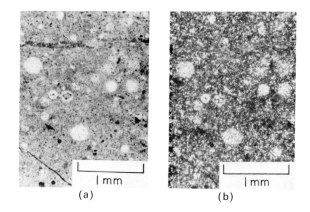

Fig. 9.5 Photomicrographs of bedded chert, composed of microquartz, with radiolarians preserved as microquartz and megaquartz. a: plane polarized light, b: crossed polars. Ordovician, Ballantrae, S.W. Scotland.

(a)　　　　　(b)

Cherts with these features have clearly been deposited by currents, probably turbidity currents derived from some nearby topographic high where the siliceous sediments were first deposited. Bedded cherts occasionally show slump folding and contemporaneous brecciation as a result of instability and mass sediment movement during sedimentation.

Many bedded cherts consist of radiolarians to the exclusion of other biogenic material. The radiolarians are generally poorly preserved, consisting of quartz-filled moulds contained in a matrix of microquartz (Fig. 9.5). Sponge spicules are common in some radiolarites. Fine clastic and carbonate sediment may be present in the cherts, and with increasing concentrations the cherts pass into siliceous shales or limestones.

Some bedded cherts are associated with volcanic rocks, others are not. Where there is a volcanic association the cherts were frequently deposited within or above pillow lavas. Lava flows and volcaniclastic sediments may be intercalated, as well as horizons of black shales and pelagic limestones. In some cases ultramafic rocks and dyke complexes are also present, so that the whole igneous-sedimentary assemblage constitutes an ophiolite suite (Coleman, 1977). Ophiolites are generally accepted as fragments of ocean floor. Classic Mesozoic examples of the volcanic-chert association occur in the Franciscan rocks of the Californian Coast Ranges (Blake & Jones, 1974), in the Ligurian Apennines of northern Italy (Abbate & Sagri, 1970; Folk & McBride, 1978) and in the Troodos Massif, Cyprus (Robertson & Hudson, 1974). Palaeozoic examples occur in the Cambrian and Ordovician of Newfoundland (Dewey & Bird, 1971) and Ballantrae, S.W. Scotland.

Bedded cherts independent of volcanics are typically associated with pelagic limestones, and siliciclastic and carbonate turbidites. Such deposits are typical of ancient passive continental margin sequences and frequently rest upon foundered platform carbonates. Examples occur in the Mesozoic of the Tethyan region, the Eastern Alps in particular (Garrison & Fischer, 1969; Schlager & Schlager, 1973; Bernoulli & Jenkyns, 1974) and the Devonian Caballos Formation of Texas (Folk & McBride, 1976). During the Lower Carboniferous, radiolarian cherts were widely deposited in basinal areas of western Europe, in S.W. England and central Germany especially.

By analogy with modern siliceous oozes, many ancient radiolarian-rich

213

bedded cherts are interpreted as very deep water in origin, having been deposited below the carbonate compensation depth (CCD), at depths of several kilometres (Garrison & Fischer, 1969). Although this may be so for some cherts, they could have been deposited at much shallower depths if there was no calcareous plankton available for sedimentation. This may well have been the case during the Palaeozoic and well into the Mesozoic, since the main calcareous planktonic organisms, coccoliths and Foraminifera, did not evolve until the Mesozoic. Variations in the CCD could also have permitted siliceous sediments to form in shallower water.

Siliceous sediments rich in diatoms are common in the Miocene and Pliocene of the circum-Pacific area and in localized regions of the Mediterranean. These diatomites, which are often bituminous and phosphatic, formed in relatively small back-arc and rifted basins which were starved of terrigenous sediment but were the sites of vigorous upwelling and thus phytoplankton productivity. Diatomites of the Monterey Formation, California, are of this type (Bramlette, 1946) and the diatom oozes of the Gulf of California are a modern analogue. In some regions, these sediments are important hydrocarbon source rocks.

The depositional environments of pelagic sediments, including cherts, of the present and past, have been reviewed by Jenkyns (1978).

9.3.2   DIAGNESIS AND THE FORMATION OF BEDDED CHERTS

The chert-igneous rock association has given rise to much discussion on the origin of cherts. The two alternative views are (i) that the cherts are entirely biogenic in origin, unrelated to igneous activity and (ii) that the cherts are a product of submarine volcanism either directly through inorganic precipitation of silica derived from subaqueous magmas or indirectly through plankton blooms induced by submarine volcanism. A better understanding of submarine volcanism in recent years, through plate tectonic theory, has made a volcanic-sedimentary origin of cherts less likely. Seafloor volcanism is restricted to oceanic ridges and localized 'hot-spots' and so is unlikely to give rise to areally-extensive cherts. In fact hydrothermal silica is precipitated close to vents, but it is quantitatively insignificant. In addition, the occurrence of radiolarian cherts in non-volcanic sequences and the dominantly biogenic origin of modern siliceous sediments, controlled by oceanographic factors, indicate that the formation of cherts is not related to contemporaneous volcanism (see reviews by Garrison, 1974 and Wise & Weaver, 1974).

Cores collected during deep sea drilling have permitted detailed studies of the siliceous ooze to chert transformation. Cores from the ocean floors of the Pacific (Heath & Moberly, 1971; Lancelot, 1973; Keene, 1975) and Atlantic (von Rad & Rosch, 1974) have encountered well-indurated cherts in Pliocene and older sections. Chert is particularly widespread in the Eocene of the North Atlantic. Contributions to silica diagenesis have also come from studies of cherts exposed on land (Robertson, 1977).

From the biogenic amorphous opal, frequently referred to as opal-A, the first diagenetic stage is the development of crystalline opal, identified by X-ray diffraction and referred to as opal-CT, also called disordered cristobalite,

214

Fig. 9.6 Lepispheres of opal-CT growing in voids in silicified Eocene chalk from the Arabian Sea. Sample from 630 m below seafloor. Prismatic crystals are of clinoptilolite (a zeolite). SEM micrograph. Courtesy of Albert Matter and Deep Sea Drilling Project, Scripps O.I., California.

alpha-cristobalite or lussatite (Jones & Segnit, 1971). Opal-CT consists of an interlayering of cristobalite and tridymite, and its disordered nature probably results from the small crystal size and incorporation of cations into the crystal lattice. Opal-CT replaces radiolarian and diatom skeletons and is precipitated as bladed crystals lining cavities and forming microspherules (5-10 μm diameter) called lepispheres (Fig. 9.6). Further diagenesis results in the metastable opal-CT being converted to quartz chert, mostly an equant mosaic of microquartz crystals but also chalcedonic quartz. This recrystallization of opal-CT to quartz obliterates the structure of many diatom and radiolarian tests.

The driving forces behind chert formation from biogenic opal-A are the solubility differences and the chemical conditions. Biogenic silica has a solubility of 120–140 ppm, cristobalite of 25–30 ppm, and quartz of 6–10 ppm in the pH ranges of marine sediment porewater (Fig. 9.7). Once the metastable opal-A dissolves the solution is saturated with respect to opal-CT and quartz. The precipitation of opal-CT in preference to quartz is probably due to the more internally-structured nature of quartz, which would require slow precipitation from less concentrated solutions. Temperature is also involved; with a rise in temperature, as through increasing depth of burial, the rate of transformation of opal-CT to quartz increases substantially (Ernst & Calvert, 1969). The formation of chert from opal-CT has been referred to as a 'maturation' process (reviewed in Wise & Weaver, 1974). A similar sequence was deduced by Bramlette, in 1946, for the Miocene Monterey Formation of California, where the term porcelanite or opaline-claystone is used for the metastable precursor to chert. Further studies have shown that the maturation of opal-A to quartz depends on the nature of the host sediment and on the chemical conditions (Lancelot, 1973; Kastner et al., 1977; Robertson, 1977). The presence of excess alkalinity in the sediments, as occurs where there is much calcareous material, favours the initial opal-CT precipitation and enhances the rate of transformation of opal-CT to quartz. Where there is much clay in the sediment, opal-CT contains abundant impurities, mainly foreign cations, which retard the maturation to quartz. This 'impurity-controlled' maturation

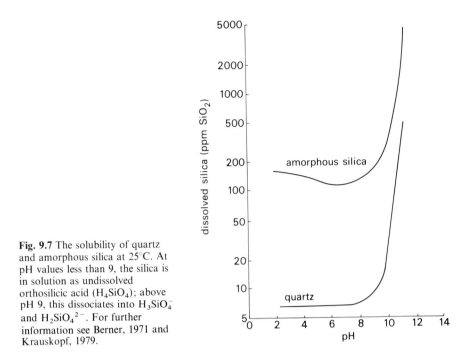

Fig. 9.7 The solubility of quartz and amorphous silica at 25°C. At pH values less than 9, the silica is in solution as undissolved orthosilicic acid (H_4SiO_4); above pH 9, this dissociates into $H_3SiO_4^-$ and $H_2SiO_4^{2-}$. For further information see Berner, 1971 and Krauskopf, 1979.

hypothesis has been put forward to explain differences in silica mineralogy and radiolarian preservation in cherts from Cyprus (Robertson, 1977). The end-product of these processes of silica solution, reprecipitation and replacement, is a mosaic of microquartz and chalcedonic quartz, with relatively few of the biogenic particles identifiable, although the original ooze was wholly composed of them.

Although most bedded cherts are now regarded as biogenic, some contain minerals thought to be volcanic alteration products, e.g. montmorillonite, palygorskite, sepiolite and clinoptilolite. The devitrification of volcanic glass and clay transformations, such as montmorillonite to illite, do liberate silica. However, in most, if not all Phanerozoic bedded cherts, there is at least some preservation of siliceous microfossils. This can be taken to indicate a dominantly biogenic origin, with most microfossils destroyed through the solution-reprecipitation of the maturation process.

9.4   Nodular cherts

Nodular cherts occur predominantly in carbonate host rocks. They are small to large, subspherical to irregular nodules, often concentrated along particular bedding planes; they may coalesce to form near-continuous layers, when they resemble bedded cherts. Nodular cherts are common in shelf limestones such as the Lower Carboniferous limestones of Britain (Orme, 1974) and North America (Meyers, 1977). They are also common in pelagic carbonates, with many examples occurring in the Cretaceous and Tertiary of the Alpine-Mediterranean-Tethys region. In the Cretaceous Chalk of western Europe and southeastern U.S.A., nodules of chert (flint) are common (Fig. 9.8) and many

216

Fig. 9.8 Chert nodules (flint) in Cretaceous chalk of Yorkshire, England. Sutured stylolites are present in the chalk.

Fig. 9.9 Silicified oolite. Ooids have been replaced by microquartz and are enclosed in a megaquartz mosaic. Crossed polars. Cambrian, Pennsylvania, U.S.A.

0·5 mm

have developed in burrow fills and nucleated around fossils. Flint pebbles derived from the Chalk are a major constituent of Tertiary and Quaternary gravels. Nodular cherts have been recovered from Miocene and older deep-sea chalks and pelagic limestones in cores from the ocean floors.

As with bedded cherts the origin of nodular cherts has also been much discussed. The older view involved the direct precipitation of silica from seawater to form blobs of silica gel on the seafloor which hardened into chert nodules. Nodular cherts in limestones, however, contain much evidence to demonstrate a replacement and thus diagenetic origin. Within the nodules, originally calcareous grains such as ooids and skeletal debris are preserved in silica (Fig. 9.9). Bedding structures such as lamination may be preserved in the nodules. The diagenetic processes involved in chert nodule formation are thought to be similar to those operating in bedded cherts. Biogenic silica disseminated in the sediments dissolves and is reprecipitated in the form of opal-CT at nodule growth points. Pore spaces are first infilled with opal-CT lepispheres and then carbonate skeletal and matrix replacement by opal-CT follows. Maturation of the opal-CT to microquartz and chalcedonic quartz takes place from the nodule centre outwards. It can often be demonstrated that the microquartz has formed by replacement of carbonate and that the chalcedonic quartz and megaquartz are dominantly pore-filling (Figs. 9.1, 9.9). From cherts in Mississippian limestones of New Mexico, it has been shown that microquartz and chalcedonic quartz formed by replacement

217

(maturation) of opal-CT and megaquartz was a later, direct precipitate (Meyers, 1977; Meyers & James, 1978).

The source of biogenic silica in shelf limestones is largely sponge spicules, whereas in deeper water pelagic limestones, the silica is largely supplied by radiolarians and diatoms. With the Cretaceous chalks of western Europe, a relatively shallow-water pelagic carbonate, silica for the flint nodules was derived mainly from sponges.

Chert nodules can also form by the replacement of evaporites, particularly anhydrite. Length-slow chalcedonic quartz is common in these occurrences (Section 5.5).

9.5   Non-marine siliceous sediments and cherts

Biogenic and inorganic siliceous sediments can form in lakes and ephemeral water bodies. Diatoms, which can occur in great abundance in lakes, form diatomaceous earths or diatomites. Such sediments are accumulating in many sediment-starved, higher latitude lakes such as Lake Lucerne, Switzerland and Lake Baikal, USSR. During the Pleistocene diatomites were deposited in many late-glacial and post-glacial lakes in Europe and North America.

Inorganic precipitation of silica can take place where there are great fluctuations in pH. Quartz, with its low solubility in most natural waters, is not affected by pH until values exceed 9, then with increasing pH the solubility increases dramatically (Fig. 9.7). In ephemeral lakes of the Coorong district of South Australia, very high pH values, greater than 10, are seasonally reached through photosynthetic activities of phytoplankton. Detrital quartz grains and clay minerals are partially dissolved at these high pH values so that the lake waters become supersaturated with respect to amorphous silica. Evaporation of lake water and a decrease in pH causes silica to be precipitated as a gel of cristobalite, which would give rise to chert on maturation (Peterson & von der Borch, 1965). Another related inorganic process has been described from East African lakes (Eugster, 1969). In very alkaline, sodium carbonate-rich lake waters, silica is leached from volcanic rocks and rock fragments. Exceptionally high concentrations of silica (up to 2500 ppm) are attained and lowering of pH by freshwater influxes causes the silica to be precipitated as magadiite, a metastable hydrated sodium silicate. This is converted to chert in a relatively short time. Cherts in ancient ephemeral lake sequences, such as the Eocene Green River Formation of Wyoming (Surdam & Wolfbauer, 1975), may well have formed via a sodium silicate precursor.

Further reading

Papers by Wise & Weaver, Rad & Rosch, Garrison, Nisbet & Price and Robertson & Hudson on modern and ancient bedded and nodular cherts in: Hsü, K.J. & Jenkyns, H.C. (Eds.) (1974) *Pelagic Sediments: on Land and under the Sea*, pp. 447. *Specl Publ. int. Ass. Sediment., 1.*
    Further papers of note include:
Robertson, A.H.F. (1977) The origin and diagenesis of cherts from Cyprus. *Sedimentology*, **24**, 11–30.
Meyers, W.J. (1977) Chertification in the Mississippian Lake Valley Formation, Sacramento Mountains, New Mexico, *Sedimentology*, **24**, 75–105.
Folk, R.L. & McBride, E.F. (1978) Radiolarites and their relation to subjacent 'oceanic crust' in Liguria, Italy. *J. sedim. Petrol* **48**, 1069–1102.

10

Volcaniclastic Sediments

10.1   Introduction

Volcaniclastic sediments are those composed chiefly of grains of volcanic origin, derived from contemporaneous volcanicity. One type, a pryoclastic deposit, is produced by explosive volcanic activity; other volcaniclastic deposits result from different processes. Sediments derived from erosion of pre-existing volcanic rocks are types of lithic sandstone (Section 2.6.1).

The term *tephra* is applied to any material, regardless of size, ejected from volcanoes. Tephra are chiefly derived from the magma itself, in which case they are usually composed of volcanic glass, but they may be crystals, if the magma had begun to crystallize before its explosive eruption. Tephra also include lithic fragments, of lava from earlier eruptions (accessory components) and of country rock (accidental components). Tephra are subdivided on particle size into volcanic dust, ash and lapilli (Table 10.1). Large fragments of magma ejected in a fluid state are bombs. Some have an elongated ellipsoidal shape as a result of rotation during flight. Large fragments of previously-solidified lava and country rock are blocks. Many tephra are in the form of pumice, highly vesicular volcanic glass which may have a porosity of more than 50%. When formed from more basic magma, the term scoria is used instead of pumice. Fragmentation of pumice gives rise to glass shards. These are a common component of many volcaniclastic sediments and can be recognized by their sickle, lunate or Y-shape, the concave surfaces representing the broken glass bubble walls (Fig. 10.1). Glass shards are more common in acidic ashes; in more basic volcaniclastics, the glass is present as droplets. In thin section, glass has a pale to yellow colour and is isotropic. Devitrified glass may show a weak

Table 10.1 Classification of volcaniclastic grains and sediments, based on grain size.

volcaniclastic grains (tephra)	volcaniclastic sediments
bombs-ejected fluid	agglomerate
blocks-ejected solid	volcanic breccia
——— 64 mm ———	
lapilli	lapilli-stone
——— 2 mm———	
ash	tuff — vitric / lithic / crystal
——— .06 mm———	
dust	

219

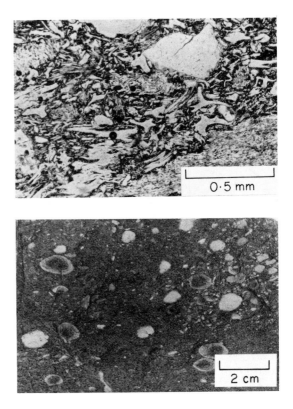

Fig. 10.1 Glass shards with characteristic broken bubble wall shape, broken crystals and streaked devitrified glass, called fiamme (lower right). Plane polarized light. Ignimbrite, Tertiary, New Zealand. Courtesy of M.H.Battey.

0·5 mm

Fig 10.2 Accretionary lapilli. Borrowdale Volcanics, Ordovician; Cumbria, England.

2 cm

birefringence if replaced by clays or zeolites (Section 10.5). Crystals ejected from volcanoes are usually euhedra or broken euhedra and they may be zoned. Quartz and feldspar crystals are common and in more basic ashes pyroxenes may occur. One particular type of lapilli consists of a spherical to ellipsoidal body composed of a nucleus, which may be a glass shard, crystal or rock fragment, that is enveloped by one or more laminae formed of volcanic ash and dust (Fig. 10.2). These accretionary lapilli, as they are called, form during both air-fall and flow processes.

The nature of volcanic explosions and eruptions depends on the volatile content (water and CO_2 especially) and viscosity of the magma. At depth, the volatiles are in solution in the magma but as the latter rises towards the Earth's surface, the gases exsolve and expand. This causes vesiculation of the magma and produces frothy and foam-like magma, which on ejection and solidification gives rise to the pumice. Acid magmas contain a higher percentage of volatiles over basic magmas and they are also more viscous, so that acid magmas give rise to more widespread tephra deposits.

Volcaniclastic sediments composed of ash are tuffs; those of lapilli are lapilli-stones, those of bombs and/or blocks are agglomerates and volcanic breccias (Table 10.1). Intermediate varieties are lapilli-tuff and tuff-breccia. On the basis of the proportions of glass, crystals and rock fragments, tuffs can be divided into vitric, crystal and lithic types. Terrigenous clastic or carbonate sediments containing volcanic ash can be described as tuffaceous, or referred to as tuffites. Volcaniclastic sediments are best discussed in terms of their modes of

Table 10.2 Main types of volcaniclastic deposit.

A)	pyroclastic-fall deposits: formed of tephra ejected from vent
B)	volcaniclastic flow deposits (and type of flow) (i) ignimbrites (fluidized ash-flows) (ii) base-surge deposits (ash-laden steam flows) (iii) lahar deposits (mudflows)
C)	hyaloclastites: fragmented and granulated lava through contact with water

formation. The principal types are pyroclastic-fall deposits, volcaniclastic flow deposits and hyaloclastites (Table 10.2).

10.2   Pyroclastic-fall deposits

These sediments are simply formed through the fallout of volcanic fragments ejected from a vent or fissure. In the majority of cases, explosive volcanoes are subaerial and so many air-fall tuffs are too, but if there are subaqueous environments nearby these will also receive the pyroclastic material. The characteristic features of air-fall deposits are a gradual decrease in both bed thickness and grain size away from the site of eruption. Blocks and bombs are deposited relatively close to the vent, whereas ash may be carried many tens of kilometres and dust thousands of kilometres away from the vent. Individual beds of air-fall material typically show normal grading of particles, although in other cases, inverse grading of pumice and lithic clasts has occurred (Self, 1976). Where deposition takes place in water quite large fragments of low density pumice may occur towards the top of an air-fall bed as a result of the pumice floating on the water surface before deposition. Particle size analysis can be used in the interpretation of pyroclastic fall deposits (Walker, 1973) and to distinguish between them and volcaniclastic flow deposits (Fig. 10.3; Walker, 1971; Bond & Sparks, 1976).

A further feature of air-fall deposits is the development of mantle bedding whereby the tephra layer follows and blankets any original topography. The

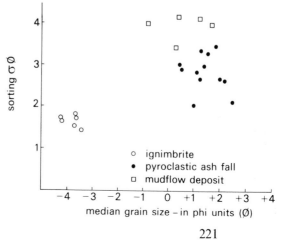

Fig. 10.3 An illustration of the use of grain size parameters to distinguish between volcaniclastic sediments of different origin. Median grain size/sorting scatter diagram for ignimbrite, pyroclastic ash fall and mudflow deposits of the Minoan eruption (1470 B.C.), Santorini, Greece. Data from Bond & Sparks (1976).

○ ignimbrite
● pyroclastic ash fall
□ mudflow deposit

impact of bombs may depress or rupture the bedding to produce 'bomb-sags'. While in the air, pyroclastic material is strongly affected by any prevailing wind so that air-fall deposits are frequently concentrated in a particular direction away from the volcanic vent. Reworking of air-fall deposits can take place in the subaerial environment by wind and rain, and in the submarine environment by waves and tidal currents. Air-fall pyroclastic deposits can be very important in stratigraphic correlation. Geologically they represent instantaneous deposition and they can cover vast areas. See for example Smith & Nash (1976).

10.3   Volcaniclastic flow deposits

In the subaerial environment, pyroclastic flows are of two principal types: those arising from fluidization by magmatic gas and giving rise to ignimbrites, and those formed through base surge mechanisms where the flows consist of ash-laden steam moving at hurricane or even greater velocities. A further type of subaerial flow is a lahar or volcanic mudflow.

10.3.1   IGNIMBRITES

With fluidized ash-flows, the fluidization (an upward movement of gas or water causing the particles to behave as a fluid) is brought about by the expansion of gases exsolved from the magma and of air caught up in the advancing flow. The flows, which are a component of the *nuée ardente* type of eruption, can travel for great distances (up to 100 km), even over flat ground. The deposits of these flows, *ignimbrites,* are characterized by a homogeneous appearance with little sorting of the finer ash particles, but if coarse lithic clasts are present they may show normal size grading and large pumice fragments are commonly reversely graded (Fig. 10.4; Sparks, 1976). There is generally a lack of internal stratification. Ignimbrites do not mantle the topography, but tend to follow valleys and low ground. Evidence for flow is provided by flattened and stretched pumice and glass fragments, termed fiamme (Fig. 10.1). Pumice clasts

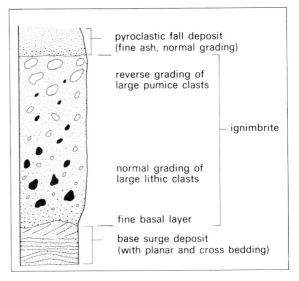

pyroclastic fall deposit
(fine ash, normal grading)

reverse grading of
large pumice clasts

ignimbrite

normal grading of
large lithic clasts

fine basal layer

base surge deposit
(with planar and cross bedding)

Fig. 10.4 Diagrammatic section through the deposits of an ignimbrite eruption. Base surges are not always associated with ignimbrite flows. After Sparks, 1976.

222

may be concentrated on the top surface of an ignimbrite. After deposition, the hot, plastic ash particles in the central part of the ignimbrite typically become welded together to form a dense rock. The term welded tuff is thus also used for ignimbrite. As a result of the more rapid heat loss, the lower and upper parts of an ignimbrite are usually non-welded and so have a higher porosity. Ignimbrite beds in a sequence may show considerable variation in degree of welding, thickness and grain size. Ignimbrites grade laterally into loose, poorly welded pumiceous tuffs. Fine air-fall ashes are frequently associated with ignimbrites (Fig. 10.4), and may be the deposits of turbulent dust clouds which are usually part of *nuée ardente*-type eruptions. Base-surge deposits (see below) may form a basal unit to an ignimbrite (Fig.10.4; Sparks, 1976).

Ignimbrites are typically derived from acid magmas and indeed they may be difficult to distinguish from rhyolite lavas. The ignimbrites can form substantial thicknesses of tuff; individual flows are frequently 10 m or more thick. Ignimbrites may also be intrusive into previously deposited pyroclastic flows. Well-developed ignimbrite sequences occur in Armenia, New Zealand and New Mexico. For discussions of ignimbrites see Smith (1960), Ross & Smith (1961) and Cook (1966).

Related ash-flow tuffs can be deposited subaqueously. This can result from submarine eruptions, such as described by Fiske & Matsuda (1964) from Japan, or from a subaerially-erupted ignimbrite flow travelling into water, as has been described from the Ordovician of North Wales (Francis & Howells, 1973). Pyroclastic material deposited in water around shallow subaqueous or emergent volcanoes may be transported into deeper water through slumping and the generation of turbidity currents to produce graded ash beds. Fiske (1963) has described such subaqueous volcaniclastic flow deposits from the Tertiary of Washington, U.S.A., and Tassé *et al.* (1978) from the Archaean of Quebec, Canada.

10.3.2 BASE-SURGE DEPOSITS

Base surges are subaerial flows formed by phreatic and phreatomagmatic eruptions (where magma comes into contact with water). The base surge is a fast-moving turbulent mixture of pyroclastic particles, water and gas (steam), The distinguishing feature of base-surge deposits is the presence of stratification: planar and cross bedding (Fig. 10.5). Antidune cross bedding, a

Fig. 10.5 Base-surge tuffs showing well-developed planar bedding and both upstream and downstream directed cross bedding (upper part) formed through antidune development. Flow from left to right. Pleistocene; Eifel, Germany. Courtesy of L. Viereck and G. Pritchard.

223

feature of high velocity flows (Section 2.3.2), is not uncommon in base-surge deposits (Schminke *et al.*, 1973). The low undulating bed forms are preserved and show up-flow directed cross bedding. With base-surge deposits the bed thickness and grain size decrease away from the source. Maximum bed thickness is generally less than 1 metre. Accretionary lapilli may develop during the surge as a result of accretion of volcanic dust and fine ash on to wet nuclei. Base surges are frequently associated with maar-type volcanoes, i.e. those which erupt in lakes.

10.3.3 LAHAR DEPOSITS

Lahars or volcanic mudflows occur on the slopes of some subaerial volcanoes. The mudflows arise chiefly from heavy rain falling on unconsolidated ash, this leads to a cold lahar, or from the eruption of ash into a crater lake which then overflows, giving a hot lahar. Lahar deposits have textures similar to those of mudflows on alluvial fans (Section 2.10.1) a lack of sorting and matrix-support fabric (Fig. 10.6).

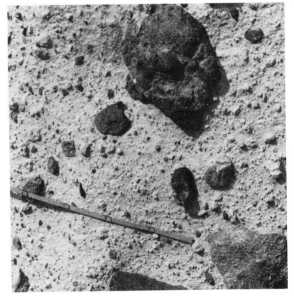

Fig. 10.6 Lahar deposit, consisting of large angular lithic clasts randomly scattered in a matrix of low density pumice and ash. From the Minoan eruption of Santorini, Greece. Courtesy of R.S.J. Sparks and Geological Society, London.

10.4 Hyaloclastites

When extruding lava comes into contact with water, then the rapid chilling and quenching causes fragmentation of the lava. The surface of the lava flow is chilled and as the flow moves, the surface rind is fragmented and granulated, allowing more magma to be chilled and fragmented. Volcaniclastic sediments produced through this process are known as hyaloclastites or aquagene tuffs (Carlisle, 1963). Such tuffs can be produced where a lava is extruded directly into water, where a subaerially-erupted lava flows into water, or where lava is erupted subglacially (often the case in Iceland). In shallow water, apart from chilling, lava fragmentation is also brought about by exsolution of gases from

Fig. 10.7 Hyaloclastite consisting of coarse, angular fragments of basaltic lava. Mid-Proterozoic, West Finnmark, Norway. Courtesy of T.C. Pharoah.

Fig. 10.8 Hyaloclastite consisting of graded units of lava fragments, deposited by density currents. Mid-Proterozoic, West Finnmark, Norway. Courtesy of T.C. Pharoah.

the magma, which causes vesiculation. The product is a hyaloclastite of small vesicular lava fragments, typically a few millimetres to centimetres in diameter (Fig. 10.7). The fragments are generally flakes and chips of glass, although in the submarine environment they are often altered to palagonite (see next section) and cemented by zeolites or calcite. Many hyaloclastites are without any apparent stratification or any degree of sorting, although in shallow water they can be reworked by waves and currents to give sedimentary structures as with any other clastic sediment. They may also be transported into deep water through slumping and turbidity currents, to give graded beds (Fig. 10.8).

In deeper water, where the hydrostatic pressure exceeds the confining pressure of gases in the magma, lava fragmentation takes place purely by the sudden chilling. The product is a volcanic breccia consisting of clasts of very variable size, from whole lava pillows to small pieces of pillow rind.

Hyaloclastites and pillow lava breccias have been recovered from many sites on the ocean floor during deep sea drilling; see for example McKelvey & Fleet (1974) and Natland (1976). They are a characteristic feature of submarine basaltic volcanism and so occur in many sequences of active continental margin-oceanic settings and in greenstone belts of the Precambrian (e.g. Carlisle, 1963; Furnes, 1972).

225

10.5 Diagenesis of volcaniclastic sediments

Volcanic glass is metastable so that excepting special conditions it is not preserved in rocks older than mid-Tertiary. Volcanic glass is readily devitrified, altered and replaced during weathering and diagenesis. Volcaniclastic sediments can be difficult to recognize as a result of this alteration. The common alteration products are clay minerals and zeolites, and in modern submarine basaltic volcaniclastic sediments, palagonite.

The clay minerals which replace volcanic glass are mainly smectites, in particular montmorillonite and saponite in more basic ashes, and kaolinite in feldspathic ashes. Chlorite may also replace basic tuffs. Smectite-rich clay beds derived from the alteration of volcanic ash are known as bentonites (Grim & Güven, 1978). Some kaolinite-rich mudrocks called tonsteins are of volcanic ash origin. Apart from clay mineralogy the presence of some glass shards or their pseudomorphs, together with euhedral or zoned phenocrysts, especially of quartz, feldspar or pyroxene, will further confirm a volcanic origin. Trace elements can also be used. Spears & Rice (1973) for example demonstrated that a tonstein of volcanic origin from the British Carboniferous had a distince trace element composition, being enriched in Ga, Pb, Sn, Zn, Th and U, and depleted in Sr, Ba, V, Cr, Mn, Co, Ni and Cu, relative to average shale. Diagenetically-altered volcaniclastic deposits include the bentonites known as fuller's earths from the Mesozoic of Britain (Hallam & Sellwood, 1968) and the Ordovician and Cretaceous in the U.S.A. (Grim & Güven, 1978).

The common zeolites formed by alteration of volcanic glass are analcime, clinoptilolite, phillipsite, laumontite and mordenite. They are usually cryptocrystalline, occasionally fibrous. Zeolites are often developed where ash has fallen into alkaline lakes. Lacustrine sequences of the Eocene Green River Formation, western U.S.A. and the Quaternary of East Africa contain such zeolitic horizons. In fact, volcanic ash is not a prerequisite; if the chemistry of a saline alkaline lake is appropriate then zeolites will be precipitated. Examples include the Triassic Lockatong Formation of eastern U.S.A. (van Houten, 1962) and the present-day Lake Natron, Tanzania (Hay, 1966). The zeolite phillipsite is an important constituent of hemipelagic muds on the floor of the Pacific Ocean, having formed from palagonite.

Many submarine basaltic tuffs are altered from sideromelane, a translucent variety of glass, to palagonite, an amorphous yellow to orange mineraloid. Pillow lavas generally possess a rind of palagonite and many hyaloclastite grains have been completely altered to palagonite (hence the older term palagonite tuffs). Processes of alteration taking place during palagonitization are hydration of the glass, oxidation of iron, a loss of some silica and oxides such as Na and Mg, and an increase in K and Fe oxides. Palagonite itself is not a mineral, but an intergrowth of montmorillonite and phillipsite.

Finally, tuffs may be replaced by silica or calcite. Silica is released on the alteration of glass to clays and zeolites and this can be precipitated as chert by replacement and as a cement. Completely silicified tuffs have been referred to as porcelanite or halleflinta. Calcite is commonly a cement of tuffs, as other sediments, but it may also replace the volcanic grains.

Further Reading

Few textbooks treat volcaniclastic sediments in any detail, but for a review see Lajoie, J. (1979) in *Facies Models* (Ed. by R.G. Walker), Geoscience Canada.

For further information on pyroclastic-fall and volcaniclastic flow deposits refer to Ross & Smith (1961)

Bond, A. & Sparks, R.S.J. (1976) The Minoan eruption of Santorini, Greece. *J. geol. Soc. Lond.* **132,** 1–16.

Ross, C.S. & Smith, R.L. (1961) Ash-flow tuffs: their origin, geologic relations and identification. *Prof. Pap. U.S. geol. Surv.* **336,** 81pp.

Schmincke, H.U., Fischer, R.V. & Waters, A.C. (1973) Antidune and chute and pool structures in base surge deposits from the Laacher See area (Germany). *Sedimentology,* **20,** 553–574.

Sparks, R.S.J., (1976) Grain size variations in ignimbrites and implications for the transport of pyroclastic flows. *Sedimentology,* **23,** 147–188.

Walker, G.P.L. (1973) Explosive volcanic eruptions—a new classification scheme. *Geol. Radsch.* **62,** 431–446.

For hyaloclastites see:

Carlisle, D. (1963) Pillow breccias and their aquagene tuffs, Quadra Island, British Columbia. *J. Geol.* **71,** 48–71.

McKelvey, B.C. & Fleet, A.J. (1974 Eocene basaltic pyroclastics at site 253, Ninetyeast Ridge. In *Initial Reports of the Deep Sea Drilling Project,* 26 (T.A. Davies, B.P. Luyendyk *et al.*) pp. 553–565. U.S. Government Printing Office, Washington D.C.

Natland, J.H. (1976) Petrology of volcanic rocks dredged from seamounts in the Line islands. In *Initial Reports of the Deep Sea Drilling Project,* 33 (S.O. Schlanger, E.D. Jackson *et al.*) pp. 749–777. U.S. Government Printing Office, Washington D.C.

References

Abbate, E. & Sagri, M. (1970) The eugeosynclinal sequences. In: *Development of the Northern Apennines Geosyncline* (Ed. by G. Sestini). *Sedim. Geol.* **4**, 251–340.

Abbott, P.L., Minch, J.A. & Peterson, G.L. (1976) Pre-Eocene Paleosol south of Tijuana, Baja California, Mexico. *J. sedim. Petrol.* **46**, 355–361.

Adams, J.E. & Rhodes, M.L. (1960) Dolomitization by seepage refluxion. *Bull. Am. Ass. Petrol. Geol.* **44**, 1912–1920.

Adams, J.K., Groot, J.J. & Hiller, N.W. (1961) Phosphatic pebbles from the Brightseat Formation of Maryland. *J. sedim. Petrol.* **31**, 546–552.

Adey, W.H. & MacIntyre, I.G. (1973) Crustose coralline algae: a re-evaluation in the geological sciences. *Bull. geol. Soc. Am.* **84**, 883–904.

Aitken, J.D. (1967) Classification and environmental significance of cryptalgal limestones and dolomites, with illustrations from the Cambrian and Ordovician of south-western Alberta. *J. sedim. Petrol.* **37**, 1163–1178.

Alexandersson, E.T. (1972) Intragranular growth of marine aragonite and Mg-calcite: evidence of precipitation from supersaturated seawater. *J. sedim. Petrol.* **42**, 441–460.

Allan, J.R. & Matthews, R.K. (1977) Carbon and oxygen isotopes as diagenetic and stratigraphic tools: surface and subsurface data, Barbados, West Indies, *Geology* **5**, 16–20.

Allen, J.R.L. (1964) Primary current lineation in the Lower Old Red Sandstone (Devonian), Anglo-Welsh Basin. *Sedimentology* **3**, 89–108.

Allen, J.R.L. (1965a) The sedimentation and palaeogeography of the Old Red Sandstone of Anglesey, North Wales. *Proc. Yorks. geol. Soc.* **35**, 139–185.

Allen, J.R.L. (1965b) A review of the origin and characteristics of Recent alluvial sediments. *Sedimentology* **5**, 89–191.

Allen, J.R.L. (1966) On bedforms and palaeocurrents. *Sedimentology* **6**, 153–190.

Allen, J.R.L. (1968) *Current Ripples*, pp. 433. North Holland, Amsterdam.

Allen, J.R.L. (1970) Studies in fluviatile sedimentation: a comparison of fining-upwards cyclothems with special reference to coarse-member composition and interpretation. *J. sedim. Petrol.* **40**, 298–323.

Allen, J.R.L. (1971) Transverse erosional marks of mud and rock: their physical basis and geological significance. *Sedim. Geol.* **5**, 167–385.

Allen, J.R.L. (1975) Studies in fluviatile sedimentation: implications of pedogenic carbonate units, Lower Old Red Sandstone, Anglo-Welsh outcrop. *Geol. J.* **9**, 181–208.

Allen, J.R.L. (1977) *Physical Processes of Sedimentation*, pp. 248. Allen & Unwin, London.

Almon, W.R., Fullerton, L.B. & Davies, D.K. (1976) Pore space reduction in Cretaceous sandstones through chemical precipitation of clay minerals. *J. sedim. Petrol.* **46**, 89–96.

Altschuler, Z.S. (1965) Precipitation and recycling of phosphate in the Florida Land-Pebble Phosphate deposits. *Prof. Pap. U.S. geol Surv.* **525-B**, 91–95.

Ames, L.L. (1959) The genesis of carbonate apatites. *Econ. Geol.* **54**, 829–841.

Amstutz, G.C. (1958) Coprolites: a review of the literature and a study of specimens from southern Washington. *J. sedim. Petrol.* **28**, 498–508.

Anderson, R.Y., Dean, W.E., Kirkland, D.W. & Snide, H.I. (1972) Permian Castile varved evaporite sequence, West Texas and New Mexico. *Bull. geol. Soc. Am.* **83**, 59–86.

Anderson, R.Y. & Kirkland, D.W. (1966) Intrabasin varve correlation. *Bull. geol. Soc. Am.* **77**, 241–256.

Anderton, R. (1976) Tidal shelf sedimentation: an example from the Scottish Dalradian. *Sedimentology* **23**, 429–458.

Arthurton, R.S. & Hemingway, J.E. (1972) The St. Bees Evaporites—a carbonate-evaporite formation of Upper Permian age in West Cumberland, England. *Proc. Yorks. geol. Soc.* **38**, 565–592.

Assereto, R.L.A.M. & Kendall, C.G.St.C. (1977) Nature, origin and classification of peritidal tepee structures and related breccias. *Sedimentology* **24**, 153–210.

Atwood, D.K. & Bubb, J.N. (1970) Distribution of dolomite in a tidal flat environment, Sugarloaf Key, Florida. *J. Geol.* **78**, 499–509.

Baas-Becking, L.G.M., Kaplan, I.R. & Moore, D. (1960) Limits of the natural environment in terms of pH and oxidation-reduction potentials. *J. Geol.* **68**, 243–284.

Badiozamani, K. (1973) The Dorag dolomitization model-application to the Middle Ordovician of Wisconsin. *J. sedim. Petrol.* **43**, 715–723.

Bagnold, R.A. (1941) *The Physics of Blown Sand and Desert Dunes*, pp. 265. Methuen, London.

Baird, G.C. (1978) Pebbly phosphorites in shale: a key to recognition of a widespread submarine discontinuity in the Middle Devonian of New York. *J. sedim. Petrol.* **48**, 545–555.

Baker, E.T. (1973) Distribution and composition of suspended sediment in the bottom waters of the Washington Continental Shelf and Slope. *J. sedim. Petrol.* **43**, 812–821.

Ball, M.M. (1967) Carbonate sand bodies of Florida and the Bahamas. *J. sedim. Petrol.* **37**, 556–591.

Banks, N.G. (1970) Nature and origin of early and late cherts in the Leadville Limestone, Colorado. *Bull. geol. Soc. Am.* **81**, 3033–3048.

Basu, A., Young, S.W., Suttner, L.J., James, W.C. & Mack, G.H. (1975) Re-evaluation of the use of undulatory extinction and polycrystallinity in detrital quartz for provenance interpretation. *J. sedim. Petrol.* **45**, 873–882.

Bathurst, R.G.C. (1959) The cavernous structure of some Mississippian *Stromatactis* reefs in Lancashire, England. *J. Geol.* **67**, 506–521.

Bathurst, R.G.C. (1964) The replacement of aragonite by calcite in the molluscan shell wall. In: *Approaches to Paleocology* (Ed. by J. Imbrie and N.D. Newell), pp. 357–376. Wiley, New York.

Bathurst, R.G.C. (1966) Boring algae, micrite envelopes and lithification of molluscan biosparites. *Geol. J.* **5**, 15–32.

Bathurst, R.G.C. (1970) Problems of lithification in carbonate rocks. *Proc. geol. Ass.* **81**, 429–440.

Bathurst, R.G.C. (1975) *Carbonate sediments and their diagenesis*, pp. 658. Elsevier, Amsterdam.

Baturin, G.N. (1970) Recent authigenic phosphorite formation on the southwest African shelf. In: *The Geology of the East Atlantic Continental Margin* **1**, pp. 90-97. *Inst. geol. Sci. Rep.* 70/13.

Bausch, W.M. (1968) Clay content and calcite crystal size of limestones. *Sedimentology* **10**, 71–75.

Beard, D.C. & Weyl, P.K. (1973) The influence of texture on porosity and permeability of unconsolidated sand. *Bull. Am. Ass. Petrol. Geol.* **57**, 349–369.

Behrens, E.W. & Land, L.S. (1972) Subtidal Holocene dolomite, Baffin Bay, Texas. *J. sedim. Petrol.* **42**, 155–161.

Bell, D.L. & Goodell, H.G. (1967) A comparative study of glauconite and the associated clay fraction in modern marine sediments. *Sedimentology*, **9**, 169–202.

Benson, L.V. & Matthews, R.K. (1971) Electron microprobe studies of magnesium distribution in carbonate cements and recrystalised skeletal grainstones from the Pleistocene of Barbados, West Indies. *J. sedim. Petrol.* **41**, 1018–1025.

Berner, R.A. (1969) Goethite stability and the origin of red beds. *Geochim. Cosmochim. Acta* **33**, 267–273.

Berner, R.A. (1970) Sedimentary pyrite formation. *Am. J. Sci.* **268**, 1–23.

Berner, R.A. (1971) *Principles of Chemical Sedimentology*, pp. 256. McGraw-Hill, New York.

Bernoulli, D. & Jenkyns, H.C. (1974) Alpine, Mediterranean and Central Atlantic Mesozoic facies in relation to the early evolution of the Tethys. In: *Modern and Ancient Geosynclinal Sedimentation* (Ed. by R.H. Dott and R.H. Shaver), pp. 129–160. *Spec. Publ. Soc. econ. Paleont. Miner.* **19**, Tulsa.

Biggarella, J.J. (1972) Eolian environments: their characteristics, recognition and importance. In: *Recognition of Ancient Sedimentary Environments* (Ed. by J.K. Rigby and W.K. Hamblin), pp. 12–62. *Spec. Publ. Soc. econ. Paleont. Miner.* **16**, Tulsa.

Birch, G.F. (1979) Phosphatic rocks on the western margin of South Africa. *J. sedim. Petrol.* **49**, 93–110.

Bjørlykke, K. (1974) Geochemical and mineralogical influence of Ordovician island arcs on epicontinental clastic sedimentation. A study of Lower Palaeozoic sedimentation in the Oslo Region, Norway. *Sedimentology* **21**, 251–272.

Black, M. (1933) The algal sediments of Andros Island, Bahamas. *Phil. Trans. Roy. Soc. London, Ser. B.* **222**, 165–192.

Blake, M.C. & Jones, D.L. (1974) Origin of Franciscan mélanges in northern California. In: *Modern and Ancient Geosynclinical Sedimentation* (Ed. by R.H. Dott and R.H. Shaver), pp. 345–357. *Spec. Publ. Soc. econ. Paleont. Miner.* **19**, Tulsa.

Blanche, J.B. & Whitaker, J.H.McD. (1978) Diagenesis of part of the Brent Sand Formation (Middle Jurassic) of the northern North Sea Basin. *Jl. geol. Soc. Lond.* **135**, 73–82.

Blatt, H. (1967) Original characteristics of clastic quartz grains. *J. sedim. Petrol.* **37**, 401–424.

Blatt, H., Middleton, G.V. & Murray, R.C. (1980) *Origin of Sedimentary Rocks* pp. 634. Prentice-Hall, New Jersey.

Blount, D.N. & Moore, C.H. (1969) Depositional and non-depositional carbonate breccias, Chiantla Quadrangle, Guatemala. *Bull. Geol. Soc. Am.* **80**, 429–442.

Bluck, B.J. (1967) Deposition of some Upper Old Red Sandstone conglomerates in the Clyde area. A study in the significance of bedding. *Scott. J. Geol.* **3**, 139–167.

Bluck, B.J. & Kelling, G. (1963) Channels from the Upper Carboniferous coal measures of South Wales. *Sedimentology* **2**, 29–53.

Bohor, B.F. & Gluskoter, H.J. (1973) Boron in illites as an indicator of paleosalinity of Illinois coals. *J. sedim. Petrol.* **43**, 945–956.

Bond, A. & Sparks, R.S.J. (1976) The Minoan eruption of Santorini, Greece. *Jl. geol. Soc. Lond.* **132**, 1–16.

Borchert, H. & Muir, R.O. (1964) *Salt deposits: the origin, metamorphism and deformation of evaporites*, pp. 338. Van Nostrand, London.

Bosellini, A. & Hardie, L.A. (1973) Depositional theme of a marginal marine evaporite. *Sedimentology* **20**, 5–27.

Bouma, A.H. (1962) *Sedimentology of some flysch deposits: A graphic approach to facies interpretation*, pp. 168, Elsevier, Amsterdam.

Bouma, A.H. (1969) *Methods for the study of sedimentary structures*, pp. 458. Interscience, New York.

Bouma, A.H. & Brouwer, A. (Eds.) (1964) *Turbidites*, pp. 264. Elsevier, Amsterdam.

Bradley, W.H. (1929) Algae reefs and oolites of the Green River Formation. *Prof. Pap. U.S. geol. Surv.* **154**, 203–233.

Bradley, W.H. (1931) Origin and microfossils of the oil shale of the Green River Formation of Colorado and Utah. *Prof. Pap. U.S. geol. Surv.* **168**, pp. 58.

Braithwaite, C.J.R. (1968) Diagenesis of phosphatic carbonate rocks on Remire, Amirantes, Indian Ocean. *J. sedim. Petrol.*, **38**, 1194–1212.

Braithwaite, C.J.R. (1973) Settling behaviour related to sieve analysis of skeletal sands. *Sedimentology*, **20**, 251–262.

Braitsch, O. (1971) *Salt deposits, their origin and composition*, pp. 297. Springer-Verlag, Berlin.

Bramlette, M.N. (1946) The Monterey Formation of California and the origin of its siliceous rocks. *Prof. Pap. U.S. geol. Surv.* **212**, pp. 57.

Brenchley, P.J. (1969) Origin of matrix in Ordovician greywackes, Berwyn Hills, North Wales. *J. sedim. Petrol.* **39**, 1297–1301.

Brenchley, P.J. & Newall, G. (1977) The significance of contorted bedding in upper Ordovician sediments in the Oslo Region, Norway. *J. sedim. Petrol.* **47**, 819–833.

Brenchley, P.J., Newall, G. & Stanistreet, I.G. (1979) A storm surge origin for sandstone beds in an epicontinental platform sequence, Ordovician, Norway. *Sedim. Geol.* **22**, 185–217.

Bricker, O.P. (Ed.) (1971) *Carbonate cements*, pp. 376. Johns Hopkins Press, Baltimore.

Bridge, J.S. (1975) Computer simulation of sedimentation in meandering streams. *Sedimentology* **22**, 3–43.

Bromley, R.G. (1975) Trace fossils at omission surfaces. In: *The study of Trace fossils* (Ed. by R.W. Frey), pp. 399–428. Springer-Verlag, New York.

Broussard, M.L. (Ed.) (1975) *Deltas; Models for Exploration*, pp. 555. Houston Geol. Soc., Houston.

Bull, W.B. (1972) Recognition of alluvial-fan deposits in the stratigraphic record. In: *Recognition of Ancient Sedimentary Environments* (Ed. by K.J. Rigby and W.K. Hamblin), pp. 68–83. *Spec. Publ. Soc. econ. Paleont. Miner.* **16**, Tulsa.

Burnett, W.C. & Gomberg, D.N. (1977) Uranium oxidation and probable subaerial weathering of phosphatized limestone from the Pourtales Terrace. *Sedimentology* **24**, 291–302.

Burst, J.F. (1958) Glauconite pellets: their mineral nature and application to stratigraphic interpretation. *Bull. Am. Ass. Petrol Geol.* **42**, 310–327.

Burst, J.F. (1969) Diagenesis of Gulf Coast clayey sediments and its possible relation to petroleum migration. *Bull. Am. Ass. Petrol. Geol.* **53**, 73–93.

Butler, G.P. (1970) Holocene gypsum and anhydrite of the Abu Dhabi sabkha, Trucial Coast: an alternative explanation of origin. In: *Third Symposium on Salt* (Ed. by J.L. Rau and L.F. Dellwig), pp. 120–152. Northern Ohio Geol. Soc., Cleveland, Ohio.

Buyce, M.R. & Friedman, G.M. (1975) Significance of authigenic K-feldspar in Cambrian-Ordovician carbonate rocks of the Proto-Atlantic shelf in North America. *J. sedim. Petrol.* **45**, 808–821.

Buzzalini, A.D., Adler, F.J. & Jodey, R.L. (Eds.) (1969) Evaporites and petroleum. *Bull. Am. Ass. Petrol. Geol.* **53**, 775–1011.

Calvert, S.E. (1966) Accumulation of diatomaceous silica in sediments of the Gulf of California. *Bull. geol. Soc. Am.* **77**, 569–596.

Cant, D.J. & Walker, R.G. (1976) Development of a braided-fluvial facies model for the Devonian Battery Point Sandstone, Quebec. *Can. J. Earth Sci.* **13**, 102–119.

Cant, D.J. & Walker, R.G. (1978) Fluvial processes and facies sequences in the sandy braided South Saskatchewan River, Canada. *Sedimentology* **25**, 625–648.

Carlisle, D. (1963) Pillow breccias and their aquagene tuffs, Quadra Island, British Columbia. *J. Geol.* **71**, 48–71.

Carroll, D. (1958) Role of clay minerals in the transportation of iron. *Geochim. Cosmochim. Acta* **14**, 1–27.

Carver, R.E. (Ed.) (1971) *Procedures in Sedimentary Petrology,* pp. 653. John Wiley, New York.

Castano, J.R. & Sparks, D.M. (1974) Interpretation of vitrinite reflectance measurements in sedimentary rocks and determination of burial history using vitrine reflectance and authigenic minerals. *Spec. Pap. geol. Soc. Am.* **153**, 31–52.

Chilingar, G.V., Mannon, R.W. & Rieke, H.H. (Eds.) (1972) *Oil and Gas Production from Carbonate Rocks,* pp. 408. Elsevier, New York.

Chilingarian, G.V., Bissell, H.J. & Wolf, K.H. (1979) Diagenesis of carbonate sediments and epigenesis (or catagenesis) of limestones. In: *Diagenesis in sediments and sedimentary rocks* (Ed. by G. Larsen & G.V. Chilingarian), pp. 249–422 *Dev. in Sed.* **25A**. Elsevier, Amsterdam.

Chilingarian, G.V. & Yen, T.F. (Eds.) (1978) *Bitumens, asphalts and tar sands,* pp. 331. Elsevier, Amsterdam.

Chilingarian, G.V., Zenger, D.H., Bissell, H.J. & Wolf, K.H. (1979) Dolomites and dolomitization. In: *Diagenesis in sediments and sedimentary rocks* (Ed. by G. Larsen & G.V. Chilingarian), pp. 423–536. *Dev. in Sed.* **25A**. Elsevier, Amsterdam.

Choquette, P.W. & Pray, L.C. (1970) Geological nomenclature and classification of porosity in sedimentary carbonates. *Bull. Am. Ass. Petrol. Geol.* **54**, 207–250.

Chowns, T.M. & Elkins, J.E. (1974) The origin of quartz geodes and cauliflower cherts through silicification of anhydrite nodules. *J. sedim. Petrol.* **44**, 885–903.

Cita, M.B. & Wright, R. (Eds.) (1979) Geodynamic and biodynamic effects of the Messinian Salinity crisis in the Mediterranean. *Palaeogeogr., Palaeoclimatol., Palaeoecol.* **29**, pp. 222.

Clemmensen, L.B. (1978) Lacustrine facies and stromatolites from the Middle Triassic of East Greenland *J. sedim. Petrol.* **48**, 1111–1128.

Cloud, P.E. (1962) Environment of calcium carbonate deposition west of Andros Island, Bahamas. *Prof. Pap. U.S. geol. Surv.* **350**, pp. 138.

Cloud, P.E. (1973) Paleoecological significance of the Banded Iron-Formation. *Econ. Geol.* **68**, 1135–1143.

Coleman, J.M. (1976) *Deltas: processes of deposition and models for exploration,* pp. 102. Continuing Education Publication Co. Ltd., Champaign, Illinois.

Coleman, R.G. (1977) *Ophiolites,* pp. 229. Springer-Verlag, Berlin.

Collinson, J.D. (1978) Alluvial Sediments. In: *Sedimentary Environments and Facies* (Ed. by H.G. Reading), pp. 15–60. Blackwell Scientific Publications, Oxford.

Conybeare, C.E.B. & Crook, K.A.W. (1968) *Manual of sedimentary structures,* pp. 327. Australian Dept. Natl. Development. *Bull. Bur. Min. Res., Geol. and Geophysics 102.*

Cook, E.F. (Ed.) (1966) *Tufflavas and ignimbrites, a survey of Soviet studies,* pp. 212. American Elsevier, New York.

Cook, H.E. & Enos, P. (Eds.) (1977) *Deep-water Carbonate Environments,* pp. 336. *Spec. Publ. Soc. econ. Paleont. Miner.* **25**, Tulsa.

Cook, P.J. & McElhinny, M.W. (1979) A re-evaluation of the spatial and temporal distribution of sedimentary phosphate deposits in the light of plate tectonics. *Econ. Geol.* **74**, 315–330.

Cotter, E. (1965) Waulsortian-type carbonate banks in the Mississippian Lodgepole Formation of central Montana. *J. Geol.* **73**, 881–888.

Crimes, T.P. & Harper, J.C. (Eds.) (1970) *Trace Fossils,* pp. 547. Seel House Press, Liverpool.

Crimes, T.P. & Harper, J.C. (Eds.) (1977) *Trace Fossils 2,* pp. 351. Seel House Press, Liverpool.

Crook, K.A.W. (1974) Lithogenesis and geotectonics: the significance of compositional variation in flysch arenites (graywackes). In: *Modern and Ancient Geosynclinal Sedimentation* (Ed. by Dott, R.H. & Shaver, R.H.), pp. 304–310. *Spec. Publ. Soc. econ. Paleont. Miner.* **19**, Tulsa.

Curray, J.R. (1956) The analysis of two-dimensional orientation data. *J. geol.* **64**, 117–131.

Curtis, C.D. & Spears, D.A. (1968) The formation of sedimentary iron minerals. *Econ. Geol.* **63**, 257–270.

Dalrymple, D.W. (1966) Calcium carbonate deposition associated with blue-green algal mats, Baffin Bay, Texas. *Inst. Marine Sci. Publ.* **10**, 187–200.

Damberger, H.H. (1974) Coalification patterns of Pennsylvanian coal basins of the eastern United States. *Spec. Pap. geol. Soc. Am.* **153**, 53–74.

Davies, D.K., Ethridge, F.G. & Berg, R.R. (1971) Recognition of barrier environments. *Bull. Am. Ass. Petrol. Geol.* **55**, 550–565.

Davies, G.R. (1970a) Carbonate bank sedimentation, eastern Shark Bay, Western Australia. *Mem. Am. Ass. Petrol. Geol.* **13**, 85–168.

Davies, G.R. (1970b) Algal-laminated sediments, Gladstone Embayment, Shark Bay, Western Australia. *Mem. Am. Ass. Petrol. Geol.* **13**, 169–205.

Davies, G.R. (1977) Former magnesian calcite and aragonite submarine cements in Upper Palaeozoic reefs of the Canadian Arctic: a summary. *Geology* **5**, 11–15.

Davies, G.R. & Ludlam, S.D. (1973) Origin of laminated and graded sediments, Middle Devonian of Western Canada. *Bull. geol. Soc. Am.* **84**, 3527–3546.

Davies, P.J., Bubela, B. & Ferguson, J. (1978) The formation of ooids. *Sedimentology*, **25**, 703–730.

Davis, J.C. (1973) *Statistics and Data Analysis in Geology*, pp. 550. John Wiley, New York.

Davis, R.A. (Ed.) (1978) *Coastal Sedimentary Environments*, pp. 420. Springer-Verlag, New York.

Davis, R.A. & Ethington, R.L. (Eds.) (1976) *Beach and Nearshore Sedimentation*, pp. 187. *Spec. Publ. Soc. econ. Paleont. Miner.* **24**, Tulsa.

Dean, W.E., Davies, G.R. & Anderson, R.Y. (1975) Sedimentological significance of nodular and laminated anhydrite. *Geology* **3**, 367–372.

Dean, W.E. & Schreiber, B.C. (Eds.) (1978) Notes for a short course on marine evaporites. *Soc. econ. Paleont. Miner.*, Tulsa, *Short Course* No. 4.

Deffeyes, K.S., Lucia, F.J. & Weyl, P.K. (1965) Dolomitization of Recent and Plio-Pleistocene sediments by marine evaporite waters on Bonaire, Netherlands Antilles. In: *Dolomitization and limestone diagenesis* (Ed. by L.C. Pray and R.C. Murray), pp. 71–88. *Spec. Publ. Soc. econ. Paleont. Miner.* **16**, Tulsa.

Degens, E.T. (1965) *Geochemistry of Sediments*, pp. 342. Prentice-Hall, New Jersey.

Dellwig, L.F. (1955) Origin of the Salina Salt of Michigan. *J. sedim. Petrol.* **25**, 83–110.

Dewey, J.F. & Bird, J.M. (1971) Origin and emplacement of the ophiolite suite: Appalachian ophiolites in Newfoundland. *J. geophys. Res.* **76**, 3179–3206.

Dickinson, W.R. (1974) Plate tectonics and sedimentation. In: *Tectonics and sedimentation* (Ed. by W.R. Dickinson), pp. 1–27. *Spec. Publ. Soc. econ. Paleont. Miner.* **22**, Tulsa.

Dickinson, W.R., Helmold, K.P. & Stein, J.A. (1979) Mesozoic lithic sandstones in central Oregon. *J. sedim. Petrol.* **49**, 501–516.

Dickson, J.A.D. (1966) Carbonate identification and genesis as revealed by staining. *J. sedim. Petrol.* **36**, 491–505.

Dickson, J.A.D. & Coleman, M.L. (1980) Changes in carbon and oxygen isotope composition during limestone diagenesis. *Sedimentology* **27**, 107–118.

Didyk, B.M., Simoneit, B.R.T., Brassell, S.C. & Eglinton, G. (1978) Organic geochemical indicators of paleoenvironmental conditions of sedimentation. *Nature* **272**, 216–222.

Dietz, R.S., Emery, K.O. & Shepard, F.P. (1942) Phosphorite deposits on the sea-floor off southern California. *Bull. geol. Soc. Am.* **53**, 815–848.

Dimroth, E. (1968) Sedimentary textures, diagenesis and sedimentary environments of certain Precambrian Ironstones. *Abh. Neues. Jahrb. Geol. Paleont.* **130**, 247–274.

Dimroth, E. (1979) Models of physical sedimentation of iron formations. Diagenetic facies of iron formation. In: *Facies Models* (Ed. by R.G. Walker), pp. 175–189. Geoscience, Canada.

Dobkins, J.E. & Folk, R.L. (1970) Shape development of Tahiti-Nui. *J. sedim. Petrol.* **40**, 1167–1203.

Donovan, R.N. (1975) Devonian lacustrine limestones at the margin of the Orcadian Basin, Scotland. *J. geol. Soc. Lond.* **131**, 489–510.

Donovan, R.N. & Foster, R.J. (1972) Subaqueous shrinkage cracks from the Caithness Flagstone Series (Middle Devonian) of northeast Scotland. *J. sedim. Petrol.* **42**, 309–317.

Dott, R.H. (1964) Wacke, graywacke and matrix—what approach to immature sandstone classification? *J. sedim. Petrol.* **34**, 625–632.

Dott, R.H. Jr. & Shaver, R.H. (Eds.) (1974) *Modern and Ancient Geosynclinal Sedimentation*, pp. 380. *Spec. Publ. Soc. econ. Paleont. Miner.* **19**, Tulsa.

Doveton, J.H. (1971) An application of Markov Chain Analysis to the Ayrshire Coal Measures succession. *Scott. Jour. Geol.* **7**, 11–27.

Dravis, J. (1979) Rapid and widespread generation of Recent oolitic hardgrounds on a high energy Bahamian Platform, Eleuthera Bank, Bahamas. *J. sedim. Petrol.* **49**, 195–207.

Drever, J.I. (1974) Geochemical model for the origin of Precambrian banded-iron formations. *Bull. geol. Soc. Am.* **85**, 1099–1106.

Duff, P.McL.D., Hallam, A. & Walton, E.K. (1967) *Cyclic Sedimentation*, pp. 280. Elsevier, Amsterdam.

Dunham, R.J. (1962) Classification of carbonate rocks according to depositional texture. In: *Classification of Carbonate Rocks* (Ed. by W.E. Ham), pp. 108–121. *Mem. Am. Ass. Petrol. Geol.* **1**, Tulsa.

Dunoyer de Segonzac, G. (1970) The transformation of clay minerals during diagenesis and low-grade metamorphism: a review. *Sedimentology* **15**, 281–346.

Dzulynski, S. & Walton, E.K. (1965) *Sedimentary Features of Flysch and Greywackes,* pp. 274. Elsevier, Amsterdam.

Edwards, M.B. (1978) Glacial Environments. In: *Sedimentary Environments and Facies* (Ed. by H.G. Reading), pp. 416–438. Blackwell Scientific Publications, Oxford.

Edwards, M.B. (1979) Late Precambrian loessites from North Norway and Svalbard. *J. sedim. Petrol.* **49,** 85–91.

Eggleston, J.R. & Dean, W.E. (1976) Freshwater stromatolitic bioherms in Green Lake, New York. In: *Stromatolites* (Ed. by M.R. Walter), pp. 479–488. Elsevier, Amsterdam.

Ekdale, A.A., Ekdale, S.F. & Wilson, J.L. (1976) Numerical analysis of carbonate microfacies in the Cupido Limestone (Neocomian–Aptian), Coahuila, Mexico. *J. sedim. Petrol.* **46,** 362–368.

Elliott, T. (1976) The morphology, magnitude and regime of a Carboniferous fluvial-distributary channel. *J. sedim. Petrol.* **46,** 70–76.

Elliot, T. (1978) Deltas. In: *Sedimentary Environments and Facies* (Ed. by H.G. Reading), pp. 97–142. Blackwell Scientific Publications, Oxford.

Embry, A.F. & Klovan, J.E. (1971) A Late Devonian reef tract on northeastern Banks Island, Northwest Territories. *Bull. Can. Petrol. Geol.* **19,** 730–781.

Emery, K.O. (1968) Relict sediments on continental shelves of the world. *Bull. Am. Ass. Petrol. Geol.* **52,** 445–464.

Enos, P. (1969) Anatomy of a flysch. *J. sedim. Petrol.* **39,** 680–723.

Ernst, W.G. & Calvert, S.E. (1969) An experimental study of the recrystallization of porcelanite and its bearing on the origin of some bedded cherts. *Am. J. Sci.* **267-A,** 114–133.

Eugster, H.P. (1969) Inorganic bedded cherts from the Magadi area, Kenya. *Contr. Mineral. and Petrol.* **22,** 1–31.

Eugster, H.P. & Chou, I-Ming (1973) The depositional environments of Precambrian Banded Iron-Formations. *Econ. Geol.* **68,** 1144–1168.

Evamy, B.D. (1967) Dedolomitization and the development of rhombohedral pores in limestones. *J. sedim. Petrol.* **37,** 1204–1215.

Evamy, B.D. (1969) The precipitational environment and correlation of some calcite cements deduced from artificial staining. *J. sedim. Petrol.* **39,** 787–793.

Fairbridge, R.W. (1967) Phases of diagenesis and authigenesis. In: *Diagenesis in Sediments* (Ed. by G. Larsen and G.V. Chilingar), pp. 18–89. Elsevier, Amsterdam.

Ferm, J. C. (1970) Allegheny deltaic deposits. In: *Deltaic Sedimentation Modern and Ancient* (Ed. by J.P. Morgan and R.H. Shaver), pp. 246–255. *Spec. Publ. Soc. econ. Paleont. Miner.* **15,** Tulsa.

Fischer, A.G. (1964) The Lofer cyclothems of the Alpine Triassic. In: *Symposium on Cyclic Sedimentation* (Ed. by D.F. Merriam), pp. 107–149. *Bull. geol. Surv. Kansas* **169.**

Fischer, A.G. & Garrison, R.E. (1967) Carbonate lithification on the sea floor. *J. Geol.* **75,** 488–496.

Fischer, A.G., Honjo, S. & Garrison, R.E. (1967) Electron micrographs of limestones and their nannofossils. *Monogr. Geol. Paleont.* Vol. 1 (Ed. by A.G. Fischer), pp. 141. Princeton University Press, Princeton.

Fiske, R.S. (1963) Subaqueous pyroclastic flows in the Ohanapecosh Formation, Washington. *Bull. geol. Soc. Am.* **74,** 391–406.

Fiske, R.S. & Matsuda, T. (1964) Submarine equivalents of ash flows in the Tokiwa Formation, Japan. *Am. J. Sci.* **262,** 76–106.

Flügel, E. (Ed.) (1977) *Fossil Algae. Recent results and Developments,* pp. 375. Springer-Verlag, Berlin.

Flügel, E. (1978) *Mikrofazielle Untersuchungs-methoden von Kalken,* pp. 454. Springer-Verlag, Berlin.

Flügel, E., Franz, H.E. & Ott, W.F. (1968) Review on electron microscope studies of limestones. In: *Recent Developments in Carbonate Sedimentology in Central Europe* (Ed. by G. Müller and G.M. Friedman), pp. 85–97. Springer-Verlag, Berlin.

Folk, R.L. (1959) Practical petrographic classification of limestones. *Bull. Am. Ass. Petrol. Geol.* **43,** 1–38.

Folk, R.L. (1962) Spectral subdivision of limestone types. In: *Classification of Carbonate Rocks* (Ed. by W.E. Ham), pp. 62–84. *Mem. Am. Ass. Petrol. Geol.* **1,** 62–84.

Folk, R.L. (1965) Some aspects of recrystallization in ancient limestones. In: *Dolomitization and limestone diagenesis* (Ed. by L.C. Pray and R.C. Murray), pp. 14–48, *Spec. Publ. Soc. econ. Paleont. Miner.* **13,** Tulsa.

Folk, R.L. (1974) *Petrology of Sedimentary Rocks,* pp. 159. Hemphill Publishing Co., Austin, Texas.

Folk, R.L. (1974) The natural history of crystalline calcium carbonate: effect of magnesium content and salinity. *J. sedim. Petrol.* **44,** 40–53.

Folk, R.L. & Land, L.S. (1975) Mg/Ca ratio and salinity: two controls over crystallization of dolomite. *Bull. Am. Ass. Petrol. Geol.* **59,** 60–68.

234

Folk, R.L. & McBride, E.F. (1976) The Caballos Novaculite revisited, Part 1: origin of novaculite members. *J. sedim. Petrol.* **46**, 659–669.

Folk, R.L. & McBride, E.F. (1978) Radiolarites and their relation to subjacent 'oceanic crust' in Liguria, Italy. *J. sedim. Petrol.* **48**, 1069–1102.

Folk, R.L. & Pittman, J.S. (1971) Length-slow chalcedony: a new testament for vanished evaporites. *J. sedim. Petrol.* **41**, 1045–1058.

Folk. R.L. & Ward, W. (1957) Brazos River bar: a study in the significance of grain size parameters. *J. sedim. Petrol.* **27**, 3–26.

Foth, H.D. (1978) *Fundamentals of Soil Science,* pp. 436. John Wiley, New York.

Francis, E.H. & Howells, M.F. (1973) Transgressive welded ash-flow tuffs among the Ordovician sediments of NE Snowdonia, N. Wales. *Jl. geol. Soc. Lond.* **129**, 621–641.

Frey, R.W. (1973) Concepts in the study of biogenic sedimentary structures. *J. sedim. Petrol.* **43**, 6–19.

Frey, R.W. (Ed.) (1975) *The Study of Trace Fossils,* pp. 562. Springer-Verlag, New York.

Friedman, G.M. (1965) Terminology of crystallization textures and fabrics in sedimentary rocks. *J. sedim. Petrol.* **35**, 643–655.

Friedman, G.M. (Ed.) (1969) *Depositional Environments in Carbonate Rocks,* pp. 209. *Spec. Publ. Soc. econ. Paleont. Miner.* **14**, Tulsa.

Friedman, G.M. (1979) Differences in size distributions of populations of particles among sands of various origins. *Sedimentology* **26**, 3–32.

Friedman, G.M. & Sanders, J.E. (1978) *Principles of Sedimentology,* pp. 792. John Wiley, New York.

Frost, S.H., Weiss, M.P. & Saunders, J.B. (1977) *Reefs and related carbonates—ecology and sedimentology. Studies in Geology, 4,* pp. 421. Am. Ass. Petrol Geol., Tulsa.

Füchtbauer, H. (1974) *Sediments and Sedimentary Rocks,* pp. 464. Schweizerbart, Stuttgart.

Furnes, H. (1972) Meta-hyaloclastite breccias associated with Ordovician pillow lavas in the Solund area, west Norway. *Norsk. geol. Tids.* **52**, 385–407.

Fürsich, F.T. (1979) Genesis, environments and ecology of Jurassic hardgrounds. *N. Jb. Geol. Paläont. Abh.* **158**, 1–63.

Gaines, A.M. (1977) (1978) Protodolomite redefined. *J. sedim. Petrol.* **47**, 543–546. Discussion: *J. sedim. Petrol.* **48**, 1004–1011.

Galloway, W.E. (1974) Deposition and diagenetic alteration of sandstone in Northeast Pacific arc-related basins: implications for graywacke genesis. *Bull. geol. Soc. Am.* **85**, 379–390.

Garrels, R.M. & Mackenzie, F.T. (1971) *Evolution of Sedimentary Rocks,* pp. 397. Norton, New York.

Garrels, R.M., Perry, E.A. & Mackenzie, F.T. (1973) Genesis of Precambrian Iron-Formations and the development of atmospheric oxygen. *Econ. Geol.* **68**, 1173–1179.

Garrison, R.E. (1974) Radiolarian cherts, pelagic limestones and igneous rocks in eugeosynclinal assemblages. In: *Pelagic Sediments: on Land and under the Sea* (Ed. by K.J. Hsü and H.C. Jenkyns), pp. 367–399. *Spec. Publ. int. Ass. Sediment.* **1**.

Garrison, R.E. & Fischer, A.G. (1969) Deep-water limestones and radiolarites of the Alpine Jurassic. In: *Depositional Environments in Carbonate Rocks* (Ed. by G.M. Friedman), pp. 20–56. *Spec. Publ. Soc. econ. Paleont. Miner.* **14**, Tulsa.

Gebelein, C.D. (1969) Distribution, morphology, and accretion rate of recent subtidal algal stromatolites, Bermuda. *J. sedim. Petrol.* **39**, 49–69.

Gebelein, C.D. (1974) Biologic control of stromatolite microstructure: implication for Precambrian time stratigraphy. *Am. J. Sci.* **274**, 575–598.

Gebelein, C.D., Steinen, R.P., Hoffman, E.J., Garrett, P. & Queen, J.M. (1979) Mixing-zone dolomite in tidal flat sediments of central-west Andros Island, Bahamas (Abs.). *Bull. Am. Ass. Petrol. Geol.* **63**, p. 457. Also in preparation as *Spec. Publ. Soc. econ. Paleont. Miner.*

Gill, W.C., Khalaf, F.L. & Massoud, M.S. (1977) Clay minerals as an index of the degree of metamorphism of the carbonate and terrigenous rocks in the South Wales coalfield. *Sedimentology* **24**, 675–691.

Ginsburg, R.N. (1975) *Tidal Deposits: A casebook of Recent Examples and Fossil Counterparts,* pp. 428. Springer-Verlag, Berlin.

Glasby, G.P. (Ed.) (1977) *Marine manganese deposits,* pp. 523. Elsevier, Amsterdam.

Glennie, K.W. (1970) *Desert Sedimentary Environments,* pp. 222. Elsevier, Amsterdam.

Glennie, K.W. (1972) Permian Rotliegendes of Northwest Europe interpreted in the light of modern desert sedimentation studies. *Bull. Am. Ass. Petrol. Geol.* **56**, 1048–1071.

Goldring, R. & Bridges, P. (1973) Sublittoral sheet sandstones. *J. sedim. Petrol.* **43**, 736–747.

Goldring, R., Boscence, D.W.J. & Blake, T. (1978) Estuarine sedimentation in the Eocene of southern England. *Sedimentology* **25**, 861–876.

Goodwin, A.M. (1973) Archaean Iron-Formations and Tectonic Basins of the Canadian Shield. *Econ. Geol.* **68**, 915–933.

Goudie, A. (1973) *Duricrusts in Tropical and Subtropical Landscapes*, pp. 174. Clarendon Press, Oxford.

Graf, D.L. & Goldsmith, J.R. (1956) Some hydrothermal syntheses of dolomite and protodolomite. *J. Geol.* **64**, 137–186.

Graham, S.A., Ingersoll, R.V. & Dickinson, W.R. (1976) Common provenance for lithic grains in Carboniferous sandstones from Ouachita Mountains and Black Warrior Basin. *J. sedim. Petrol.* **46**, 620–632.

Greensmith, J.T. (1978) *Petrology of the Sedimentary Rocks*, pp. 241. George Allen & Unwin, London.

Greer, S.A. (1975) Sandbody geometry and sedimentary facies at the estuary-marine transition zone, Ossabaw Sound, Georgia: a stratigraphic model. *Senckenberg. Mar.* **7**. 105–135.

Griffin, J.J., Windom, H. & Goldberg, E.D. (1968) The distribution of clay minerals in the world ocean. *Deep-Sea Res.* **15**, 433–459.

Griffiths, J.C. (1967) *Scientific Method in Analysis of Sediments*, pp. 508. McGraw-Hill, New York.

Grim, R.E. (1968) *Clay Mineralogy*, pp. 596. McGraw-Hill, New York.

Grim, R.E. & Güven, N. (1978) *Bentonites*, pp. 256. Elsevier, Amsterdam.

Grover, G. & Read, J.F. (1978) Fenestral and associated vadose diagenetic fabrics of tidal flat carbonates, Middle Ordovician New Market Limestone, Southwestern Virginia. *J. sedim. Petrol.* **48**, 453–473.

Gulbrandsen, R.A. (1969) Physical and chemical factors in the formation of marine apatite. *Econ. Geol.* **64**, 365–382.

Hallam, A. (1964) Origin of the limestone-shale rhythm in the Blue Lias of England: a composite theory. *J. Geol.* **72**, 157–169.

Hallam, A. (1966) Depositional environment of British Liassic ironstones considered in the context of their facies relationships. *Nature* **209**, 1306–1309.

Hallam, A. & Sellwood, B.W. (1968) Origin of Fuller's Earth in the Mesozoic of southern England. *Nature* **220**, 1193–1195.

Halley, R.B. (1977) Ooid fabric and fracture in the Great Salt Lake and the geologic record. *J. sedim. Petrol.* **47**, 1099–1120.

Hallimond, A.F. (1925) Iron ores: bedded ores of England and Wales. Petrography and chemistry. *Mem. geol. Surv. spec. Rep. Miner. Resour. Gt. Br.* **29**, 139 pp.

Hancock, N.J. (1978) Possible causes of Rotliegend sandstone diagenesis in northern West Germany. *Jl. geol. Soc. Lond.* **135**, 35–40.

Hancock, N.J. & Taylor, A.M. (1978) Clay mineral diagenesis and oil migration in the Middle Jurassic Brent Sand Formation. *Jl. geol. Soc. Lond.* **135**, 69–72.

Hanor, J.S. (1978) Precipitation of beachrock cements: mixing of marine and meteoric waters vs. CO_2-degassing. *J. sedim. Petrol.* **48**, 489–501.

Hanshaw, B.B., Black, W. & Dieke, R.G. (1971) A geochemical hypothesis for dolomitization by ground-water. *Econ. Geol.* **66**, 710–724.

Harbaugh, J.W. & Bonham-Carter, G. (1970) *Computer Simulation in Geology*, pp. 575. John Wiley, New York.

Hardie, L.A. (1967) The gypsum-anhydrite equilibrium at one atmosphere pressure. *Am. Miner.* **52**, 171–200.

Hardie, L.A. (Ed.) (1977) *Sedimentation on the Modern Carbonate Tidal Flats of Northwestern Andros Island, Bahamas*. Johns Hopkins Press, Baltimore.

Hardie, L.A. & Eugster, H.P. (1971) The depositional environment of marine evaporites: a case for shallow clastic accumulation. *Sedimentology* **16**, 187–220.

Hardie, L.A., Smoot, J.P. & Eugster, H.P. (1978) Saline lakes and their deposits: a sedimentological approach. In: *Modern and Ancient Lake Sediments* (Ed. by A. Matter and M.E. Tucker), pp. 7–41. *Spec. Publs int. Ass. Sediment.* **2**.

Harland, W.B. & Hambrey, M. (Eds.) (1980) *Pre-Pleistocene Tillites*. Cambridge University Press.

Harms, J.C., Southard, J.B., Spearing, D.R. & Walker, R.G. (1975). Depositional environments as interpreted from primary sedimentary structures and stratification sequences. *Soc. econ. Paleont. Miner., Short Course 2*, pp. 161. Dallas, Texas.

Hawkins, P.J. (1978) Relationship between diagenesis, porosity reduction, and oil emplacement in late Carboniferous sandstone reservoirs, Bothamsall Oilfield, E. Midlands. *Jl. geol. Soc. Lond.* **135**, 7–24.

Hay, R.L. (1966) Zeolites and zeolitic reactions in sedimentary rocks. *Spec. Pap. geol. Soc. Am.* **85**, 130 pp.

Heath, G.R. & Moberly, R. (1971) Cherts from the Western Pacific, Leg 7, Deep Sea Drilling Project. In: *Initial Reports of the Deep Sea Drilling Project, 7* (Ed. by E.L. Winterer *et al.*), pp. 991–1007. U.S. Government Printing Office, Washington, D.C.

Heckel, P.H. (1972a) Possible inorganic origin for stromatactis in calcilutite mounds in the Tully Limestone, Devonian of New York. *J. sedim. Petrol.* **42**, 7–18.

236

Heckel, P.H. (1972b) Recognition of ancient shallow marine environments. In: *Recognition of Ancient Sedimentary Environments* (Ed. by J.K. Rigby and W.K. Hamblin), pp. 226–296. *Spec. Publ. Soc. econ. Paleont. Miner.* **16**, Tulsa.

Heckel, P.H. (1974) Carbonate buildups in the geologic record: a review. In: *Reefs in Time and Space* (Ed. by L.F. Laporte), pp. 90–154. *Spec. Publ. Soc. econ. Paleont. Miner.* **18**, Tulsa.

Helwig, J. (1970) Slump folds and early structures, northeastern Newfoundland Appalachians. *J. Geol.* **78**, 172–187.

Hendry, H.E. & Stauffer, M.R. (1975) Penecontemporaneous recumbent folds in trough cross-bedding of Pleistocene sands in Saskatchewan, Canada. *J. sedim. Petrol.* **45**, 932–943.

Heward, A.P. (1978) Alluvial fan and lacustrine sediments from the Stephanian A and B (La Magdalena, Cinera-Matallana and Sabero) coalfields, northern Spain. *Sedimentology* **25**, 451–488.

Hine, A.C. (1977) Lily Bank, Bahamas: history of an active oolite sand shoal. *J. sedim. Petrol.* **47**, 1554–1582.

Hiscott, R.N. (1979) Clastic sills and dykes associated with deep-water sandstones, Tourelle Formation, Ordovician, Quebec. *J. sedim. Petrol.* **49**, 1–9.

Ho, C. & Coleman, J.M. (1969) Consolidation and cementation of Recent sediments in the Atchafalaya Basin. *Bull. geol. Soc. Am.* **80**, 183–192.

Hobday, D.K. & Eriksson, K.A. (Eds.) (1977) Tidal Sedimentation, with particular reference to S. African examples. *Sedim. Geol.* **18**, 1–287.

Hobson, G.D. & Tiratsoo, E.N. (1975) *Introduction to Petroleum Geology,* pp. 300. Scientific Press, Beaconsfield.

Hoffman, P. (1974) Shallow and deepwater stromatolites in Lower Proterozoic platform-basin facies change, Great Slave Lake, Canada. *Bull. Am. Ass. Petrol. Geol.* **58**, 856–867.

Hoffman, P. (1976) Stromatolite morphogenesis in Shark Bay, Western Australia. In: *Stromatolites* (Ed. by M.R. Walter), pp. 261–272. Elsevier, Amsterdam.

Holliday, D.W. (1970) The petrology of secondary gypsum rocks: a review. *J. sedim. Petrol.* **40**, 734–744.

Holliday, D.W. (1973) Early diagenesis in nodular anhydrite rocks. *Trans. Inst. Min. Metall.* **82**, B81–84.

Holser, W.T. (1966a) Bromide geochemistry of salt rocks. In: *Second Symposium on Salt* (Ed. by J.L. Rau), v. 1, pp. 248–275. Northern Ohio Geol. Soc., Cleveland, Ohio.

Holser, W.T. (1966b) Diagenetic polyhalite in Recent salt from Baja California. *Am. Miner.* **51**, 99–109.

Hood, A., Gutjahr, C.C.M. & Heacock, C.L. (1975) Organic metamorphism and the generation of petroleum. *Bull. Am. Ass. Petrol. Geol.* **59**, 986–996.

Horowitz, A.S. & Potter, P.E. (1971) *Introductory Petrography of Fossils,* pp. 302. Springer-Verlag, Berlin.

Howard, J.D. & Frey, R.W. (1973) Characteristic physical and biogenic sedimentary structures in Georgia estuaries. *Bull. Am. Ass. Petrol. Geol.* **57**, 1169–1184.

Hower, J., Eslinger, E.V., Hower, M.E. & Perry, E.A. (1976) Mechanism of burial metamorphism of argillaceous sediment: 1. Mineralogical and chemical evidence. *Bull. geol. Soc. Am.* **87**, 725–737.

Hsü, K.J. (1972) Origin of saline giants: a critical review after the discovery of the Mediterranean Evaporite. *Earth Sci. Rev.* **8**, 371–396.

Hsü, K.J. & Jenkyns, H.C. (Eds.) (1974) *Pelagic Sediments: on Land and under the Sea. Spec. Publ. int. Ass. Sediment.* **1**, pp. 447.

Hsü, K.J. & Siegenthaler, C. (1969) Preliminary experiments on hydrodynamic movement induced by evaporation and their bearing on the dolomite problem. *Sedimentology* **12**, 11–25.

Hsü, K.J. *et al.* (1977) History of the Mediterranean salinity crisis. *Nature* **267**, 399–403.

Hsü, K.J. *et al.* (1978) History of the Mediterranean salinity crisis. In: *Initial Reports of the Deep Sea Drilling Project, 42,* part 1 (K.J. Hsü, L. Montadert *et al.*) pp. 1053–1078. U.S. Government Printing Office, Washington, D.C.

Hubert, J.F. & Reed, A.A. (1978) Red-bed diagenesis in the East Berline Formation, Newark Group, Connecticut Valley. *J. sedim. Petrol.* **48**, 175–184.

Hudson, J.D. (1962) Pseudo-pleochroic calcite in recrystallized shell-limestones. *Geol. Mag.* **99**, 492–500.

Hudson, J.D. (1975) Carbon isotopes and limestone cement. *Geology* **3**, 19–22.

Hudson, J.D. (1977) Stable isotopes and limestone lithification. *Jl. geol. Soc. Lond.* **133**, 637–660.

Hudson, J.D. (1978) Concretions, isotopes, and the diagenetic history of the Oxford Clay (Jurassic) of central England. *Sedimentology* **25**, 339–369.

Hunt, C.B. (1972) *Geology of soils,* pp. 344. Freeman, San Francisco.

Hunter, R.E. (1970) Facies of iron sedimentation in the Clinton Group. In: *Studies of Appalachian*

Geology (Ed. by G.W. Fisher, F.J. Pettijohn, J.C. Reed & K.N. Weaver), pp. 101–124. Wiley-Interscience, New York.

Hunter, R.E. (1977) Basic types of stratification in small eolian dunes. *Sedimentology* **24,** 361–388.

Hutchison, C.S. (1974) *Laboratory handbook of petrographic techniques,* pp. 527. John Wiley, New York.

Illing, L.V., Wells, A.J. & Taylor, J.C.M. (1965) Penecontemporary dolomite in the Persian Gulf. In: *Dolomitization and Limestone Diagenesis* (Ed. by L.C. Pray and R.C. Murray), pp. 89–111. *Spec. Publ. Soc. econ. Paleont.* **13,** Tulsa.

Ingersoll, R.V. (1978) Petrofacies and petrologic evolution of the late Cretaceous fore-arc basin, northern and central California. *J. Geol.* **86,** 335–352.

Irwin, M.L. (1965) General theory of epeiric clear water sedimentation. *Bull. Am. Ass. Petrol. Geol.* **49,** 445–459.

Jacobs, M.B. (1974) Clay mineral changes in Antarctic deep-sea sediments and Cenozoic climatic events. *J. sedim. Petrol.* **44,** 1079–1086.

James, H.E. & Van Houten, F.B. (1979) Miocene goethitic and chamosite oolites, northeastern Columbia. *Sedimentology* **26,** 125–133.

James, H.L. (1954) Sedimentary Facies of Iron-Formation. *Econ. Geol.* **49,** 235–293.

James, N.P. (1972) Holocene and Pleistocene calcareous crust (caliche) profiles: criteria for subaerial exposure. *J. sedim. Petrol.* **42,** 817–836.

James, N.P. (1974) Diagenesis of scleractinian corals in the subaerial vadose environment. *J. Paleont.* **48,** 785–799.

James, N.P., Ginsburg, R.N., Marszalek, D.S. & Choquette, P.W. (1976) Facies and fabric specificity of early subsea cements in shallow Belize (British Honduras) reefs. *J. sedim. Petrol.* **46,** 523–544.

James, W.C. & Oaks, R.Q. (1977) Petrology of the Kinnikinic Quartzite (Middle Ordovician), east-central Idaho. *J. sedim. Petrol.* **47,** 1491–1511.

Jeans, C.V. (1978) The origin of the Triassic clay assemblages of Europe with special reference to the Keuper Marl and Rhaetic of parts of England. *Phil. Trans. Roy. Soc. Lond.,* Ser. A, **289,** 549–639.

Jenkyns, H.C. (1974) Origin of red nodular limestones (Ammonitico Rosso, Knollenkalke) in the Mediterranean Jurassic: a diagenetic model. In: *Pelagic Sediments: on Land and under the Sea* (Ed. by K.J. Hsü and H.C. Jenkyns), pp. 249–271. *Spec. Publ. int. Ass. Sediment.* **1.**

Jenkyns, H.C. (1978) Pelagic Environments. In: *Sedimentary Environments and Facies* (Ed. by H.G. Reading), pp. 314–371. Blackwell Scientific Publications, Oxford.

Jenkyns, H.C. (1980) Cretaceous anoxic events: from continents to oceans, *Jl. geol. Soc. Lond.* **137,** 171–188.

Johnson, H.D. (1975) Tide- and wave-dominated inshore and shoreline sequences from the late Precambrian, Finnmark, north Norway. *Sedimentology* **22,** 45–73.

Johnson, H.D. (1977) Shallow marine and bar sequences: an example from the late Precambrian of North Norway. *Sedimentology* **24,** 245–270.

Johnson, H.D. (1978) Shallow Siliciclastic Seas. In: *Sedimentary Environments and Facies* (Ed. by H.G. Reading), pp. 207–258. Blackwell Scientific Publications, Oxford.

Johnson, J.H. (1961) *Limestone—building algae and algal limestones,* pp. 297. Colorado School of Mines.

Jones, B. & Dixon, O.A. (1976) Storm deposits in the Read Bay Formation (Upper Silurian), Somerset Island, Arctic Canada (an application of Markov chain analysis). *J. sedim. Petrol.* **46,** 393–401.

Jones, J.B. & Segnit, E.R. (1971) The nature of Opal. Nomenclature and constituent phases. *J. geol. Soc. Aust.* **18,** 57–68.

Kahle, C.F. (1965) Possible roles of clay minerals in the formation of dolomite. *J. sedim. Petrol.* **35,** 448–453.

Kastner, M., Keene, J.B. & Gieskes, J.M. (1977) Diagenesis of siliceous oozes—1. Chemical controls on the rate of opal-A to opal-CT transformation—an experimental study. *Geochim. Cosmochim. Acta* **41,** 1041–1059.

Keene, J.B. (1975) Cherts and porcelanites from the North Pacific, D S D P Leg 32. In: *Initial Reports of the Deep Sea Drilling Project, 32* (R.L. Larson, R. Moberly *et al.*), pp. 429–507. U.S. Government Printing Office, Washington, D.C.

Keller, W.D. (1970) Environmental aspects of clay minerals. *J. sedim. Petrol.* **40,** 788–813.

Kendall, A.C. (1979) Continental and supratidal (sabkha) evaporites. Subaqueous evaporites. In: *Facies Models* (Ed. by R.G. Walker), pp. 145–174. Geoscience Canada.

Kendall, A.C. & Broughton, P.W. (1978) Origin of fabrics in speleothems composed of columnar calcite crystals. *J. sedim. Petrol.* **48,** 519–538.

Kendall, A.C. & Tucker, M.E. (1973) Radiaxial fibrous calcite: a replacement after acicular carbonate. *Sedimentology* **20,** 365–390.

238

Kendall, C.G.St.C. & Skipwith, P.A.d'E. (1968) Recent algal mats of a Persian Gulf lagoon. *J. sedim. Petrol.* **38,** 1040–1058.

Kennedy, W.J. & Garrison, R.E. (1975a) Morphology and genesis of nodular chalks and hardgrounds in the Upper Cretaceous of southern England. *Sedimentology* **22,** 311–386.

Kennedy, W.J. & Garrison, R.E. (1975b) Morphology and genesis of nodular phosphates in the Cenomanien Glauconitic Marl of southeast England. *Lethaia* **8,** 339–360.

Kerr, S.D. & Thomson, A. (1963) Origin of nodular and bedded anhydrite in Permian shelf sediments, Texas and New Mexico. *Bull. Am. Ass. Petrol. Geol.* **47,** 1726–1732.

Kidd, R.B., Cita, M.B. & Ryan, W.B.F. (1978) Stratigraphy of eastern Mediterranean sapropel sequences recovered during DSDP Leg 42A and their palaeoenvironmental significance. In: *Initial Reports of the Deep Sea Drilling Project, 42,* part 1 (K.J. Hsü, L. Montadert *et al.*), pp. 421–443. U.S. Government Printing Office, Washington, D.C.

Kimberley, M.M. (1979) (1980) Origin of oolitic iron formation *J. sedium. Petrol.* **49.** 111–131. Discussion: *J. sedim. Petrol.* **50,** 295–302, 1001–1004.

Kinsman, D.J.J. (1969) Modes of formation, sedimentary associations and diagnostic features of shallow-water and supratidal evaporites. *Bull. Am. Ass. Petrol. Geol.* **53,** 830–840.

Kinsman, D.J.J. (1975) Salt floors to geosynclines. *Nature* **255,** 375–378.

Kinsman, D.J.J. & Holland, H.D. (1969) The co-precipitation of cations with $CaCO_3$ IV. The co-precipitation of Sr^{+2} with aragonite between $16°$ and $96°C$. *Geochim. cosmochim. Acta* **33,** 1–17.

Kirkland, D.W. & Evans, R. (Eds.) (1973) *Marine evaporites: origins, diagenesis and geochemistry,* pp. 426. Dowden, Hutchinson & Ross, Stroudsburg.

Klein, G.deV. (1977) *Clastic tidal facies,* pp. 149. CEPCO, Champaign, Illinois.

Klemme, H.D. (1975) Geothermal gradients, heat flow, and hydrocarbon recovery. In: *Petroleum and global tectonics* (Ed. by A.G. Fischer and S. Judson), pp. 251–304. Princeton Univ. Press, Princeton.

Klovan, J.E. (1974) Developments of western Canadian Devonian reefs and comparison with Holocene analogues. *Bull. Am. Ass. Petrol. Geol.* **58,** 787–799.

Knebel, H.J., Conomos, T.J. & Commeau, J.A. (1977) Clay-mineral variability in the suspended sediments of the San Francisco Bay system, California. *J. sedim. Petrol.* **47,** 229–236.

Kocurko, M.J. (1979) Dolomitization by spray-zone brine-seepage, San Andres, Columbia. *J. sedim. Petrol.* **49,** 209–214.

Kolodny, Y. & Kaplan, I.R. (1970) Uranium isotopes in sea floor phosphorites. *Geochim. cosmochim. Acta* **34,** 3–34.

Krauskopf, K.B. (1979) *Introduction to Geochemistry,* pp. 617. McGraw-Hill, New York.

Krebs, W. (1974) Devonian carbonate complexes of Central Europe. In: *Reefs in Time and Space* (Ed. by L.F. Laporte), pp. 155–208. *Spec. Publ. Soc. econ. Paleont. Miner.* **18,** Tulsa.

Krinsley, D.H. & Doornkamp, J.C. (1973) *Atlas of sand surface textures,* pp. 91. Cambridge University Press, Cambridge.

Kuenen, Ph.H. & Migliorini, C.I. (1950) Turbidity currents as a cause of graded bedding. *J. Geol.* **58,** 91–127.

La Berge, G.L. (1973) Possible biological origin of Precambrian iron-formations. *Econ. Geol.* **68,** 1098–1109.

Lajoie, J. (Ed.) (1970) *Flysch Sedimentology in North America,* pp. 242. Geol. Assoc. Canada, Spec. Pap. **7.**

Lajoie, J. (1979) Volcaniclastic rocks. In: *Facies Models* (Ed. by R.G. Walker), pp. 191–200. Geoscience Canada.

Lancelot, Y. (1973) Chert and silica diagenesis in sediments from the central Pacific. In: *Initial Reports of the Deep Sea Drilling Project, 17* (E.L. Winterer, J.I. Ewing *et al.*), pp. 377–405. U.S. Government Printing Office, Washington, D.C.

Land, L.S., Behrens, E.W. & Frishman, S.A. (1979) The ooids of Baffin Bay, Texas. *J. sedim. Petrol.* **49,** 1269–1277.

Land, L.S., Mackenzie, F.T. & Gould, S.J. (1967) Pleistocene history of Bermuda. *Bull. geol. Soc. Am.* **78,** 993–1006.

Laporte, L.F. (1967) Carbonate deposition near mean sea-level and resultant facies mosaic: Manlius Formation (Lower Devonian) of New York State. *Bull. Am. Ass. Petrol. Geol.* **51,** 73–101.

Laporte, L.F. (Ed.) (1974) *Reefs in time and space,* pp. 256. *Spec. Publ. Soc. econ. Paleont. Miner.* **18,** Tulsa.

Larsen, G. & Chilingarian, G.V. (Eds.) (1979) *Diagenesis in Sediments and sedimentary rocks,* pp. 579. Elsevier, Amsterdam.

Larsen, V. & Steel, R.J. (1978) The sedimentary history of a debris-flow dominated Devonian alluvial fan—a study of textural inversion. *Sedimentology* **25,** 37–59.

Lauff, F.H. (Ed.) (1967) *Estuaries,* pp. 757. Am. Ass. Adv. of Science, Washington, D.C.

239

Leeder, M.R. (1973) Sedimentology and palaeogeography of the Upper Old Red Sandstone in the Scottish Border Basin. *Scott. J. Geol.* **9**, 117–144.

Lees, A. (1964) The structure and origin of the Waulsortian (Lower Carboniferous) "Reefs" of west-central Eire. *Phil. Trans. Roy. Soc. London, Ser. B.* **247**, 483–531.

Lees, A. (1975) Possible influences of salinity and temperature on modern shelf carbonate sedimentation. *Mar. Geol.* **19**, 159–198.

Lemoalle, J. & Dupont, B. (1973) Iron-bearing oolites and the present conditions of iron sedimentation in Lake Chad (Africa). In: *Ores in Sediments* (Ed. by G.C. Amstutz and A.J. Bernard), pp. 167–178. Springer-Verlag, Berlin.

Lepp, H. (Ed.) (1975) *Geochemistry of Iron*, pp. 464. Dowden, Hutchinson & Ross, Stroudsburg.

Levorsen, A.I. (1967) *Geology of Petroleum*, pp. 724. Freeman & Co., San Francisco.

Lewan, M.D. (1978) Laboratory classification of very fine grained sedimentary rocks. *Geology* **6**, 745–748.

Lindholm, R.C. (1979) Utilization of programmable calculators in sedimentology. *J. sedim. Petrol.* **49**, 615–620.

Link, M.H. & Osborne, R.H. (1978) Lacustrine facies in the Pliocene Ridge Basin Group: Ridge Basin, California. In: *Modern and Ancient Lake Sediments* (Ed. by A. Matter and M.E. Tucker), pp. 167–189. *Spec. Publ. int. Ass. Sediment.* **2**.

Lippmann, F. (1973) *Sedimentary Carbonate Minerals*, pp. 228. Springer-Verlag, Berlin.

Lisitzin, A.P. (1972) *Sedimentation in the World Ocean*, pp. 218. *Spec. Publ. econ. Paleont. Miner.* **17**, Tulsa.

Logan, B.W. (1974) Inventory of diagenesis in Holocene-Recent carbonate sediments, Shark Bay, Western Australia. *Mem. Am. Ass. Petrol. Geol.* **22**, 195–249.

Logan, B.W., Harding, J.L., Ahr, W.M., Williams, J.D. & Snead, R.G. (1969) Carbonate sediments and reefs, Yucatan shelf, Mexico. *Mem. Am. Ass. Petrol. Geol.* **11**, 1–198.

Logan, B.W., Hoffman, P. & Gebelein, C.F. (1974) Algal mats, cryptalgal fabrics and structures, Hamelin Pool, Western Australia. *Mem. Am. Ass. Petrol. Geol.* **22**, 140–194.

Logan, B.W., Rezak, R. & Ginsburg, R.N. (1964) Classification and environmental significance of algal stromatolites. *J. Geol.* **72**, 62–83.

Lohmann, K.C. & Meyers, W.J. (1977) Microdolomite inclusions in cloudy prismatic calcites: A proposed criterion for former high magnesium calcites. *J. sedim. Petrol.* **47**, 1078–1088.

Longman, M.W. (1977) Factors controlling the formation of microspar in the Bromide Formation. *J. sedim. Petrol.* **47**, 347–350.

Longman, M.W. (1980) Carbonate diagenetic textures from nearsurface diagenetic environments. *Bull. Am. Ass. Petrol. Geol.* **64**, 461–487.

Loreau, J.P. & Purser, B.H. (1973) Distribution and ultrastructure of Holocene ooids in the Persian Gulf. In: *The Persian Gulf* (Ed. by B.H. Purser), pp. 279–328. Springer-Verlag, Berlin.

Lowe, D.R. (1975) Water escape structures in coarse-grained sediments. *Sedimentology* **22**, 157–204.

Lowe, D.R. (1976) Subaqueous liquified and fluidized sediment flows and their deposits. *Sedimentology* **23**, 285–308.

Macintyre, I.G. (1977) Distribution of submarine cements in a modern Caribbean fringing reef, Galeta Point, Panama. *J. sedim. Petrol.* **47**, 503–516.

Mack, G.H. (1978) The survivability of labile light-mineral grains in fluvial, aeolian and littoral marine environments: the Permian Cutler and Cedar Mesa Formations, Moab, Utah. *Sedimentology* **25**, 587–604.

Majewske, O.P. (1969) *Recognition of invertebrate fossil fragments in rocks and thin sections*, pp. 101. Brill, Leiden.

Manheim, F., Rowe, G.T. & Jipa, D. (1975) Marine phosphorite formation off Peru. *J. sedim. Petrol.* **45**, 243–251.

Marlowe, J.I. (1971) Dolomite, phosphorite and carbonate diagenesis on a Caribbean seamount. *J. sedim. Petrol.* **41**, 809–827.

Marshall, J.F. & Davies, P.J. (1975) High-magnesium calcite ooids from the Great Barrier Reef. *J. sedim. Petrol.* **45**, 285–291.

Masson, P.H. (1955) An occurrence of gypsum in Texas. *J. sedim. Petrol.* **25**, 72–77.

Matter, A. (1967) Tidal flat deposits in the Ordovician of Western Maryland. *J. sedim. Petrol.* **37**, 601–609.

Matter, A. (1974) Burial diagenesis of pelitic and carbonate deep-sea sediments from the Arabian Sea. In: *Initial Reports of the Deep Sea Drilling Project, 23* (R.B. Whitmarsh, O.E. Weser, D.A. Ross *et al.*), pp. 421–469. U.S. Government Printing Office, Washington.

Matter, A. & Tucker, M.E. (Eds.) (1978) *Modern and Ancient Lake Sediments. Spec. Publ. int. Ass. Sediment.* **2**, pp. 290.

Matthews, R.K. (1966) Genesis of Recent lime mud in Southern British Honduras. *J. sedim. Petrol.* **36**, 428–454.

240

McKee, E.D. (1966) Dune structures. *Sedimentology* **7**, 3–69.
McKelvey, B.C. & Fleet, A.J. (1974) Eocene basaltic pyroclastics at site 253, Ninetyeast Ridge. In: *Initial Reports of the Deep Sea Drilling Project, 26* (T.A. Davies, B.P. Luyendyk *et al.*), pp. 553–565. U.S. Government Printing Office, Washington, D.C.
McKelvey, V.E., Williams, J.S., Sheldon, R.P., Cressman, E.R., Cheney, T.M. & Swanson, R.S. (1959) The Phosphoria, Park City, and Shedborn formations in the western phosphate field. *Prof. Pap. U.S. geol. Surv.* **313-A**, 1–47.
McRae, S.G. (1972) Glauconite. *Earth-Sci. Rev.* **8**, 397–440.
Meischner, K.D. (1964) Allodapische Kalke, Turbidite in Riff-Nahen Sedimentations-Becken. In: *Turbidites* (Ed. by A.H. Bouma and A. Brouwer), pp. 156–191. Elsevier, Amsterdam.
Merriam, D.F. (Ed.) (1964) Symposium on Cyclic Sedimentation, pp. 636. *Bull. geol. Surv. Kansas,* **169.**
Meyers, W.J. (1974) Carbonate cement stratigraphy of the Lake Valley Formation (Mississippian), Sacramento Mountains, New Mexico. *J. sedim. Petrol.* **44**, 837–861.
Meyers, W.J. (1977) Chertification in the Mississippian Lake Valley Formation, Sacramento Mountains, New Mexico. *Sedimentology* **24**, 75–105.
Meyers, W.J. (1978) Carbonate cements: their regional distribution and interpretation in Mississippian limestones of southwestern New Mexico. *Sedimentology* **25**, 371–400.
Meyers, W.J. & James, A.T. (1978) Stable isotopes of cherts and carbonate cements in the Lake Valley Formation (Mississippian), Sacramento Mts., New Mexico. *Sedimentology* **25**, 105–124.
Miall, A.D. (1973) Markov Chain analysis applied to an ancient alluvial plain succession. *Sedimentology* **20**, 347–364.
Miall, A.D. (1976) Palaeocurrent and palaeohydraulic analysis of some vertical profiles through a Cretaceous braided stream deposit, Banks Island, Arctic Canada. *Sedimentology,* **23**, 459–483.
Miall, A.D. (1977) A review of the braided-river depositional environment. *Earth-Sci. Rev.* **13**, 1–62.
Miall, A.D. (Ed.) (1978) *Fluvial Sedimentology,* pp. 859. Canadian Society of Petroleum Geologists, Calgary, Alberta.
Middleton, G.V. (1976) Hydraulic interpretations of sand size distributions. *J. Geol.* **84**, 405–426.
Middleton, G.V. & Hampton, M.A. (1976) Subaqueous sediment transport and deposition by sediment gravity flows. In: *Marine Sediment Transport and Environmental Management* (Ed. by D.J. Stanley and D.J.P. Swift), pp. 197–218. John Wiley, New York.
Milliken, K.L. (1979) The silicified evaporite syndrome—two aspects of silicification history of former evaporite nodules from southern Kentucky and northern Tennessee. *J. sedim. Petrol.* **49**, 245–256.
Milliman, J.D. (1974) *Marine Carbonates,* pp. 375. Springer-Verlag, Berlin.
Milliman, J.D. & Barretto, H.T. (1975) Relict magnesian calcite oolite and subsidence of the Amazon shelf. *Sedimentology* **22**, 137–145.
Milliman, J.D. & Müller, J. (1973) Precipitation and lithification of magnesian calcite in the deep-sea sediments of the eastern Mediterranean Sea. *Sedimentology* **20**, 29–46.
Milliman, J.D., Ross, D.A. & Ku, T.H. (1969) Precipitation and lithification of deep-sea carbonates in the Red Sea. *J. sedim. Petrol.* **39**, 724–736.
Millot, G. (1970) *Geology of Clays,* pp. 429. Springer-Verlag, New York.
Misik, M. (1971) Observations concerning calcite veinlets in carbonate rocks. *J. sedim. Petrol.* **41**, 450–460.
Mitchell, A.H.G. & Reading, H.G. (1969) Continental margins, geosynclines and ocean floor spreading. *J. Geol.* **77**, 629–646.
Mitterer, R.M. (1971) Influence of natural organic matter on $CaCO_3$ precipitation. In: *Carbonate Cements* (Ed. by O.P. Bricker), pp. 252–258. Johns Hopkins, Baltimore.
Monseur, G. & Pel, J. (1973) Reef environment and stratiform ore deposits. In: *Ores in Sediments* (Ed. by G.C. Amstutz and A.J. Bernard), pp. 195–207. Springer-Verlag, Heidelberg.
Monty, C.L.V. (1967) Distribution and structure of Recent stromatolitic algal mats, eastern Andros Island, Bahamas. *Annls. Soc. geol. Belg.* **90**, 57–93.
Monty, C.V.L. & Hardie, L.A. (1976) The geological significance of the fresh-water blue-green algal calcareous marsh. In: *Stromatolites* (Ed. by M.R. Walter), pp. 447–477. Elsevier, Amsterdam.
Moore, J.C. (1974) Turbidite and terrigenous muds, DSDP leg 25. In: *Initial Reports of the Deep Sea Drilling Project 25* (E.S.W. Simpson, R. Schlich *et al.*), pp. 441–479. U.S. Government Printing Office, Washington, D.C.
Moore, L.R. (1964) Microbiology, mineralogy and genesis of a tonstein. *Proc. Yorks. geol. Soc.* **34**, 235–291.
Morgan, J.P. (Ed.) (1970) *Deltaic Sedimentation, modern and ancient,* pp. 312. *Spec. Publ. Soc. econ. Paleont. Miner.* **15**, Tulsa.
Morris, K.A. (1979) A classification of Jurassic marine shale sequences: an example from the

241

Toarcian (Lower Jurassic) of Great Britain. *Palaeogeogr., Palaeoclimatol., Palaeoecol.* **26,** 117–126.

Müller, G. (1967) *Methods in Sedimentary Petrography,* pp. 283. Hafner Pub. Co., New York.

Müller, G., Irion, G. & Förstner, U. (1972) Formation and diagenesis of inorganic Ca–Mg-carbonates in the lacustrine environment. *Naturwissenschaften* **59,** 158–164.

Müller, J. & Fabricius, F. (1974) Magnesium-calcite nodules in the Ionian deep sea: an actualistic model for the formation of some nodular limestones. In: *Pelagic Sediments: On Land and Under the Sea* (Ed. by K.J. Hsü and H.C. Jenkyns), pp. 249–271. *Spec. Publ. Int. Ass. Sediment.* **1.**

Müller-Jungblath, W.V. (1968) Sedimentary petrologic investigation of the Upper Triassic "Hauptdolomit" of the Lechthaler Alps, Tyrol, Austria. In: *Recent Developments in Carbonate Sedimentology in Central Europe* (Ed. by G. Müller and G.M. Friedman), pp. 228–239. Springer-Verlag, Berlin.

Multer, H.G. & Hoffmeister, J.E. (1968) Subaerial laminated crusts of the Florida Keys. *Bull. geol. Soc. Am.* **79,** 183–192.

Murray, R.C. (1960) Origin of porosity in carbonate rocks. *J. sedim. Petrol.* **30,** 59–84.

Natland, J.H. (1976) Petrology of volcanic rocks dredged from seamounts in the Line islands. In: *Initial Reports of the Deep Sea Drilling Project, 33* (S.O. Schlanger, E.D. Jackson *et al.*), pp. 749–777. U.S. Government Printing Office, Washington, D.C.

Nelson, C.S. (1978) Temperate shelf carbonate sediments in the Cenozoic of New Zealand. *Sedimentology* **25,** 737–771.

Neumann, A.C. & Land, L.S. (1975) Lime mud deposition and calcareous algae in the Bight of Abaco, Bahamas: a budget. *J. sedim. Petrol.* **45,** 763–786.

Newell, N.D., Rigby, J.K., Fischer, A.G., Whiteman, A.J., Hickox, J.E. & Bradley, J.S. (1953) *The Permian Reef Complex of the Guadalupe Mountain Region, Texas and New Mexico,* pp. 236. W.H. Freeman, San Francisco.

Newton, R.S. (1968) Internal structure of wave-formed ripple marks in the nearshore-zone. *Sedimentology* **11,** 275–292.

Nisbet, E.G. & Price, I. (1974) Siliceous turbidites: bedded cherts as redeposited, ocean ridge-derived sediments. In: *Pelagic Sediments: on Land and under the Sea* (Ed. by K.J. Hsü and H.C. Jenkyns), pp. 351–366. *Spec. Publ. int. Ass. Sediment.* **1.**

Oldershaw, A.E. & Scoffin, T.P. (1967) The source of ferroan and non-ferroan calcite cements in the Halkin and Wenlock Limestones. *J. Geol.* **5,** 309–320.

Orme, G.R. (1974) Silica in the Viséan limestones of Derbyshire, England. *Proc. Yorks. geol. Soc.* **40,** 63–104.

Page, H.G. (1955) Phi-millimeter conversion table. *J. sedim. Petrol.* **25,** 285–292.

Pannella, G. (1976) Geophysical inferences from stromatolite lamination. In: *Stromatolites* (Ed. by M.R. Walter), pp. 673–686. Elsevier, Amsterdam.

Park, R. (1976) A note on the significance of lamination in stromatolites. *Sedimentology* **23,** 379–393.

Parkash, B. & Middleton, G.V. (1970) Downcurrent textural changes in Ordovician turbidite greywackes. *Sedimentology* **14,** 259–293.

Parker, R.J. (1975) The petrology and origin of some glauconitic and glauco-conglomeratic phosphorites from the South African continental margin. *J. sedim. Petrol.* **45,** 230–242.

Perry, E.D. & Hower, J. (1970) Burial diagenesis in Gulf Coast pelitic sediments. *Clays Clay Miner.* **18,** 165–177.

Peterson, M.N.A. & von der Borch, C.C. (1965) Chert: modern inorganic deposition in a carbonate-precipitating locality. *Science* **149,** 1501–1503.

Pett, J.W. & Walker, R.G. (1971) Relationship of flute cast morphology to internal sedimentary structures in turbidites. *J. sedim. Petrol.* **41.** 114–128.

Pettijohn, F.J. (1975) *Sedimentary Rocks,* pp. 628. Harper & Row, New York.

Pettijohn, F.J. & Potter, P.E. (1964) *Atlas and Glossary of Primary Sedimentary Structures,* pp. 370. Springer-Verlag, Berlin.

Pettijohn, F.J., Potter, P.E. & Siever, R. (1973) *Sand and Sandstone,* pp. 617. Springer-Verlag, Berlin.

Pevear, D.R. (1966) The estuarine formation of U.S. Atlantic coastal plain phosphorite. *Econ. Geol.* **61,** 251–256.

Picard, M.D. (1971) Classification of fine-grained sedimentary rocks. *J. sedim. Petrol.* **41,** 179–195.

Picard, M.D. & High, L.R. (1972) Criteria for recognizing lacustrine rocks. In: *Recognition of Ancient Sedimentary Environments* (Ed. by J.K. Rigby and W.K. Hamblin), pp. 108–145. *Spec. Publs. Soc. econ. Paleont. Miner.* **16,** Tulsa.

Pingitore, N.E. (1976) Vadose and phreatic diagenesis: processes, products and their recognition in corals. *J. sedim. Petrol.* **46,** 985–1006.

Playford, P.E. & Cockbain, A.E. (1976) Modern algal stromatolites at Hamelin Pool, a hypersaline barred basin in Shark Bay, Western Australia. In: *Stromatolites* (Ed. by M.R. Walter), pp. 389–412. Elsevier, Amsterdam.

Playford, P.E. & Lowry, D.C. (1966) Devonian reef complexes of the Canning basin, Western Australia. *Bull. Geol. Surv. West. Austr.* **118.**

Porrenga, D.H. (1967) Glauconite and chamosite as depth indicators in the marine environment. *Mar. Geol.* **5,** 495–501.

Potter, P.E., Maynard, J.B. & Pryor, W.A. (1980) *Sedimentology of Shale,* pp. 270. Springer-Verlag, Berlin.

Potter, P.E. & Pettijohn, F.J. (1977) *Paleocurrents and Basin Analysis,* pp. 425. Springer-Verlag, Berlin.

Powers, M.C. (1967) Fluid release mechanisms in compacting marine mudrocks and their importance in oil exploration. *Bull. Am. Ass. Petrol. Geol.* **51,** 1240–1254.

Pray, L.C. (1958) Fenestrate bryozoan core facies, Mississippian bioherms, Southwestern United States. *J. sedim. Petrol.* **28,** 261–273.

Pryor, W.A. (1973) Permeability—porosity patterns and variations in some Holocene sand bodies. *Bull. Am. Ass. Petrol. geol.* **57,** 162–189.

Purdy, E.G. (1974) Reef configurations: cause and effect. In: *Reefs in Time and Space* (Ed. by L.F. Laporte), pp. 9–76. *Spec. Publ. Soc. econ. Paleont. Miner.* **18,** Tulsa.

Purser, B.H. (1969) Syn-sedimentary marine lithification of Middle Jurassic limestones in the Paris Basin. *Sedimentology* **12,** 205–230.

Purser, B.H. (1978) Early diagenesis and the preservation of porosity in Jurassic limestones. *J. Petrol. Geol.* **1,** 83–94.

Raaf, J.F.M. De, Boersma, J.R. & Gelder, A. Van (1977) Wave-generated structures and sequences from a shallow marine succession, Lower Carboniferous, County Cork, Ireland. *Sedimentology* **24,** 451–483.

Rad, U. & Rosch, H. (1974) Petrography and diagenesis of deep-sea cherts from the central Atlantic. In: *Pelagic Sediments: on Land and under the Sea* (Ed. by K.J. Hsü and H.C. Jenkyns), pp. 327–347. *Spec. Publ. int. Ass. Sediment.* **1.**

Raiswell, R. (1971) The growth of Cambrian and Liassic concretions. *Sedimentology* **17,** 147–171.

Ramsay, A.T.S. (1974) The distribution of calcium carbonate in deepsea sediments. In: *Studies in Paleo-oceanography* (Ed. by W.W. Hay), pp. 58–76. *Spec. Publ. Soc. econ. Paleont. Miner.* **20,** Tulsa.

Rao, C.P. & Naqvi, I.H. (1977) Petrography, geochemistry and factor analysis of a Lower Ordovician subsurface sequence, Tasmania, Australia. *J. sedim. Petrol.* **47,** 1036–1055.

Rateev, M.A., Gorbunova, Z.N., Lisitzyn, A.P. & Nosov, G.L. (1969) The distribution of clay minerals in the oceans. *Sedimentology* **13,** 21–43.

Raup, O.B. & Miesch, A.T. (1957) A new method for obtaining significant average directional measurements in cross-stratification studies. *J. sedim. Petrol.* **27,** 313–321.

Read, J.F. (1974) Calcrete deposits and Quaternary sediments, Edel Provinces, Shark Bay, Western Australia. *Mem. Am. Ass. Petrol. Geol.* **12,** 250–282.

Reading, H.G. (1972) Global tectonics and the genesis of flysch successions. *24th Int. geol. Cong. Proc. Sect.* **6,** 59–66.

Reading, H.G. (Ed.) (1978) *Sedimentary Environments and Facies,* pp. 557. Blackwell Scientific Publications, Oxford.

Reeves, C.C. (1976) *Caliche; origin, classification, morphology and uses.* Estacado Books, Lubbock, Texas.

Reineck, H.E. (1972) Tidal flats. In: *Recognition of Ancient Sedimentary Environments* (Ed. by J.K. Rigby and W.K. Hamblin), pp. 146–159. *Spec. Publ. Soc. econ. Paleont. Miner.* **16,** Tulsa.

Reineck, H.E. & Singh, I.B. (1973) *Depositional Sedimentary Environments—With Reference to Terrigenous Clastics,* pp. 439. Springer-Verlag, Berlin.

Richter, D.K. & Füchtbauer, H. (1978) Ferroan calcite replacement indicates former magnesian calcite skeletons. *Sedimentology* **25,** 843–860.

Richter-Bernberg, G. (1950) Zur Frage der absoluten Geschwindigkeit geologischer Vorgänge. *Die Naturwissenschaften* **37,** 1–8.

Rieke, H.H. & Chilingarian, G.V. (1974) *Compaction of argillaceous sediments,* pp. 424. Elsevier, Amsterdam.

Riggs, S.R. (1979) Petrology of the Tertiary phosphorite system of Florida. *Econ. Geol.* **74,** 195–220.

Riley, J.P. & Skirrow, G. (1975) *Chemical Oceanography,* 2nd Edition, 5 volumes. Academic Press, London.

Robertson, A.H.F. (1977) The origin and diagenesis of cherts from Cyprus. *Sedimentology* **24,** 11–30.

Robertson, A.H.F. & Hudson, J.D. (1974) Pelagic sediments in the Cretaceous and Tertiary history of the Troodos Massif, Cyprus. In: *Pelagic Sediments: on Land and under the Sea* (Ed. by K.J.

243

Hsü and H.C. Jenkyns), pp. 403–436. *Spec. Publ. int. Ass. Sediment.* **1**.

Rohrlich, V., Price, N.B. & Calvert, S.E. (1969) Chamosite in Recent sediments of Loch Etive, Scotland. *J. sedim. Petrol.* **39**, 624–631.

Rooney, T.P. & Kerr, P.F. (1967) Mineralogic nature and origin of phosphorite, Beaufort County, North Carolina. *Bull. geol. Soc. Am.* **78**, 731–748.

Ross, C.S. & Smith, R.L. (1961) Ash-flow tuffs: their origin, geologic relations and identification. *Prof. Pap. U.S. geol. Surv.* **366**, 81 pp.

Ross, R.J., Jaanusson, V. & Friedman, I. (1975) Lithology and origin of Middle Ordovician calcareous mudmounds at Meiklejohn Peak, southern Nevada. *Prof. Pap. U.S. geol. Surv.* **871**.

Rupke, N.A. (1978) Deep Clastic Seas. In: *Sedimentary Environments and Facies* (Ed. by H.G. Reading), pp. 372–415. Blackwell Scientific Publications, Oxford.

Sandberg, P.A. (1975) New interpretations of Great Salt Lake ooids and of ancient non-skeletal carbonate mineralogy. *Sedimentology* **22**, 497–538.

Schäfer, A. & Stapf, K.R.G. (1978) Permian Saar-Nahe Basin and Recent Lake Constance (Germany): two environments of lacustrine algal carbonates. In: *Modern and Ancient Lake Sediments* (Ed. by A. Matter and M.E. Tucker), pp. 83–107. *Spec. Publ. int. Ass. Sediment.* **2**.

Schenk, P.E. (1969) Carbonate-sulphate-redbed facies and cyclic sedimentation of the Windsorian stage (middle Carboniferous), Maritime Provinces. *Can. J. Earth Sci.* **6**, 1037–1066.

Schlager, W. & Bolz, H. (1977) Clastic accumulation of sulphate evaporites in deep water. *J. sedim. Petrol.* **47**, 600–609.

Schlager, W. & James, N.P. (1978) Low-magnesian calcite limestones forming at the deep-sea floor, Tongue of the Ocean, Bahamas. *Sedimentology* **25**, 675–702.

Schlager, W. & Schlager, M. (1973) Clastic sediments associated with radiolarites (Tauglboden Schichten, Upper Jurassic, Eastern Alps). *Sedimentology* **20**, 65–89.

Schlanger, S.O. & Douglas, R.G. (1974) The pelagic ooze-chalk-limestone transition and its implications for marine stratigraphy. In: *Pelagic Sediments: on Land and under the Sea* (Ed. by K.J. Hsü and H.C. Jenkyns), pp. 117–148. *Spec. Publ. int. Ass. Sediment.* **1**.

Schlanger, S.O. & Jenkyns, H.C. (1976) Cretaceous oceanic anoxic events: causes and consequences. *Geol. Mijnb.*, **55**, 179–184.

Schmalz, R.F. (1969) Deep-water evaporite deposition: a genetic model. *Bull. Am. Ass. Petrol. Geol.* **53**, 798–823.

Schmincke, H.U., Fischer, R.V. & Waters, A.C. (1973) Antidune and chute and pool structures in base surge deposits from the Laacher See area (Germany). *Sedimentology* **20**, 553–574.

Scholle, P.A. (1971) Sedimentology of fine-grained deep water carbonate turbidites, Monte Antola flysch (Upper Cretaceous), northern Apennines, Italy. *Bull. geol. Soc. Am.* **82**, 629–658.

Scholle, P.A. (1978) A colour illustrated guide to Carbonate Rock Constituents, Textures, Cements and Porosities. *Mem. Am. Ass. Petrol. Geol.* **27**, 241.

Scholle, P.A. (1979) A colour illustrated guide to Constituents, Textures, Cements and Porosities of Sandstones and Associated Rocks. *Mem. Am. Ass. Petrol. Geol.* **28**, 201.

Scholle, P.A. & Kinsman, D.J.J. (1974) Aragonitic and high-Mg calcite caliche from the Persian Gulf—a modern analogy for the Permian of Texas and New Mexico. *J. sedim. Petrol.* **44**, 904–916.

Scholle, P.A. & Kling, S.A. (1972) Southern British Honduras: lagoonal coccolith ooze. *J. sedim. Petrol.* **42**, 195–204.

Scholle, P.A. & Schluger, P.R. (Eds.) (1979) *Aspects of diagenesis,* pp. 400. *Spec. Publ. Soc. econ. Paleont. Miner.* **26**, Tulsa.

Schreiber, B.C., Friedman, G.M., Decima, A. & Schreiber, E. (1976) The depositional environments of the Upper Miocene (Messinian) evaporite deposits of the Sicilian Basin. *Sedimentology* **23**, 729–760.

Schroeder, J.H. (1972) Fabrics and sequences of submarine carbonate cements in Holocene Bermuda cup reefs. *Geol. Rdsch.* **61**, 708–730.

Schwartz, M.L. (1973) *Barrier Islands,* pp. 451. Dowden, Hutchinson and Ross, Stroudsburg.

Seilacher, A. (1964) Biogenic sedimentary structures. In: *Approaches to Palaeoecology* (Ed. by J. Imbrie and N. Newell), pp. 296–316. John Wiley, New York.

Seilacher, A. (1977) Evolution of trace fossil communities. In: *Patterns of Evolution: illustrated by the Fossil Record* (Ed. by A. Hallam), pp. 359–376. Elsevier, North Holland, New York.

Self, S. (1976) The Recent volcanology of Terceira, Azores. *Jl. geol. Soc. Lond.* **132**, 645–666.

Selley, R.C. (1968) A classification of palaeocurrent models. *J. Geol.* **76**, 99–110.

Selley, R.C. (1976) *An Introduction to Sedimentology,* pp. 408. Academic Press, London.

Selley, R.C. (1978a) *Ancient Sedimentary Environments,* pp. 287. Chapman and Hall, London.

Shelley, R.C. (1978b) Porosity gradients in North Sea oil-bearing sandstones. *Jl. geol. Soc. Lond.* **135**, 119–132.

Sellwood, B.W. (1978) Shallow-water carbonate environments. In: *Sedimentary Environments and Facies* (Ed. by H.G. Reading). pp. 259–313. Blackwell Scientific Publications, Oxford.

244

Shearman, D.J. (1966) Origin of evaporites by diagenesis. *Trans. Inst. Min. Metall.* **75**, B208–215.

Shearman, D.J. (1970) Recent halite rock, Baja California, Mexico. *Trans. Inst. Min. Metall.* **79**, B155–162.

Shearman, D.J. & Fuller, J.G. (1969) Anhydrite diagenesis, calcitisation and organic laminites, Winnepegosis Formation, Middle Devonian, Saskatchewan. *Bull. Canad. Petrol. Geol.* **17**, 469–525.

Shearman, D.J., Khouri, J. & Taha, S. (1961) On the replacement of dolomite by calcite in some Mesozoic limestones from the French Jura. *Proc. geol. Ass.* **72**, 1–12.

Shearman, D.J., Mossop, G., Dunsmore, H. & Martin, M. (1972) Origin of gypsum veins by hydraulic fracture. *Trans. Inst. Min. Metall.* **81**, B149–155.

Shearman, D.J., Tyman, J., Kandkarimi, M. (1970) The genesis and diagenesis of oolites. *Proc. geol. Ass.* **81**, 1905–1914.

Sheldon, R.P. (1964) Exploration for phosphorite in Turkey—a case history. *Econ. Geol.* **59**, 1159–1175.

Shelton, J.W. & Mack, D.E. (1970) Grain orientation in determination of paleocurrents and sandstone trends. *Bull. Am. Ass. Petrol. geol.* **54**, 1108–1119.

Shinn, E.A. (1968) Practical significance of birdseye structures in carbonate rocks. *J. sedim. Petrol.* **38**, 215–223.

Shinn, E.A. (1969) Submarine lithification of Holocene carbonate sediments in the Persian Gulf. *Sedimentology* **12**, 109–144.

Shinn, E.A., Ginsburg, R.N. & Lloyd, R.M. (1965) Recent supratidal dolomite from Andros Island, Bahamas. In: *Dolomitization and Limestone Diagenesis* (Ed. by L.C. Pray and R.C. Murray), pp. 112–123. *Spec. Publ. Soc. econ. Paleont. Miner.* **13**, Tulsa.

Shinn, E.A., Halley, R.B., Hudson, J.H. & Lidz, B.H. (1977) Limestone compaction: An enigma. *Geology* **5**, 21–24.

Shinn, E.A., Lloyd, R.M. & Ginsburg, R.N. (1969) Anatomy of a modern carbonate tidal flat, Andros Island, Bahamas. *J. sedim. Petrol.* **39**, 1202–1228.

Sibley, D.F. & Blatt, H. (1976) Intergranular pressure solution and cementation of the Tuscarora Quartzite. *J. sedim. Petrol.* **46**, 881–896.

Simons, D.B., Richardson, E.V. & Nordin, C.F. (1965) Sedimentary structures generated by flow in alluvial channels. In: *Primary Sedimentary Structures and their Hydrodynamic Interpretation* (Ed. by G.V. Middleton), pp. 34–52. *Spec. Publ. Soc. econ. Paleont. Miner.* **12**, Tulsa.

Sippel, R.F. (1968) Sandstone petrology, evidence from luminescence petrography. *J. sedim. Petrol.* **38**, 530–554.

Skipper, K. & Bhattacharjee, S.B. (1978) Backset bedding in turbidites: a further example from the Cloridorme Formation (Middle Ordovician), Gaspé, Quebec. *J. sedim. Petrol.* **48**, 193–202.

Smalley, I.J. (1976) *Loess Lithology and Genesis*, pp. 429. Dowden, Hutchinson & Ross, Stroudsburg.

Smith, D.B. (1971) Possible displacive halite in the Permian Upper Evaporite Group of Northeast Yorkshire. *Sedimentology* **17**, 221–232.

Smith, D.B. (1973) The origin of the Permian Middle and Upper Potash deposits of Yorkshire: an alternative hypothesis. *Proc. Yorks. geol. Soc.* **39**, 327–346.

Smith, N.D. (1970) The braided stream depositional environment: Comparison of the Platte River with some Silurian clastic rocks, North-Central Appalachians. *Bull. geol. Soc. Am.* **81**, 2993–3014.

Smith, R.P. & Nash, W.P. (1976) Chemical correlation of volcanic ash deposits in the Salt Lake Group, Utah, Idaho and Nevada. *J. sedim. Petrol.* **46**, 930–939.

Smith, R.L. (1960) Ash-flows. *Bull. geol. Soc. Am.* **71**, 795–842.

Smosna, R. & Warshaver, S.M. (1979) A scheme for multivariate analysis in carbonate petrology with an example from the Silurian Tonoloway Limestone. *J. sedim. Petrol.* **49**, 257–272.

Sneed, E.D. & Folk, R.L. (1958) Pebbles in the lower Colorado River, Texas, a study in particle morphogenesis. *J. Geol.* **66**, 114–150.

Spackman, W., Riegel, W.L. & Dolsen, C.P. (1969) Geological and biological interactions in the swamp-marsh complex of southern Florida. In: *Environments of coal deposition* (Ed. by E.C. Dapples and M.E. Hopkins), pp. 1–35. *Spec. Pap. geol. Soc. Am.* **114**.

Sparks, R.S.J. (1976) Grain size variations in ignimbrites and implications for the transport of pyroclastic flows. *Sedimentology* **23**, 147–188.

Spears, D.A. (1980) Towards a classification of shales. *Jl. geol. Soc. Lond.* **137**, 125–129.

Spears, D.A. & Rice, C.M. (1973) An Upper Carboniferous tonstein of volcanic origin. *Sedimentology* **20**, 281–294.

Stablein, N.K. & Dapples, E.C. (1977) Feldspars of the Tunnel City Group (Cambrian), western Wisconsin. *J. sedim. Petrol.* **47**, 1512–1538.

Stach, E., Mackowsky, M.-Th., Teichmüller, M., Taylor, G.H., Chandra, D. & Teichmüller, R. (1975) *Textbook of Coal Petrology*, pp. 428. Gebrüder Borntraeger, Stuttgart.

Stalder, P.J. (1979) Organic and inorganic metamorphism in the Taveyannaz Sandstone of the Swiss Alps and equivalent sandstones in France and Italy. *J. sedim. Petrol.* **49**, 463–482.

Staub, J.R. & Cohen, A.D. (1978) Kaolinite-enrichment beneath coals: a modern analogue, Snuggedy Swamp, South Carolina. *J. sedim. Petrol.* **48**, 203–210.

Steel, R.J. (1974) Cornstone (fossil caliche)—its origin, stratigraphic and sedimentary importance in the New Red Sandstone, western Scotland. *J. Geol.* **82**, 351–369.

Steinen, R.P. (1978) On the diagenesis of lime-mud: scanning electron microscopic observations of subsurface material from Barbados, W.I. *J. sedim. Petrol.* **48**, 1139–1148.

Steinen, R.P. & Matthews, R.K. (1973) Phreatic vs. vadose diagenesis: stratigraphy and mineralogy of a cored borehole on Barbados, W.I. *J. sedim. Petrol.* **43**, 1012–1020.

Steinmetz, R. (1962) Analysis of a vectorial data. *J. sedim. Petrol.* **32**, 801–812.

Stewart, F.H. (1963) Marine evaporites. *Prof. pap. U.S. geol. Surv.* 440-Y.

Stockman, K.W., Ginsburg, R.N. & Shinn, E.A. (1967) The production of lime mud by algae in South Florida. *J. sedim. Petrol.* **37**, 663–648.

Stoddart, D.R. (1969) Ecology and morphology of Recent coral reefs. *Biol. Rev.* **44**, 433–498.

Stopes, M.C. (1919) On the four visible ingredients in banded bituminous coal. Studies in the composition of coal. *Proc. Roy. Soc. Lond. Ser. B* **90**, 470–487.

Stopes, M.C. (1935) On the petrology of banded bituminous coal. *Fuel, London* **14**, 4–13.

Stow, D.A.V. & Lovell, J.P.B. (1979) Contourites: their recognition in modern and ancient sediments. *Earth Sci. Rev.* **14**, 251–291.

Sturm, M. & Matter, A. (1978) Turbidites and varves in Lake Brienz (Switzerland): deposition of clastic detritus by density currents. In: *Modern and Ancient Lake Sediments* (Ed. by A. Matter and M.E. Tucker), pp. 147–168. *Spec. Publ. int. Ass. Sediment.* **2**.

Surdam, R.C. & Wolfbauer, C.A. (1975) Green River Formation, Wyoming: a playa-lake complex. *Bull. geol. Soc. Am.* **86**, 335–345.

Surdam, R.C. & Wray, J.L. (1976) Lacustrine stromatolites, Eocene Green River Formation, Wyoming. In: *Stromatolites* (Ed. by M.R. Walter), pp. 535–541. Elsevier, Amsterdam.

Swift, D.J.P. (1976) Continental shelf sedimentation. In: *Marine sediment transport and environmental management* (Ed. by D.J. Stanley and D.J.P. Swift), pp. 311–350. John Wiley, New York.

Talbot, M.R. (1974) Ironstones in the Upper Oxfordian of southern England. *Sedimentology* **21**, 433–450.

Tanner, W.F. (1967) Ripple mark indices and their uses. *Sedimentology* **9**, 89–104.

Tassé, N., Lajoie, J. & Dimroth, E. (1978) The anatomy and interpretation of an Archean volcaniclastic sequence, Noranda region, Quebec. *Can. J. Earth Sci.* **15**, 874–888.

Taylor, J.C.M. & Illing, L.V. (1969) Holocene intertidal calcium carbonate cementation, Qatar, Persian Gulf. *Sedimentology* **12**, 69–107.

Taylor, J.H. (1949) Petrology of the Northampton Sand Ironstone Formation. *Mem. geol. Surv. U.K.*, 111 pp.

Teichmuller, M. & Teichmuller, R. (1968) Geological aspects of coal metamorphism. In: *Coal and Coal-Bearing Strata* (Ed. by D. Murchison and T.S. Westoll), pp. 233–267. Oliver & Boyd, Edinburgh.

Teichmüller, M. & Teichmüller, R. (1979) Diagenesis of coal (coalification). In: *Diagenesis of sediments and sedimentary rocks* (Ed. by G. Larsen and G.V. Chilingarian), pp. 207–246. *Dev. in Sed. 25A.* Elsevier, Amsterdam.

Thompson, D.B. (1969) Dome-shaped aeolian dunes in the Frodsham Member of the so-called 'Keuper' Sandstone Formation (Scythian-?Anislan: Triassic) at Frodsham, Cheshire (England). *Sedim. Geol.* **3**, 263–289.

Till, R. (1974) *Statistical methods for the Earth scientist: an introduction*, pp. 154. MacMillan, London.

Tissot, B. (1977) The application of the results of organic geochemical studies in oil and gas exploration. In: *Developments in Petroleum Geology—1* (Ed. by G.D. Hobson), pp. 53–82. Applied Science Publishers, London.

Tissot, B. (1979) Effects of prolific petroleum source rocks and major coal deposits caused by sea-level changes. *Nature* **277**, 463–465.

Tissot, B.P. & Welte, D.H. (1978) *Petroleum Formation and Occurrence*, pp. 538. Springer-Verlag, Berlin.

Tooms, J.S., Summerhayes, C.P. & Cronan, D.S. (1969) Geochemistry of phosphate and manganese deposits. *Oceanogr. Mar. Biol. Ann. Rev.* **7**, 49–100.

Trendall, A.F. (1973) Varve cycles in the Weeli Wolli Formation of the Precambrian Hamersley Group, Western Australia. *Econ. Geol.* **68**, 1089–1097.

Tucker, M.E. (1969) Crinoidal turbidites from the Devonian of Cornwall and their palaeogeographic significance. *Sedimentology* **13**, 281–290.

Tucker, M.E. (1974) Sedimentology of Palaeozoic pelagic limestones: the Devonian Griotte

(Southern France) and Cephalopodenkalk (Germany). In: *Pelagic Sediments: on Land and under the Sea* (Ed. by K.J. Hsü and H.C. Jenkyns), pp. 71–92. *Spec. Publ. int. Ass. Seciment* **1**.

Tucker, M.E. (1976) Replaced evaporites from the Late Precambrian of Finnmark, Arctic Norway. *Sed. Geol.* **16**, 193–204.

Tucker, M.E. (1978) Triassic lacustrine sediments from South Wales: shore-zone clastics, evaporites and carbonates. In: *Modern and Ancient Lake Sediments* (Ed. by A. Matter and M.E. Tucker), pp. 205–224. *Spec. Publ. int. Ass. Sediment.* **2**.

Tucker, M.E. (1981) *Field description of sedimentary rocks*, pp. 128. Open Univ. Enterprises, Milton Keynes, England.

Tucker, M.E. & Kendall, A.C. (1973) The diagenesis and low-grade metamorphism of Devonian styliolinid-rich pelagic carbonates from West Germany: possible analogues of Recent pteropod oozes. *J. sedim. Petrol.* **43**, 672–687.

Turner, P. & Archer, R. (1977) The role of biotite in the diagenesis of red beds from the Devonian of northern Scotland. *Sedim. Geol.* **19**, 241–251.

Turmel, R.J. & Swanson, R.G. (1976) The development of Rodriguez Bank, a Holocene mudbank in the Florida reef tract. *J. sedim. Petrol.* **46**, 497–518.

Van Houten, F.B. (1962) Cyclic sedimentation and the origin of analcime-rich Upper Triassic Lockatong Formation, West-Central New Jersey and adjacent Pennsylvania. *Am. J. Sci.* **260**, 561–576.

Van Houten, F.B. (1968) Iron oxides in red beds. *Bull. geol. Soc. Am.* **79**, 399–416.

Van Houten, F.B. (1972) Iron and clay in tropical savanna alluvium, northern Columbia: a contribution to the origin of red beds. *Bull. geol. Soc. Am.* **83**, 2761–2772.

Velde, B. (1977) *Clays and clay minerals in natural and synthetic systems*, pp. 218. Elsevier, Amsterdam.

Von der Borch, C.C. (1976) Stratigraphy and formation of Holocene dolomitic carbonate deposits of the Coorong area, South Australia. *J. sedim. Petrol.* **46**, 952–966.

Walkden, G.M. (1974) Palaeokarstic surfaces in Upper Viséan (Carboniferous) Limestones of the Derbyshire Block, England. *J. sedim. Petrol.* **44**, 1232–1247.

Walker, G.P.L. (1971) Grain size characteristics of pyroclastic deposits. *J. Geol.* **79**, 696–714.

Walker, G.P.L. (1973) Explosive volcanic eruptions—a new classification scheme. *Geol. Rdsch.* **62**, 431–446.

Walker, R.G. (1965) The origin and significance of the internal sedimentary structures of turbidites. *Proc. Yorks. geol. Soc.* **35**, 1–32.

Walker, R.G. (1967) Turbidite sedimentary structures and their relationship to proximal and distal depositional environments. *J. sedim. Petrol.* **37**, 25–43.

Walker, R.G. (1973) Mopping-up the turbidite mess. In: *Evolving Concepts in Sedimentology* (Ed. by R.N. Ginsburg), pp. 1–37. Johns Hopkins Univ. Press, Baltimore.

Walker, R.G. (Ed.) (1979) *Facies Models*, pp. 211. Geoscience Canada.

Walker, R.G. & Middleton, G.V. (1979) Eolian sands. In: *Facies Models* (Ed. by R.G. Walker), pp. 33–41. Geoscience Canada.

Walker, R.G. & Mutti, E. (1973) Turbidite facies and facies associations. In: *Turbidites and Deep Water Sedimentation*, pp. 119–157. *Soc. econ. Paleont. Miner. Pacific Section*, Short Course Anaheim.

Walker, T.R. (1967) Formation of red beds in modern and ancient deserts. *Bull. geol. Soc. Am.* **78**, 353–368.

Walker, T.R. (1974) Formation of red beds in moist tropical climates: A hypothesis. *Bull. geol. Soc. Am.* **85**, 633–638.

Walker, T.R. & Honea, R.M. (1969) Iron content of modern deposits in the Sonoran desert: a contribution to the origin of red beds. *Bull. geol. Soc. Am.* **80**, 535–544.

Walker, T.R., Waugh, B. & Crone, A.J. (1978) Diagenesis in first-cycle desert alluvium of Cenozoic age, southwestern United States and northwestern Mexico. *Bull. geol. Soc. Am.* **89**, 19–32.

Walls, R.A., Harris, W.B. & Nunan, W.E. (1975) Calcareous crust (caliche) profiles and early subaerial exposure of Carboniferous carbonates, northeastern Kentucky. *Sedimentology* **22**, 417–440.

Walter, M.R. (Ed.) (1976) *Stromatolites*, pp. 790. Elsevier, Amsterdam.

Wanless, H.R. (1979) Limestone response to stress: pressure solution and dolomitization. *J. sedim. Petrol.* **49**, 437–462.

Wardlaw, N.C. (1962) Aspects of diagenesis in some Irish Carboniferous limestones. *J. sedim. Petrol.* **32**, 776–780.

Wardlaw, N.C. & Schwerdtner, W.M. (1966) Halite–anhydrite seasonal layers in the Middle Devonian Prairie Evaporite Formation, Saskatchewan, Canada. *Bull. geol. Soc. Am.* **77**, 331–342.

Waugh, B. (1970) Petrology, provenance and silica diagenesis of the Penrith Sandstone (Lower Permian) of northwest England. *J. sedim. Petrol.* **40**, 1226–1240.

247

Waugh, B. (1978) Authigenic K-feldspar in British Permo-Triassic sandstones. *Jl. geol. Soc. Lond.* **135**, 51–56.

Weaver, C.E. & Pollard, L.D. (1973) *The chemistry of clay minerals*, pp. 213. Elsevier, Amsterdam.

Wendt, J. (1971) Genese und Fauna submariner sedimentärer Spaltenfüllungen im mediterranen Jura. *Palaeontographica, A* **136**, 122–192.

West, I.M. (1964) Evaporite diagenesis in the Lower Purbeck Beds of Dorset. *Proc. Yorks. geol. Soc.* **34**, 315–330.

West, I.M., Ali, Y.A. & Hilmy, M.E. (1979) Primary gypsum nodules in a modern sabkha on the Mediterranean coast of Egypt. *Geology* **7**, 354–358.

Westoll, T.S. (1968) Sedimentary rhythms in coal-bearing Strata. In: *Coal and Coal-Bearing Strata* (Ed. by D. Murchison and T.S. Westoll), pp. 71–103. Oliver & Boyd, Edinburgh.

Wilkinson, B.H. & Landing, E. (1978) "Eggshell diagenesis" and primary radial fabric in calcite ooids. *J. sedim. Petrol.* **48**, 1129–1137.

Williams, G.E. (1968) Torridonian weathering and its bearing on Torridonian palaeoclimate and source. *Scott. J. Geol.* **4**, 164–184.

Williams, P.F. & Rust, B.R. (1969) The sedimentology of a braided river. *J. sedim. Petrol.* **39**, 649–679.

Wilson, I.G. (1972) Aeolian bedforms—their development and origins. *Sedimentology* **19**, 173–210.

Wilson, J.L. (1975) *Carbonate Facies in Geologic History*, pp. 471. Springer-Verlag, Berlin.

Wilson, M.D. & Pittman, E.D. (1977) Authigenic clays in sandstones: recognition and influence on reservoir properties and paleoenvironmental analysis. *J. sedim. Petrol.* **47**, 3–31.

Wilson, M.J. (1965) The origin and geological significance of the South Wales underclays. *J. sedim. Petrol.* **35**, 91–99.

Wise, S.W. & Weaver, F.M. (1974) Chertification of oceanic sediments. In: *Pelagic Sediments: on Land and under the Sea* (Ed. by K.J. Hsü and H.C. Jenkyns), pp. 301–326. *Spec. Publ. int. Ass. Sediment.* **1**.

Wood, G.V. & Wolfe, M.J. (1969) Sabkha cycles in Arab/Darb Formation of the Trucial Coast of Arabia. *Sedimentology* **12**, 165–191.

Woodcock, N.H. (1976) Ludlow series slumps and turbidites and the form of the Montgomery Trough, Powys, Wales. *Proc. Geol. Ass.* **87**, 169–182.

Wright, A.E. & Moseley, F. (1975) *Ice Ages: Ancient and Modern*, pp. 320. *Geol. J. Spec. Issue 6*. Seel House Press, Liverpool.

Yen, T.F. & Chilingarian, G.V. (1976) *Oil Shale*, pp. 292. Elsevier, Amsterdam.

Zankl, H. (1969) Der Hohe Göll. Aufbau und Lebensbild eines Dachsteinkalk-Riffes in der Obertrias der nördlichen Kalkalpen. *Abh. Senckenberg. naturforsch. Ges.* **519**, 1–123.

Zeff, M.L. & Perkins, R.D. (1979) Microbial alteration of Bahamian deep-sea carbonates. *Sedimentology* **26**, 175–201.

Index

1614